ADVANCED HIGHER MATHS

CONTENTS

Introduction

About this book

This book provides a resource to practise and assess your understanding of the mathematics covered for the Advanced Higher qualification. There are separate chapters for all of the skills, knowledge and understanding required for full coverage of the Advanced Higher specification. All the chapters use the same set of features to help you progress. You will find a range of worked examples to show you how to tackle problems, and an extensive set of exercises to help you develop the whole range of operational and reasoning skills needed for your Advanced Higher examination.

You should not work through the book from page 1 to the end. Your teacher will choose a range of topics throughout the school year and teach them in the order they think works best for your class, so you will use different parts of the book at different parts of the year.

Features

CHAPTER TITLE

The chapter title shows the general mathematical areas covered in the chapter.

8 Matrices

THIS CHAPTER WILL SHOW YOU HOW TO:

Each chapter opens with a list of topics covered in the chapter, and tells you what you should be able to do when you have worked your way through the chapter.

This chapter will show you how to:

- understand and use matrix algebra
- calculate the determinant of a 2×2 and a 3×3 matrix
- determine the inverse of a 2×2 and a 3×3 matrix
- perform geometric transformations
- use Gaussian elimination to solve a 3×3 system of linear equations.

YOU SHOULD ALREADY KNOW:

After the list of topics covered in the chapter, there is a list of topics you should already know before you start the chapter. Some of these topics will have been covered before in Higher Maths, and others will depend on preceding chapters in this book.

You should already know:

- the correct order of operations
- how to do fraction calculations
- how to solve simultaneous equations with two unknowns.

EXAMPLES

Each new topic is demonstrated with at least one worked Example, which shows how to go about tackling the questions in the following Exercise. Each Example breaks the question and solution down into steps, so you can see what calculations are involved, what kind of rearrangements are needed and how to work out the best way of answering the question. Most Examples have comments, which help explain the processes.

Example 2.1

Find $f'(x)$ of $3x^2$

Let $f(x) = 3x^2$

$$f'(x) = \lim_{h \to 0} \left[\frac{f(x + h) - f(x)}{h} \right]$$
— Start with the formula to find $f'(x)$.

$$= \lim_{h \to 0} \left[\frac{3(x + h)^2 - 3x^2}{h} \right]$$
— Substitute $f(x) = 3x^2$ and $f(x + h) = 3(x + h)^2$

$$= \lim_{h \to 0} \left[\frac{3x^2 + 6xh + 3h^2 - 3x^2}{h} \right] = \left[\frac{6xh + 3h^2}{h} \right]$$
— Multiply out the brackets in the numerator and simplify.

$$= \lim_{h \to 0} \left[\frac{h(6x + 3h)}{h} \right]$$
— Cancel out the common factor of h.

$$= \lim_{h \to 0} \left[(6x + 3h) \right] = 6x$$
— As h tends to zero, $3h$ also tends to zero and $f'(x)$ tends to $6x$.

EXERCISES

The most important parts of the book are the Exercises. The questions in the Exercises are carefully graded in difficulty, so you should be developing your skills as you work through an Exercise. If you find the questions difficult, look back at the Example for ideas on what to do. Use Key questions, marked with a star, to assess how confident you can feel about a topic.

Exercise 6C

★ 1 Determine the equations of any oblique asymptotes of the graphs of the following functions.

a $f(x) = \dfrac{2x^2 - 2}{x}$

b $f(x) = \dfrac{3 - x^2}{x}$

c $f(x) = \dfrac{4x^2}{x + 1}$

d $f(x) = \dfrac{x^3}{2x^2 - 1}$

e $f(x) = \dfrac{x - 2x^3}{x^2}$

f $f(x) = \dfrac{1 - 2x^2}{3x + 1}$

HINTS

Where appropriate, Hints are given in the text and in Exercises, to help give extra support.

3 Consider the function $h(\theta) = \sin\theta - \cos\theta$. Identify all stationary points of $h(\theta)$ and use the second derivative test to determine which are maxima and which are minima.

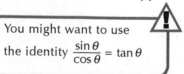
You might want to use the identity $\dfrac{\sin\theta}{\cos\theta} = \tan\theta$

HIGHLIGHT BOXES

Tinted highlight boxes are used to show key results.

$i = \sqrt{-1}$ and since $i = \sqrt{-1}$ then $i^2 = -1$ $(i \notin \mathbb{R})$

CHAPTER REVIEW

The last exercise in each chapter is a review exercise, with questions which draw on the skills and knowledge developed throughout the whole chapter. These review questions are ideal exam preparation.

Chapter review

1 Given that $z_1 = 3 - 4i$ and $z_2 = 6 + i$, calculate the following, giving your answers in the form $a + ib$ where $a, b \in \mathbb{R}$.

   a $z_1 + z_2$ **b** $3z_2 - z_1$ **c** $z_1 z_2$

   d $\dfrac{z_2}{z_1}$ **e** $(z_2)^2$ **f** $\sqrt{z_1}$

2 Solve $(2a + 3i)(5 + bi) = 16 + 11i$ for a and b where $a, b \in \mathbb{R}$.

3 Draw $z_1 = 5 + 3i$ and $z_2 = -3 + 4i$ on an Argand diagram.

4 Express $z = \sqrt{3} - 3i$ in polar form.

END-OF-CHAPTER SUMMARY

Each chapter closes with a summary of learning statements showing what you should be able to do when you complete the chapter. The summary identifies the Key questions for each learning statement. You can use the End-of-chapter summary and the Key questions to check you have a good understanding of the topics covered in the chapter.

- I understand and can use matrix algebra. ★ Exercise 8A Q4, Q5 ★ Exercise 8B Q1, Q5

- I can calculate the determinant of a 2 × 2 and a 3 × 3 matrix. ★ Exercise 8C Q1, Q3

- I can determine the inverse of a 2 × 2 and a 3 × 3 matrix. ★ Exercise 8D Q1, Q5 ★ Exercise 8E Q3

- I can use matrices to describe geometric transformations. ★ Exercise 8F Q1, Q2

- I can use Gaussian elimination to solve a 3 × 3 system of linear equations. ★ Exercise 8G Q1

ANSWERS

Answers to all Exercise questions are provided online at

www.collins.co.uk/pages/Scottish-curriculum-free-resources

Exam formulae list

Standard derivatives	
$f(x)$	$f'(x)$
$\sin^{-1}x$	$\dfrac{1}{\sqrt{1-x^2}}$
$\cos^{-1}x$	$-\dfrac{1}{\sqrt{1-x^2}}$
$\tan^{-1}x$	$\dfrac{1}{1+x^2}$
$\tan x$	$\sec^2 x$
$\cot x$	$-\operatorname{cosec}^2 x$
$\sec x$	$\sec x \tan x$
$\operatorname{cosec} x$	$-\operatorname{cosec} x \cot x$
$\ln x$	$\dfrac{1}{x}$
e^x	e^x

Standard integrals	
$f(x)$	$\int f(x)\,dx$
$\sec^2(ax)$	$\dfrac{1}{a}\tan(ax)+c$
$\dfrac{1}{\sqrt{a^2-x^2}}$	$\sin^{-1}\left(\dfrac{x}{a}\right)+c$
$\dfrac{1}{a^2+x^2}$	$\dfrac{1}{a}\tan^{-1}\left(\dfrac{x}{a}\right)+c$
$\dfrac{1}{x}$	$\ln|x|+c$
e^{ax}	$\dfrac{1}{a}e^{ax}+c$

Summations

(Arithmetic series) $\qquad S_n = \frac{1}{2}n\big[2a+(n-1)d\big]$

(Geometric series) $\qquad S_n = \dfrac{a\left(1-r^n\right)}{1-r},\ r \neq 1$

$$\sum_{r=1}^{n} r = \frac{n(n+1)}{2}, \quad \sum_{r=1}^{n} r^2 = \frac{n(n+1)(2n+1)}{6}, \quad \sum_{r=1}^{n} r^3 = \frac{n^2(n+1)^2}{4}$$

Binomial theorem

$$(a + b)^n = \sum_{r=0}^{n} \binom{n}{r} a^{n-r} b^r \quad \text{where} \quad \binom{n}{r} = {}^n C_r = \frac{n!}{r!(n-r)!}$$

Maclaurin expansion

$$f(x) = f(0) + f'(0)x + \frac{f''(0)x^2}{2!} + \frac{f'''(0)x^3}{3!} + \frac{f^{iv}(0)x^4}{4!} + \dots$$

De Moivre's theorem

$$\left[r(\cos\theta + i\sin\theta) \right]^n = r^n (\cos n\theta + i\sin n\theta)$$

Vector product

$$\mathbf{a} \times \mathbf{b} = |\mathbf{a}||\mathbf{b}| \sin\theta\, \hat{\mathbf{n}} = \begin{vmatrix} \mathbf{i} & \mathbf{j} & \mathbf{k} \\ a_1 & a_2 & a_3 \\ b_1 & b_2 & b_3 \end{vmatrix} = \mathbf{i} \begin{vmatrix} a_2 & a_3 \\ b_2 & b_3 \end{vmatrix} - \mathbf{j} \begin{vmatrix} a_1 & a_3 \\ b_1 & b_3 \end{vmatrix} + \mathbf{k} \begin{vmatrix} a_1 & a_2 \\ b_1 & b_2 \end{vmatrix}$$

Matrix transformation

Anti-clockwise rotation through an angle, θ, about the origin, $\begin{bmatrix} \cos\theta & -\sin\theta \\ \sin\theta & \cos\theta \end{bmatrix}$

1 Algebra

Expressing rational functions as a sum of partial fractions

Previously, you have learned how to add or subtract algebraic fractions to give a single fraction. Expressing a rational function as the sum of partial fractions is the opposite of this process. It is used in other areas of the Advanced Higher Maths course, including calculus and binomial expansion.

The way a fraction is dealt with is determined by the elements in the denominator:

- **linear factors**

  $\dfrac{11x + 7}{(x - 1)(x + 5)}$ can be expressed in the form $\dfrac{A}{x - 1} + \dfrac{B}{x + 5}$, where A and B are constants

- **an irreducible quadratic**

  $\dfrac{x^2 + 10x - 22}{(x - 4)(x^2 + 1)}$ can be expressed in the form $\dfrac{A}{x - 4} + \dfrac{Bx + C}{x^2 + 1}$, where A, B and C are constants

- **repeated linear factors**

  $\dfrac{6x^2 + 15x + 1}{(x - 7)(x + 3)^2}$ can be expressed in the form $\dfrac{A}{x - 7} + \dfrac{B}{x + 3} + \dfrac{C}{(x + 3)^2}$, where A, B and C are constants.

Linear factors in the denominator

A **rational function** is a function of the form $\dfrac{p(x)}{q(x)}$, where p and q are polynomials in x. This can be **decomposed** (broken down) into partial fractions.

Example 1.1

Express $\dfrac{11x + 7}{(x - 1)(x + 5)}$ as a sum of partial fractions.

Method 1

$\dfrac{11x + 7}{(x - 1)(x + 5)} = \dfrac{A}{x - 1} + \dfrac{B}{x + 5}$

> Express in partial fractions form.

$= \dfrac{A(x + 5) + B(x - 1)}{(x - 1)(x + 5)}$

> Multiply through by the denominator to combine the two fractions.

$A(x + 5) + B(x - 1) = 11x + 7$

> The denominators are the same so you can equate the numerators.

Let $x = -5$, $-6B = -48$
                  $B = 8$

> Solve for B. By letting $x = -5$, the first term on the LHS side of the identity becomes 0 so A is eliminated.

Let $x = 1$, $6A = 18$
                $A = 3$

> Solve for A. It does not matter whether you solve for A or B first.

So $\dfrac{11x + 7}{(x - 1)(x + 5)} = \dfrac{3}{x - 1} + \dfrac{8}{x + 5}$

> State your final answer.

Method 2 (comparing coefficients)

$$\frac{11x + 7}{(x - 1)(x + 5)} = \frac{A}{x - 1} + \frac{B}{x + 5}$$

$$= \frac{A(x + 5) + B(x - 1)}{(x - 1)(x + 5)}$$

$A(x + 5) + B(x - 1) = 11x + 7$

$Ax + 5A + Bx - B = 11x + 7$ ●————— Multiply out the brackets on the LHS of equation.

$(A + B)x + (5A - B) = 11x + 7$ ●————— Collect like terms.

so $A + B = 11$ and $5A - B = 7$ ●————— Compare coefficients on RHS and LHS of equation.

so $A = 3$ and $B = 8$ ●————— Solve for A and B using simultaneous equations.

So $\dfrac{11x + 7}{(x - 1)(x + 5)} = \dfrac{3}{x - 1} + \dfrac{8}{x + 5}$

Exercise 1A

Express the algebraic fractions in Questions 1–4 in partial fractions.

1 a $\dfrac{4x + 2}{(x + 2)(x - 4)}$ b $\dfrac{7x + 17}{(x + 1)(x + 3)}$ c $\dfrac{5x + 1}{(x - 1)(x + 2)}$

   d $\dfrac{9x - 5}{x(x - 1)}$ e $\dfrac{2x - 3}{(x + 1)(x - 4)}$ f $\dfrac{x - 7}{(x + 5)(x + 1)}$

   g $\dfrac{x - 6}{x(x - 3)}$ h $\dfrac{10x}{(x - 2)(x + 3)}$

★ 2 a $\dfrac{2x + 5}{x^2 + 5x + 6}$ b $\dfrac{2x + 2}{x^2 + 2x - 3}$ c $\dfrac{x + 1}{x^2 + 7x + 12}$

   d $\dfrac{17 - x}{2x^2 + 9x + 4}$ e $\dfrac{x + 15}{6x^2 - 7x - 3}$ f $\dfrac{2}{4x^2 - 1}$

3 a $\dfrac{2x + 3}{x(x + 2)}$ b $\dfrac{2x + 1}{x^2 + 4x + 3}$ c $\dfrac{x}{2x^2 + 15x + 18}$

4 a $\dfrac{2x^2 + 12x - 10}{(x - 1)(2x - 1)(x + 3)}$ b $\dfrac{2x + 5}{(x + 1)(x + 2)(x + 3)}$ c $\dfrac{2x^2 + x - 3}{2x^3 + 3x^2 - 2x}$

Irreducible quadratic in the denominator

An expression may have a linear factor and an irreducible quadratic (a quadratic that cannot be factorised) in the denominator. This can be decomposed into partial fractions.

Example 1.2

Express $\dfrac{x^2 + 10x - 22}{(x - 4)(x^2 + 1)}$ as a sum of partial fractions.

Method 1

$$\frac{x^2 + 10x - 22}{(x - 4)(x^2 + 1)} = \frac{A}{x - 4} + \frac{Bx + C}{x^2 + 1}$$

$$= \frac{A(x^2 + 1) + (Bx + C)(x - 4)}{(x - 4)(x^2 + 1)}$$

$A(x^2 + 1) + (Bx + C)(x - 4) = x^2 + 10x - 22$

Let $x = 4$, $\qquad 17A = 34$

$\qquad\qquad\qquad A = 2$

> Solve for A by letting $x = 4$ to eliminate $(Bx + C)$.

Let $x = 0$, $\quad 2 - 4C = -22$

$\qquad\qquad\qquad C = 6$

> Solve for C. Once you have found A, use $x = 0$ to find C.

Let $x = 1$, $\quad 4 - 3B - 18 = -11$

$\qquad\qquad\qquad B = -1$

> Solve for B. Once you have found A and C, to find B pick any value for x. If you haven't used $x = 1$ to find A, then that is the best value of x to use.

So $\dfrac{x^2 + 10x - 22}{(x - 4)(x^2 + 1)} = \dfrac{2}{x - 4} + \dfrac{-x + 6}{x^2 + 1}$

or $\dfrac{x^2 + 10x - 22}{(x - 4)(x^2 + 1)} = \dfrac{2}{x - 4} + \dfrac{6 - x}{x^2 + 1}$

or $\dfrac{x^2 + 10x - 22}{(x - 4)(x^2 + 1)} = \dfrac{2}{x - 4} - \dfrac{x - 6}{x^2 + 1}$

> The final answer in this example can be written in different forms.

Method 2

$$\frac{x^2 + 10x - 22}{(x - 4)(x^2 + 1)} = \frac{A}{x - 4} + \frac{Bx + C}{x^2 + 1}$$

$$= \frac{A(x^2 + 1) + (Bx + C)(x - 4)}{(x - 4)(x^2 + 1)}$$

$A(x^2 + 1) + (Bx + C)(x - 4) = x^2 + 10x - 22$

$Ax^2 + A + Bx^2 + Cx - 4Bx - 4C = x^2 + 10x - 22$

$(A + B)x^2 + (C - 4B)x + (A - 4C) = x^2 + 10x - 22$

so $A + B = 1$, $C - 4B = 10$ and $A - 4C = -22$

$\Rightarrow A = 2$, $B = -1$ and $C = 6$ ●———— Solve simultaneously for A, B and C.

$$\frac{x^2 + 10x - 22}{(x - 4)(x^2 + 1)} = \frac{2}{x - 4} + \frac{-x + 6}{x^2 + 1}$$

Exercise 1B

Express the algebraic fractions in Questions 1–3 in partial fractions.

1 a $\dfrac{2x^2 + 4x + 3}{(x + 1)(x^2 + x + 1)}$ b $\dfrac{2x^2 + 7}{(x + 2)(x^2 - 5x + 1)}$ c $\dfrac{4x^2 + 6x + 11}{(x + 2)(x^2 + 1)}$

  d $\dfrac{-7 - x^2}{(x + 3)(x^2 + 2x + 5)}$ e $\dfrac{5x^2 + 7x + 6}{(2x + 1)(x^2 - x + 3)}$ f $\dfrac{4x^2 + 14x + 5}{(4x + 3)(2x^2 + 5x + 1)}$

  g $\dfrac{3x^2 + 2x + 14}{(2x + 1)(x^2 + x + 3)}$ h $\dfrac{x^2 + 6x + 15}{(2x + 3)(x^2 + x + 2)}$ i $\dfrac{1 - 5x - 2x^2}{(x - 1)(x^2 + 2)}$

  j $\dfrac{2x^2 + 7x - 7}{(x - 3)(2x^2 - x + 1)}$

★ 2 a $\dfrac{2x}{(x + 3)(2x^2 + 4x + 6)}$ b $\dfrac{3x^2 + 7x + 12}{(x + 1)(3x^2 + 6x + 15)}$ c $\dfrac{3x^2 + 11x + 22}{(x + 2)(2x^2 + 4x + 8)}$

★ 3 a $\dfrac{5x^2 - 3x - 2}{x^3 - x^2 - 3x - 1}$ b $\dfrac{x^2 - 7x + 6}{x^3 - 5x^2 + 7x - 2}$ c $\dfrac{-5x^2 + 8x - 13}{x^3 - 2x^2 + 2x - 15}$

Repeated linear factor in the denominator

An expression may have a factor in the denominator that is repeated. For example, $(x + 3)^2$ repeats the factor $(x + 3)$. This can be decomposed into partial fractions.

Example 1.3

Express $\dfrac{6x^2 + 15x + 1}{(x - 7)(x + 3)^2}$ as a sum of partial fractions.

$$\frac{6x^2 + 15x + 1}{(x - 7)(x + 3)^2} = \frac{A}{x - 7} + \frac{B}{x + 3} + \frac{C}{(x + 3)^2}$$

The alternative method of comparing coefficients shown in Examples 1.1 and 1.2 could be used here, but can be cumbersome and time consuming.

$$= \frac{A(x + 3)^2 + B(x - 7)(x + 3) + C(x - 7)}{(x - 7)(x + 3)^2}$$

$$A(x + 3)^2 + B(x - 7)(x + 3) + C(x - 7) = 6x^2 + 15x + 1$$

Let $x = 7$, $\quad 100A = 400$

$$A = 4$$

Let $x = -3$, $\quad -10C = 10$

$$C = -1$$

Let $x = 0$, $\quad 9A - 21B - 7C = 1$

$$-21B = -42$$

$$B = 2$$

Solve for B. Once you have found A and C, pick any value for x to find B. If you haven't used $x = 0$ to find A or C, then that is the best value of x to use.

So $\dfrac{6x^2 + 15x + 1}{(x - 7)(x + 3)^2} = \dfrac{4}{x - 7} + \dfrac{2}{x + 3} - \dfrac{1}{(x + 3)^2}$

Exercise 1C

Express the algebraic fractions in Questions 1 and 2 in partial fractions.

1 a $\dfrac{5x^2 + 22x + 29}{(x + 1)(x + 3)^2}$

   b $\dfrac{5x^2 + 2x}{(2x + 1)(x + 1)^2}$

   c $\dfrac{x^2 + 3x - 13}{(x - 1)(x + 2)^2}$

   d $\dfrac{6 + x - x^2}{x^2(x - 2)}$

   e $\dfrac{16x^2 + 3}{x(x + 1)^2}$

   f $\dfrac{5}{(2 - x)(x + 1)^2}$

★ 2 a $\dfrac{2}{x^3 - 3x^2}$

   b $\dfrac{3x + 15}{x^3 + 3x^2 - 4}$

   c $\dfrac{x^2 - 2x + 1}{x^3 - 7x^2 + 16x - 12}$

   d $\dfrac{2 + x}{3x^3 + 8x^2 + 7x + 2}$

   e $\dfrac{x^2 + 4x + 7}{x^3 + 9x^2 + 27x + 27}$

   f $\dfrac{x^2 + 10x + 5}{x^3 + x^2 - x - 1}$

Division of algebraic fractions

A fraction is called **proper** when the degree of the numerator is **less than** the degree of the denominator. Otherwise it is called an **improper fraction**. This idea follows for a rational function which is one expressed in fractional form (whose numerator and denominator are polynomials).

For example, $\dfrac{x+3}{x^2+1}$, $\dfrac{2x}{(x+3)(x+1)}$ and $\dfrac{3}{x^2+2x+5}$ are all **proper** rational functions

whereas $\dfrac{x^3+2}{x^2+1}$, $\dfrac{x}{x+8}$ and $\dfrac{5x^2}{(x+3)(x+1)}$ are all **improper** rational functions.

Improper rational functions can be simplified by algebraic division.

Example 1.4

Simplify $\dfrac{x^2-3x-4}{x-2}$ by algebraic division.

Method 1

$$
\begin{array}{r}
x - 1 \\
x-2\,\overline{\big)\,x^2-3x-4} \\
-\,(x^2-2x) \\
\hline
-x-4 \\
-(-x+2) \\
\hline
-6
\end{array}
$$

① The leading term in the divisor (denominator) needs to be multiplied by a term to get the leading term in the numerator.

② Multiply the leading term in the divisor by x to get the leading term in the numerator. Write the x on top line (in line with the x terms) and multiply the divisor by x.

③

④ Subtract ② from ①.

The leading term in the divisor needs to be multiplied by -1 to get the leading term in $(-x-4)$. Write the -1 on the top line beside the x (in line with the numerical terms) and multiply the divisor by -1.

So $\dfrac{x^2-3x-4}{x-2} = x-1+\dfrac{-6}{x-2}$

$\qquad\qquad = x-1-\dfrac{6}{x-2}$

Subtract ④ from ③ to find your final remainder. Stop when the degree of the remainder (-6) is less than the degree of the divisor ($x-2$).

Rewriting $\dfrac{x^2-3x-4}{x-2}$ in this way is similar to rewriting $\dfrac{13}{4} = 3$ remainder $1 = 3 + \dfrac{1}{4}$

(continued)

Method 2 (using synthetic division)

Using synthetic division works if you are dividing by a linear expression.

Coefficients of polynomial in decreasing powers of x.

$$
\begin{array}{c|ccc}
2 & 1 & -3 & -4 \\
  & \downarrow & 2 & -2 \\
\hline
  & 1 & -1 & -6
\end{array}
$$

Remainder

Coefficients of quotient.

Quotient $= x - 1$

Remainder $= -6$

So $\dfrac{x^2 - 3x - 4}{x - 2} = x - 1 + \dfrac{-6}{x - 2}$

$\qquad\qquad = x - 1 - \dfrac{6}{x - 2}$

Method 3 (nesting)

$$\frac{x^2 - 3x - 4}{x - 2} = \frac{x(x - 2) + 2x - 3x - 4}{x - 2}$$

Nest the denominator into the leading term of the numerator remembering to $+2x$ to compensate for the $-2x$ if you were to multiply out the brackets.

$$= \frac{x(x - 2) - x - 4}{x - 2}$$

Simplify numerator.

$$= \frac{x(x - 2) - 1(x - 2) - 2 - 4}{x - 2}$$

Nest the denominator into the numerator. Remember to -2 to compensate for the $+2$ if you were to multiply out the brackets.

$$= \frac{x(x - 2) - 1(x - 2) - 6}{x - 2}$$

Simplify numerator.

$$= x - 1 - \frac{6}{x - 2}$$

Divide through by denominator.

Example 1.5

Simplify $\dfrac{x^3 + 4x^2 - x + 2}{x^2 + x}$

Method 1

$$
\begin{array}{r}
x + 3 \\
x^2 + x \,\big|\, \overline{x^3 + 4x^2 - x\ \ + 2} \\
\underline{-(x^3 + \ x^2)} \\
3x^2 - x\ \ + 2 \\
\underline{-(3x^2 + 3x)} \\
-4x + 2
\end{array}
$$

So $\dfrac{x^3 + 4x^2 - x + 2}{x^2 + x} = x + 3 + \dfrac{-4x + 2}{x^2 + x}$

$$= x + 3 - \dfrac{4x - 2}{x^2 + x} \quad \text{or} \quad x + 3 + \dfrac{2 - 4x}{x^2 + x}$$

Method 2 (nesting)

$$\dfrac{x^3 + 4x^2 - x + 2}{x^2 + x} = \dfrac{x(x^2 + x) - x^2 + 4x^2 - x + 2}{x^2 + x}$$

$$= \dfrac{x(x^2 + x) + 3x^2 - x + 2}{x^2 + x}$$

$$= \dfrac{x(x^2 + x) + 3(x^2 + x) - 3x - x + 2}{x^2 + x}$$

$$= \dfrac{x(x^2 + x) + 3(x^2 + x) - 4x + 2}{x^2 + x}$$

$$= x + 3 + \dfrac{-4x + 2}{x^2 + x} \quad \text{or} \quad x + 3 + \dfrac{2 - 4x}{x^2 + x}$$

Exercise 1D

Simplify the algebraic fractions in Questions 1–3.

1 a $\dfrac{3x + 4}{x + 1}$ b $\dfrac{4x - 5}{x - 2}$ c $\dfrac{7 - x}{x - 5}$

   d $\dfrac{4x - 3}{2x - 1}$ e $\dfrac{15x - 41}{3x - 8}$

★ 2 a $\dfrac{2x^2 - x + 2}{2x - 3}$ b $\dfrac{2x^2 + 7x - 1}{x + 4}$ c $\dfrac{3x^2 + 4x - 4}{x + 1}$

   d $\dfrac{x^3 + 3x^2 + 10x + 1}{x + 1}$ e $\dfrac{2x^3 + 5x^2 - 8x - 8}{x + 3}$

3 a $\dfrac{3x^2 + 7x + 10}{x^2 + 2x + 3}$ b $\dfrac{8x^2 + 2x + 14}{2x^2 + x + 3}$ c $\dfrac{3x^3 + 13x^2 + 5x + 7}{3x^2 + x + 2}$

Division of algebraic fractions and partial fractions

We often have to make a rational algebraic fraction proper before we can write it as a sum of partial fractions.

Example 1.6

Express $\dfrac{x^2 - x + 6}{x^2 + x - 2}$ as a sum of partial fractions.

> ⚠ The degree of the numerator is not less than the degree of the denominator, so this is an improper fraction and you need to make the fraction proper first.

$$
\begin{array}{r}
1 \\
x^2 + x - 2 \overline{\big)\; x^2 - x + 6} \\
-(x^2 + x - 2) \\
\hline
-2x + 8
\end{array}
$$

So $\dfrac{x^2 - x + 6}{x^2 + x - 2} = 1 + \dfrac{8 - 2x}{x^2 + x - 2}$

> You now have a proper fraction which can be expressed in partial fractions form.

$\dfrac{8 - 2x}{x^2 + x - 2} = \dfrac{8 - 2x}{(x + 2)(x - 1)}$

> Consider the proper fraction.

$\dfrac{8 - 2x}{(x + 2)(x - 1)} = \dfrac{A}{x + 2} + \dfrac{B}{x - 1}$

> Express in partial fractions form.

$\phantom{\dfrac{8 - 2x}{(x + 2)(x - 1)}} = \dfrac{A(x - 1) + B(x + 2)}{(x + 2)(x - 1)}$

> Multiply through by the denominator to combine the two fractions.

$A(x - 1) + B(x + 2) = 8 - 2x$

Let $x = 1$, $\quad 2B = 4$

> Solve for A and B.

$ B = 2$

Let $x = -2$, $\quad -3A = 12$

$ A = -4$

> Express as a sum of partial fractions.

So $\dfrac{8 - 2x}{(x + 2)(x - 1)} = \dfrac{-4}{x + 2} + \dfrac{2}{x - 1}$

Hence:

$$
\begin{aligned}
\frac{x^2 - x + 6}{x^2 + x - 2} &= 1 + \frac{8 - 2x}{x^2 + x - 2} \\
&= 1 + \frac{-4}{x + 2} + \frac{2}{x - 1} \\
&= 1 - \frac{4}{x + 2} + \frac{2}{x - 1}
\end{aligned}
$$

Exercise 1E

Express the algebraic fractions in Questions 1–4 in partial fractions.

★ **1** **a** $\dfrac{2x^2 + 11x + 11}{(x + 1)(x + 3)}$

    b $\dfrac{15x^2 + 45x + 15}{(x + 2)(3x + 1)}$

  c $\dfrac{3x^2 + 7x - 32}{x^2 + 2x - 8}$

    d $\dfrac{-x^2 + 7x - 2}{x^2 - 4x + 3}$

2 **a** $\dfrac{3x^3 + 13x^2 + 30x + 38}{x^3 + 5x^2 + 11x + 15}$

  b $\dfrac{2x^3 + x^2 - 4x - 26}{x^3 + x^2 - 2x - 8}$

  c $\dfrac{x^3 - x^2 - 1}{x^3 + x^2 + x}$

3 **a** $\dfrac{x^3 - x^2 - 5x + 1}{x^2 - 2x - 3}$

  b $\dfrac{x^3 + 4x^2 - x + 1}{x^2 + x}$

  c $\dfrac{x^3 - 7x^2 + 18x - 21}{x^2 - 5x + 6}$

4 **a** $\dfrac{x^4 + 2x^2 - 2x + 1}{x^3 + x}$

  b $\dfrac{x^4 - 4x^3 + 9x^2 - 17x + 12}{x^3 - 4x^2 + 4x}$

An algebraic fraction can be written as a sum of two (or more) simpler fractions to give **partial fractions**. Methods for splitting algebraic fractions into partial fractions depend on the denominator:

- two (or more) distinct linear factors in denominator

$$\frac{4x + 3}{(x - 1)(x + 2)(x + 3)} = \frac{A}{x - 1} + \frac{B}{x + 2} + \frac{C}{x + 3}$$

- irreducible quadratic factor in denominator

$$\frac{3x^2 + 2x + 1}{(x - 1)(x^2 + 2x + 2)} = \frac{A}{x - 1} + \frac{Bx + C}{x^2 + 2x + 2}$$

- repeated linear factor in denominator

$$\frac{2}{x(x + 1)^2} = \frac{A}{x} + \frac{B}{(x + 1)} + \frac{C}{(x + 1)^2}$$

or $\dfrac{2}{x(x + 1)^3} = \dfrac{A}{x} + \dfrac{B}{(x + 1)} + \dfrac{C}{(x + 1)^2} + \dfrac{D}{(x + 1)^3}$

The constants can then be found by substituting values for x that will cancel out one of the constants or by using simultaneous equations.

If the degree of the numerator is greater than or equal to the degree of the denominator then you must divide through to make a **proper** rational function **before** attempting to split the algebraic fraction into partial fractions.

Permutations and combinations

Permutations and **combinations** are ways of selecting members of a set. A **permutation** is an **ordered** arrangement of r objects from n objects. A **combination** is an **unordered** arrangement of r objects from n objects. There are mathematical methods which can be used to calculate the numbers of permutations and combinations which can be drawn from a given set.

Permutations

A football manager has 8 defenders in her squad, called A, B, C, D, E, F, G and H. She wants to pick 3 defenders for her team. How many different ways are there of doing this?

She could list them:

    ABC   ABD   ABE   … etc.

Or she could select one player from the 8 available, with the next selection made from 7 players, and the next selection made from the remaining 6. This means that there are $8 \times 7 \times 6$ (= 336) different selections when the order of selections matters. This is a permutation.

A permutation is an ordered arrangement of objects. In the case of the football team, the manager lists a permutation of 3 players from 8. The total number of permutations of 3 objects from 8 unlike objects is $8 \times 7 \times 6$.

Calculations like this can be written in terms of **factorials**. For example, $5 \times 4 \times 3 \times 2 \times 1 = 120$ can be written as 5! and read as '5 factorial'.

$n!$ can be defined as $n \times (n-1) \times (n-2) \times \dots \times 3 \times 2 \times 1$. It can be calculated on your calculator.

> Know your calculator! Your calculator will probably have a factorial button labelled $x!$, but you may need to use the SHIFT button to use it. Take some time to find it and learn how to use it – don't wait until the exam.

Hence, using factorial notation, the total number of permutations of 3 objects from 8 (which we know is $8 \times 7 \times 6$) can be written as:

$$8 \times 7 \times 6 = \frac{8 \times 7 \times 6 \times 5 \times 4 \times 3 \times 2 \times 1}{5 \times 4 \times 3 \times 2 \times 1} = \frac{8!}{5!} = \frac{8!}{(8-3)!}$$

Notation

The notation ^nP_r is used to show the total number of permutations of r objects from n objects when the order matters. In the football example, this is $^8P_3 = \frac{8!}{(8-3)!}$. The number of permutations of a set of numbers can be calculated using your calculator.

$$^nP_r = \frac{n!}{(n-r)!}$$

> Your calculator will probably have an ^nP_r button labelled nPr, but you may need to use the SHIFT button to use it.

Combinations

If we consider the selections of the football manager, selecting the defenders ABC is the same as selecting ACB (as it is the same as BAC, BCA, CAB and CBA), so the selection of the player A, player B and player C can occur in 6 different ways. Using factorial notation, there are $3!$ permutations of the players A, B and C. We know that the total number of permutations of 3 defenders from 8 is $\frac{8!}{5!}$ and we also know that within that total of 336 there are repeated combinations (i.e. the same combination can occur 6 times in the list of all possible selections). Hence the number of different combinations, when the order of selection does not matter, is $\frac{8!}{5!} \div 6$ or, because $6 = 3!$, we could write it as $\frac{8!}{5!} \div 3!$

So the number of possible selections of 3 defenders from the 8 is:

$$\frac{8!}{5!} \div 3! = \frac{8!}{5!3!} = \frac{8!}{(8-3)!3!}$$

Notation

The notation $^8C_3$ is used to show the total number of ways of selecting 3 objects from 8 when the order of selection does not matter. Here the word 'combination' is used for the process of selecting r objects from n objects where the order of selection does not matter, and we use the notation:

$$^nC_r = \binom{n}{r}$$

The number of combinations of a set of numbers can be calculated using your calculator.

$$^nC_r = \binom{n}{r} = \frac{n!}{r!(n-r)!}$$

Your calculator will probably have an ^nC_r button labelled nCr, but you may need to use the SHIFT button to use it.

If we consider $^nC_r = \binom{n}{r} = \frac{n!}{r!(n-r)!}$ without using factorial notation, we can write:

$$\binom{n}{r} = \frac{n!}{r!(n-r)!}$$

Write without using factorial notation.

$$= \frac{n(n-1)(n-2)(n-3)...(n-r+1)(n-r)(n-r-1)(n-r-2)... \times 3 \times 2 \times 1}{(r(r-1)(r-2)... \times 4 \times 3 \times 2 \times 1) \times ((n-r)(n-r-1)(n-r-2)... \times 3 \times 2 \times 1)}$$

$$= \frac{n(n-1)(n-2)(n-3)...(n-r+1)}{r(r-1)(r-2)... \times 4 \times 3 \times 2 \times 1}$$

Cancel terms on numerator and denominator.

For example:

$$\binom{10}{7} = \frac{10!}{7!(10-7)!}$$

$$= \frac{10!}{7!3!}$$

$$= \frac{10 \times 9 \times 8 \times 7 \times 6 \times 5 \times 4 \times 3 \times 2 \times 1}{(7 \times 6 \times 5 \times 4 \times 3 \times 2 \times 1) \times (3 \times 2 \times 1)}$$

$$= \frac{10 \times 9 \times 8}{3 \times 2 \times 1}$$

$$= 120$$

Similarly:

$$\binom{n}{2} = \frac{n \times (n-1) \times (n-2) \times (n-3) \times \ldots \times 3 \times 2 \times 1}{(2 \times 1)((n-2) \times (n-3) \times \ldots \times 3 \times 2 \times 1)}$$

$$= \frac{n(n-1)}{2}$$

Hobson's choice

Thomas Hobson (1544–1631) was a stable owner who offered customers the choice of either taking the horse in the stall nearest the door – that is, the one he offered them – or taking no horse at all. Understanding this in the context of combinations, the question asked is: how many ways can you pick one horse when there is only one horse to pick from, so $^{1}C_1$?

We know that there is one choice of horse so:

$$^{1}C_1 = 1$$

$$\frac{1!}{1!(1-1)!} = 1$$

$$\frac{1!}{1!0!} = 1$$

For this calculation to make sense, we can conclude that **0! = 1** and **1! = 1**.

The domain of the factorial function is $\mathbb{W}$, the set of whole numbers.

This textbook uses the notation $\binom{n}{r}$ rather than $^{n}C_r$. However, both forms of notation should be understood.

Exercise 1F

1 a Evaluate, using a calculator:

    i $^{7}C_4$ ii $^{3}C_2$ iii $^{10}C_5$ iv $^{5}C_1$

  b Calculate, without a calculator:

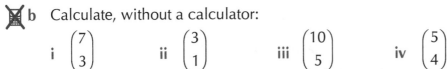

    i $\binom{7}{3}$ ii $\binom{3}{1}$ iii $\binom{10}{5}$ iv $\binom{5}{4}$

  c Calculate $\binom{6}{4}$ and $\binom{6}{2}$. What do you notice? Explain why this is the case.

2 a How many different hands can be dealt from an ordinary pack of 52 cards when playing rummy, in which each player requires 7 cards?

  b A football manager has a squad of 12 players. The manager has to select a five-a-side team, and each selection must include one substitute (so six players in total). How many different teams could the manager select?

c A teacher has to group a class of 15 into groups of 3. How many different groups could the teacher select?

d When the National Lottery started in November 1994, 6 numbers were selected from the numbers 1–49. In October 2015, the rules changed and 6 numbers are now selected from the numbers 1–59. How many more selections exist in draws made after October 2015?

3 Bookmakers offer a full cover bet which consists of doubles (when 2 selections are made), trebles (when 3 selections are made) and accumulators (when 4 or more selections are made). For example, a Trixie involves selecting 3 teams and consists of 4 bets – 3 doubles and a treble. How many bets are there in the following:

a a Yankee (4 teams are selected)

b a Super Yankee (5 teams are selected)

c a Super Heinz (7 teams are selected)

d a Goliath (8 teams are selected)?

Pascal's triangle and the binomial coefficient

Pascal's triangle is a triangular array of numbers known as the **binomial coefficients**. It is named after the French mathematician Blaise Pascal who lived in the seventeenth century. However, many other mathematicians in India, Persia, China, Germany, and Italy had studied the array of numbers many hundreds of years before Pascal.

Pascal's triangle

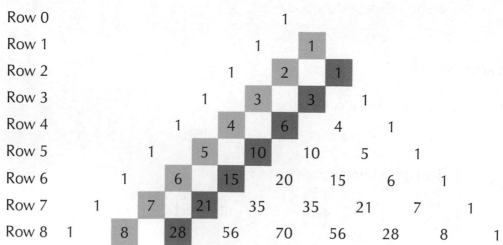

Row 0									1								
Row 1								1		1							
Row 2							1		2		1						
Row 3						1		3		3		1					
Row 4					1		4		6		4		1				
Row 5				1		5		10		10		5		1			
Row 6			1		6		15		20		15		6		1		
Row 7		1		7		21		35		35		21		7		1	
Row 8	1		8		28		56		70		56		28		8		1

Many patterns and features are found in Pascal's triangle:

- **each term in Pascal's triangle comes from adding together the two terms directly above it**. For example, in row 6, the third entry of 15 comes from adding together the two terms directly above it: $5 + 10 = 15$

- the diagonal coloured yellow in the diagram contains the counting (natural) numbers

- the diagonal coloured orange in the diagram contains the triangular numbers

- if the numbers in each row are added together, the result is a power of 2 (where the power is the row number); for example:

$$\text{row 4: } 1 + 4 + 6 + 4 + 1 = 16 \ (= 2^4)$$
$$\text{row 6: } 1 + 6 + 15 + 20 + 15 + 6 + 1 = 64 \ (= 2^6)$$

Notation

Any number in Pascal's triangle can be identified by its row (n) and number of entry in that row (r), with the first entry being $r = 0$. Each term is then expressed as $\binom{n}{r}$, for example, $\binom{5}{3}$ is the entry in row 5 and $r = 3$.

Using a calculator, $\binom{5}{3} = {}^5C_3 = 10$

or without a calculator, $\binom{5}{3} = \dfrac{5!}{3!2!} = \dfrac{5 \times 4 \times 3 \times 2 \times 1}{3 \times 2 \times 1 \times 2 \times 1} = \dfrac{5 \times 4}{2 \times 1} = 10$

So Pascal's triangle can also be written as:

$$\binom{0}{0}$$

$$\binom{1}{0} \quad \binom{1}{1}$$

$$\binom{2}{0} \quad \binom{2}{1} \quad \binom{2}{2}$$

$$\binom{3}{0} \quad \binom{3}{1} \quad \binom{3}{2} \quad \binom{3}{3}$$

$$\binom{4}{0} \quad \binom{4}{1} \quad \binom{4}{2} \quad \binom{4}{3} \quad \binom{4}{4}$$

$$\binom{5}{0} \quad \binom{5}{1} \quad \binom{5}{2} \quad \binom{5}{3} \quad \binom{5}{4} \quad \binom{5}{5} \text{ ... and so on.}$$

Properties

Looking at row 4 of Pascal's triangle, the second entry and the second last entry are both 4, so:

$$\binom{4}{1} = \binom{4}{3}$$

Similarly, looking at row 6, the third entry and the third last entry are both 15, so:

$$\binom{6}{2} = \binom{6}{4}$$

In general:

$$\binom{n}{r} = \binom{n}{n-r}$$

Proof

The result $\binom{n}{r} = \binom{n}{n-r}$ can be proved as shown here.

$$\binom{n}{n-r} = \frac{n!}{(n-r)!(n-(n-r))!} \qquad \text{By definition}$$

$$= \frac{n!}{(n-r)!(n-n+r)!}$$

$$= \frac{n!}{(n-r)!\,r!}$$

$$= \frac{n!}{r!(n-r)!}$$

$$= \binom{n}{r}$$

Properties

Given that each term in Pascal's triangle comes from adding together the two terms directly above it, then, for example:

$$\binom{5}{2} + \binom{5}{3} = \binom{6}{3}$$

In general:

$$\binom{n}{r-1} + \binom{n}{r} = \binom{n+1}{r}$$

Proof

The result $\binom{n}{r-1} + \binom{n}{r} = \binom{n+1}{r}$ can be proved as shown here.

$$\binom{n}{r-1} + \binom{n}{r} = \frac{n!}{(r-1)!(n-(r-1))!} + \frac{n!}{r!(n-r)!}$$

By definition of $\binom{n}{r}$

$$= \frac{n!}{(r-1)!(n-r+1)!} + \frac{n!}{r!(n-r)!}$$

Simplify denominator in first fraction.

$$= \frac{n!r}{r!(n-r+1)!} + \frac{n!(n-r+1)}{r!(n-r+1)!}$$

To get the same denominator in both fractions, multiply the first fraction by $\frac{r}{r}$ and the second fraction by $\frac{n-r+1}{n-r+1}$

Remember that $r \times (r-1)! = r!$ and $(n-r+1) \times (n-r)! = (n-r+1)!$

$$= \frac{n!r + n!(n-r+1)}{r!(n-r+1)!}$$

Denominators of both fractions are the same so add the numerators.

$$= \frac{n!(r+n-r+1)}{r!(n-r+1)!}$$

Common factor of $n!$

$$= \frac{n!(n+1)}{r!(n-r+1)!}$$

Simplify numerator.

$$= \frac{(n+1)!}{r!((n+1)-r)!}$$

Rearrange denominator into required form and use $n! \times (n+1) = (n+1)!$

$$= \binom{n+1}{r}$$

Example 1.7

Solve $\binom{n}{2} = 6$

$$\binom{n}{2} = 6$$

$$\frac{n!}{2!(n-2)!} = 6$$

By definition of $\binom{n}{r}$

$$\frac{n!}{2(n-2)!} = 6$$

Since $2! = 2 \times 1 = 2$

$$\frac{n!}{(n-2)!} = 12$$

$$\frac{n \times (n-1) \times (n-2) \times (n-3) \times \dots \times 3 \times 2 \times 1}{(n-2) \times (n-3) \times \dots \times 3 \times 2 \times 1} = 12$$

By definition of $n!$

$$n(n - 1) = 12$$

Cancel terms on the LHS.

$$n^2 - n - 12 = 0$$

$$(n - 4)(n + 3) = 0$$

Solve the quadratic.

$$n = -3 \text{ or } 4$$

So $n = 4$ (since $n \in \mathbb{N}$)

Example 1.8

Solve $\begin{pmatrix} n \\ 1 \end{pmatrix} + \begin{pmatrix} n \\ 2 \end{pmatrix} = 28$

$$\begin{pmatrix} n \\ 1 \end{pmatrix} + \begin{pmatrix} n \\ 2 \end{pmatrix} = 28$$

$$\begin{pmatrix} n + 1 \\ 2 \end{pmatrix} = 28$$

Since $\begin{pmatrix} n \\ r - 1 \end{pmatrix} + \begin{pmatrix} n \\ r \end{pmatrix} = \begin{pmatrix} n + 1 \\ r \end{pmatrix}$

$$\frac{(n + 1)!}{2!(n - 1)!} = 28$$

By definition of $\begin{pmatrix} n \\ r \end{pmatrix}$

$$\frac{(n + 1)!}{2(n - 1)!} = 28$$

Since $2! = 2 \times 1 = 2$

$$\frac{(n + 1)!}{(n - 1)!} = 56$$

$$\frac{(n + 1) \times n \times (n - 1) \times (n - 2) \times \ldots \times 3 \times 2 \times 1}{(n - 1) \times (n - 2) \times \ldots \times 3 \times 2 \times 1} = 56$$

By definition of $n!$

$$(n + 1)n = 56$$

Cancel terms on the LHS.

$$n^2 + n - 56 = 0$$

$$(n + 8)(n - 7) = 0$$

Solve the quadratic.

$$n = -8 \text{ or } 7$$

So $n = 7$ (since $n \in \mathbb{N}$)

Exercise 1G

1   a   Calculate $^5C_2$ and hence state the value of $^5C_3$.

   b   Calculate $^{10}C_2$ and hence state the value of $^{10}C_8$.

   c   Calculate $^{12}C_4$ and hence state the value of $^{12}C_8$.

   d   Copy and complete the statement $\dbinom{n}{r} = \dbinom{n}{\cdots}$

   e   Copy and complete the statement $\dbinom{2n}{r} = \dbinom{\cdots}{\cdots}$

 2   Solve these equations:

   a $\dbinom{n}{2} = 3$ b $\dbinom{n}{2} = 10$ c $\dbinom{n}{2} = 21$ d $\dbinom{n}{2} = 55$

   e $\dbinom{2n}{2} = 6$ f $\dbinom{2n}{2} = 15$ g $\dbinom{n}{3} = 10$ h $\dbinom{n}{3} = 35$

 3   Solve these equations.

   a $\dbinom{n}{n-2} = 3$ b $\dbinom{n}{n-2} = 28$ c $\dbinom{n}{n-2} = 66$

$$\dbinom{n}{r} = \dbinom{n}{n-r}$$

 4   Solve these equations.

   a $\dbinom{n}{1} + \dbinom{n}{2} = 6$ b $\dbinom{n}{1} + \dbinom{n}{2} = 15$

$$\dbinom{n}{r-1} + \dbinom{n}{r} = \dbinom{n+1}{r}$$

   c $\dbinom{2n-1}{1} + \dbinom{2n-1}{2} = 45$ d $\dbinom{n+1}{1} + \dbinom{n+1}{2} = 10$

 5   Using $\dbinom{n}{r} = \dfrac{n!}{r!(n-r)!}$, show that:

   a $\dbinom{n}{2} + \dbinom{n}{3} = \dbinom{n+1}{3}$ b $\dbinom{n+1}{2} - \dbinom{n}{1} = \dbinom{n}{2}$ c $\dbinom{n^2}{n^2 - 1} = n^2$

   d $\dbinom{n+1}{3} - \dbinom{n}{3} = \frac{1}{2}n(n-1)$ e $\dbinom{n+2}{3} + \dbinom{n+1}{3} = \frac{1}{6}n(n+1)(2n+1)$

The binomial theorem

The **binomial theorem** describes the algebraic expansion of powers of a binomial expression.

The polynomial $(x + y)^n$ can be expanded into a sum involving terms of the form $ax^b y^c$, where the exponents b and c are non-negative integers with $b + c = n$, and a is the **coefficient**.

For example:

$$(x + y)^3 = x^3 + 3x^2 y + 3xy^2 + y^3$$

The coefficient a in the term of $ax^b y^c$ is the binomial coefficient $\binom{n}{r}$

Exploring the binomial theorem

Expand $(1 + x)^n$:

$n = 0 \quad (1 + x)^0 = 1$

$n = 1 \quad (1 + x)^1 = 1 + x$

$n = 2 \quad (1 + x)^2 = 1 + 2x + x^2$

$n = 3 \quad (1 + x)^3 = 1 + 3x + 3x^2 + x^3$

$n = 4 \quad (1 + x)^4 = 1 + 4x + 6x^2 + 4x^3 + x^4$

$n = 5 \quad (1 + x)^5 = 1 + 5x + 10x^2 + 10x^3 + 5x^4 + x^5$

$n = 6 \quad (1 + x)^6 = 1 + 6x + 15x^2 + 20x^3 + 15x^4 + 6x^5 + x^6$

Can you spot any patterns?

What about the coefficients? Have you seen them before?

What would the expansion be for $n = 7$?

Similarly, expand $(x + y)^n$:

$n = 0 \quad (x + y)^0 = 1$

$n = 1 \quad (x + y)^1 = x + y$

$n = 2 \quad (x + y)^2 = x^2 + 2xy + y^2$

$n = 3 \quad (x + y)^3 = x^3 + 3x^2 y + 3xy^2 + y^3$

$n = 4 \quad (x + y)^4 = x^4 + 4x^3 y + 6x^2 y^2 + 4xy^3 + y^4$

$n = 5 \quad (x + y)^5 = x^5 + 5x^4 y + 10x^3 y^2 + 10x^2 y^3 + 5xy^4 + y^5$

$n = 6 \quad (x + y)^6 = x^6 + 6x^5 y + 15x^4 y^2 + 20x^3 y^3 + 15x^2 y^4 + 6xy^5 + y^6$

Can you spot the patterns in this set of expansions?

What about the coefficients? Have you seen them before?

- The coefficients in the expansions are from Pascal's triangle. This can be written using $\begin{pmatrix} n \\ r \end{pmatrix}$ notation. The row of Pascal's triangle that is used corresponds to the power of the bracket that is being expanded.
- As the powers of x **decrease**, the powers of y **increase**.
- The sum of the powers of each term equals the value of the power of the bracket being expanded, i.e. n.

Consider:

$$(x + y)^5 = x^5 + 5x^4y + 10x^3y^2 + 10x^2y^3 + 5xy^4 + y^5$$

We can write this as:

$$(x + y)^5 = \begin{pmatrix} 5 \\ 0 \end{pmatrix}x^5y^0 + \begin{pmatrix} 5 \\ 1 \end{pmatrix}x^4y^1 + \begin{pmatrix} 5 \\ 2 \end{pmatrix}x^3y^2 + \begin{pmatrix} 5 \\ 3 \end{pmatrix}x^2y^3 + \begin{pmatrix} 5 \\ 4 \end{pmatrix}x^1y^4 + \begin{pmatrix} 5 \\ 5 \end{pmatrix}x^0y^5$$

In general (and since $b^0 = 1$ and $a^0 = 1$):

$$(a + b)^n = \begin{pmatrix} n \\ 0 \end{pmatrix}a^n + \begin{pmatrix} n \\ 1 \end{pmatrix}a^{n-1}b^1 + \begin{pmatrix} n \\ 2 \end{pmatrix}a^{n-2}b^2 + ... + \begin{pmatrix} n \\ r \end{pmatrix}a^{n-r}b^r + ... \begin{pmatrix} n \\ n \end{pmatrix}b^n$$

This can be written using sigma notation (Σ):

$$(a + b)^n = \sum_{r=0}^{n} \begin{pmatrix} n \\ r \end{pmatrix}a^{n-r}b^r$$

The symbol Σ is the Greek capital letter sigma. The symbol means 'the sum of'. The terms that are added together are generated when $r = 0$, $r = 1$, $r = 2$,..., $r = n - 1$, $r = n$ are substituted into the general term, which is the expression after the Σ symbol.

Proof of the binomial theorem (using proof by induction)

For further details on proof by induction, see Chapter 11.

Base case

Consider $n = 1$:

$$\text{LHS} = (x + y)^1 = x + y \qquad \text{RHS} = \binom{1}{0}x^1 + \binom{1}{1}y^1 = x + y$$

LHS = RHS

So the statement is true when $n = 1$.

Inductive step

Assume that the theorem is true for some value $n = k$, $k \in \mathbb{N}$:

$$(x + y)^k = \binom{k}{0}x^k + \binom{k}{1}x^{k-1}y^1 + \binom{k}{2}x^{k-2}y^2 + \ldots + \binom{k}{r}x^{k-r}y^r + \ldots + \binom{k}{k}y^k$$

We need to prove the theorem is true for $n = k + 1$, so consider when $n = k + 1$:

$$(x + y)^{k+1} = (x + y)(x + y)^k$$

$$= (x + y)\left(\binom{k}{0}x^k + \binom{k}{1}x^{k-1}y^1 + \binom{k}{2}x^{k-2}y^2 + \ldots + \binom{k}{r}x^{k-r}y^r + \ldots + \binom{k}{k}y^k\right)$$

$$= \binom{k}{0}x^{k+1} + \binom{k}{1}x^k y^1 + \binom{k}{2}x^{k-1}y^2 + \ldots + \binom{k}{r}x^{k-r+1}y^r + \ldots + \binom{k}{k}xy^k$$

$$+ \binom{k}{0}x^k y + \binom{k}{1}x^{k-1}y^2 + \binom{k}{2}x^{k-2}y^2 + \ldots + \binom{k}{r-1}x^{k-r+1}y^r$$

$$+ \binom{k}{r}x^{k-r}y^{r+1} + \ldots + \binom{k}{k}y^{k+1}$$

$$= \binom{k}{0}x^{k+1} + \left(\binom{k}{1} + \binom{k}{0}\right)x^k y + \left(\binom{k}{2} + \binom{k}{1}\right)x^{k-1}y^2 + \ldots$$

$$+ \left(\binom{k}{r} + \binom{k}{r-1}\right)x^{k-r+1}y^r + \ldots + \binom{k}{k}y^{k+1}$$

Given that $\binom{n}{r-1} + \binom{n}{r} = \binom{n+1}{r}$, then:

$$(x + y)^{k+1} = \binom{k+1}{0}x^{k+1} + \binom{k+1}{1}x^{(k+1)-1}y^1 + \binom{k+1}{2}x^{(k+1)-2}y^2 + \ldots$$

$$+ \binom{k+1}{r}x^{(k+1)-r}y^r + \ldots + \binom{k+1}{k+1}y^{k+1}$$

as required (since this is the binomial theorem for $n = k + 1$).

So if the theorem is true for $n = k$ then it is true for $n = k + 1$ but since it is also true for $n = 1$, then by induction, it is true for all $n \in \mathbb{N}$.

Example 1.9

Expand $(1 + x)^4$ using the binomial theorem.

$$(1 + x)^4 = \binom{4}{0}x^4 + \binom{4}{1}x^3y^1 + \binom{4}{2}x^2y^2 + \binom{4}{3}xy^3 + \binom{4}{4}y^4$$

$$= x^4 + 4x^3y + 6x^2y^2 + 4xy^3 + y^4$$

Example 1.10

Expand $(3x - 2y)^3$ using the binomial theorem.

$$(3x + (-2y))^3 = \binom{3}{0}(3x)^3 + \binom{3}{1}(3x)^2(-2y)^1 + \binom{3}{2}(3x)^1(-2y)^2 + \binom{3}{3}(-2y)^3$$

$$= 27x^3 + 3(9x^2)(-2y) + 3(3x)(4y^2) + (-8y^3)$$

$$= 27x^3 - 54x^2y + 36xy^2 - 8y^3$$

Example 1.11

What is the coefficient of x^6 in the expansion of $(3 - x)^9$?

You only need to consider the term with x^6 in it so there is no need to do the whole expansion. In this case, you use the fact that each term in a binomial expansion has the form $\binom{n}{r}x^{n-r}y^r$

$$\binom{9}{6}3^{9-6}(-x)^6 = \binom{9}{6}3^3(-1)^6x^6$$

$n = 9$ (the power of the bracket in the expansion); $r = 6$ (the power of the value of x for which the coefficient is needed)

$$= \binom{9}{6}3^3x^6$$

The coefficient of x^6 is

$$\frac{9!}{6!3!} \times 3^3 = 84 \times 27$$

$$= 2268$$

Example 1.12

Write down and simplify the general term in the expansion of $\left(4x^2 + \dfrac{2}{x}\right)^9$

Hence, obtain the term independent of x.

> The word 'Hence' means that you are expected to use your work on the general term in the first part of the question to answer the second part of the question. If you were to find the term independent of x by expanding the expression fully, you would not gain any marks in an exam.

$$\binom{9}{r}\left(4x^2\right)^{9-r}\left(\frac{2}{x}\right)^r = \binom{9}{r}(4)^{9-r}\left(x^2\right)^{9-r}\left(\frac{2^r}{x^r}\right)$$

> A term in the expansion of $(x+y)^n$ has the form
> $$\binom{n}{r}x^{n-r}y^r$$

$$= \binom{9}{r}(4)^{9-r}\left(2^r\right)\left(x^{18-2r}\right)\left(\frac{1}{x^r}\right)$$

> Collect numerical terms together and algebraic terms together.

$$= \binom{9}{r}(4)^{9-r}\left(2^r\right)\left(x^{18-3r}\right)$$

> Simplify algebraic terms.

$$= \binom{9}{r}\left(2^2\right)^{9-r}\left(2^r\right)\left(x^{18-3r}\right)$$

> Start to simplify numerical terms.

$$= \binom{9}{r}\left(2^{18-2r}\right)\left(2^r\right)\left(x^{18-3r}\right)$$

$$= \binom{9}{r}2^{18-r}x^{18-3r}$$

> Simplify fully.

$$18 - 3r = 0$$

> The term independent of x occurs at x^0

$$r = 6$$

Hence, the term independent of x is:

$$\binom{9}{6}\left(2^{18-6}\right)\left(x^{18-3\times6}\right) = \binom{9}{6}\left(2^{12}\right)\left(x^0\right)$$

$$= 84 \times 4096$$

$$= 344\,064$$

Example 1.13

Expand $(1 - 2x + x^2)^3$ using the binomial theorem.

> The binomial theorem is used to expand brackets with two terms in them, so to use it in this example, you need to rewrite the expression as $((1 - 2x) + x^2)^3$

$$\left(1 - 2x + x^2\right)^3 = \left((1 - 2x) + x^2\right)^3$$

$$= \binom{3}{0}(1 - 2x)^3 + \binom{3}{1}(1 - 2x)^2\left(x^2\right)^1 + \binom{3}{2}(1 - 2x)^1\left(x^2\right)^2 + \binom{3}{3}\left(x^2\right)^3$$

$$= \left(1 - 6x + 12x^2 - 8x^3\right) + 3\left(1 - 4x + 4x^2\right)x^2 + 3(1 - 2x)x^4 + x^6$$

$$= 1 - 6x + 15x^2 - 20x^3 + 15x^4 - 6x^5 + x^6$$

Exercise 1H

★ 1   Expand and simplify these expressions:

a $(a + b)^4$
b $(a - b)^3$
c $(1 + 2x)^5$
d $(2 - 3m)^4$

e $(2a + 3b)^4$
f $(x^2 - y^3)^4$
g $(5 - 3x)^5$
h $(5f + 2g)^3$

2   Expand these expressions:

a $\left(x + \dfrac{1}{x}\right)^3$
b $\left(2m - \dfrac{1}{m}\right)^4$
c $\left(z + \dfrac{1}{z}\right)^6$

d $\left(3x - \dfrac{1}{2x}\right)^6$
e $\left(x^2 - \dfrac{3}{x}\right)^5$

★ 3   In each of these expressions:

i   find and simplify the general term

ii   find the coefficient of the term in square brackets.

a $(x + 5)^4 \left[x^3\right]$
b $(4 + x)^{12} \left[x^8\right]$
c $(5 + 3x)^8 \left[x^3\right]$

d $(1 - 2x)^7 \left[x^4\right]$
e $(3 + 2x)^{18} \left[x^{12}\right]$

4   Find the term independent of x in each expression.

a $\left(2x - \dfrac{1}{x}\right)^{10}$
b $\left(3x^2 - \dfrac{2}{x}\right)^{15}$
c $\left(\dfrac{4}{x^2} - 2x\right)^9$

5   a   Expand $(1 + x)(2 + x)^2$

b   Expand $(1 + x)(2 + x)^5$

c   State the coefficient of x^3 in the expansion of $(3 + 2x)(x + 3)^4$

6   Expand $(1 + x - x^2)^3$

> $(1 + x - x^2)$ can be rewritten as $((1 + x) - x^2)$; you can now apply the binomial theorem because there are two terms in the bracket.

Approximations

The binomial theorem can be used to find approximate values of functions or calculations.

Example 1.14

Determine the first four terms of the expansion $(1 + 3x)^{12}$ in ascending powers of x. Use this expansion to find an approximation for $1 \cdot 003^{12}$ and $0 \cdot 997^{12}$.

$$(1 + 3x)^{12} = 1 + \binom{12}{1}(3x)^1 + \binom{12}{2}(3x)^2 + \binom{12}{3}(3x)^3 + \dots$$

$$= 1 + (12)(3x) + (66)(9x^2) + (220)(27x^3) + \dots$$

$$= 1 + 36x + 594x^2 + 5940x^3 + \dots$$

Let $x = 0 \cdot 001$:

> Select the value of x so that $1 + 3x = 1 \cdot 003$

then $1 \cdot 003^{12} = (1 + 3(0 \cdot 001))^{12}$

$$= 1 + 36(0 \cdot 001) + 594(0 \cdot 001)^2 + 5940(0 \cdot 001)^3 + \dots$$

$$\approx 1 \cdot 036\,599\,94$$

Let $x = -0 \cdot 001$:

> Select the value of x so that $1 + 3x = 0 \cdot 997$

then $0 \cdot 997^{12} = (1 + 3(-0 \cdot 001))^{12}$

$$= 1 + 36(-0 \cdot 001) + 594(-0 \cdot 001)^2 + 5940(-0 \cdot 001)^3 + \dots$$

$$\approx 0 \cdot 964\,588\,06$$

Small values of x

When x is small, then the greater the power of x, the smaller its value and hence its contribution to the total sum becomes negligible. This means that the sum of the first few terms in the expansion will give a good approximation for the required number.

When deciding how many terms are required in an expansion to obtain a good approximation:

- first, consider the value of x: the smaller the value the fewer terms are needed to obtain a good approximation

- second, consider the specified degree of accuracy: if the number is to be calculated to a certain number of decimal places or significant figures, then an answer correct to 3 decimal places requires fewer terms than an answer that requires an answer correct to 5 decimal places.

Exercise 1I

★ 1 For each expansion:

    i write down the first four terms

    ii substitute an appropriate value of x to find an approximation to the given term.

    a **i** $(1 + x)^{20}$ **ii** $(1·01)^{20}$

    b **i** $(1 + 2x)^{8}$ **ii** $(1·02)^{8}$

    c **i** $(1 + 2x)^{12}$ **ii** $(1·02)^{12}$

    d **i** $(1 - 4x)^{10}$ **ii** $(0·96)^{10}$

    e **i** $(1 + x)^{10}$ **ii** $(0·998)^{10}$

    f **i** $\left(1 - \dfrac{x}{10}\right)^{7}$ **ii** $(0·99)^{7}$

    g **i** $\left(2 + \dfrac{x}{5}\right)^{9}$ **ii** $(2·1)^{9}$

2 Using the first four terms of the appropriate binomial expansion, find an approximation for:

    a $(1·01)^{10}$ **b** $(0·98)^{10}$

Chapter review

1 Express these algebraic fractions in partial fractions:

    a $\dfrac{5}{(x + 2)(x + 3)}$ **b** $\dfrac{x + 3}{(x + 1)^{2}(2x + 1)}$ **c** $\dfrac{2x + 1}{(x + 1)(x^{2} + 2x + 5)}$

    d $\dfrac{9 - 5x}{(x - 1)(x - 2)(x - 3)}$ **e** $\dfrac{3x^{2} - 10x - 24}{x^{3} - 4x}$ **f** $\dfrac{4x^{2} - 3x + 2}{2x^{2} - x - 1}$

    g $\dfrac{x^{3} - x^{2} - x - 3}{x^{2} - x}$

2 **a** Solve $\dbinom{n}{2} = 21$ **b** Solve $\dbinom{n}{3} = 10$ **c** Solve $\dbinom{n}{1} + \dbinom{n}{2} = 15$

    d Show that $\dbinom{n}{3} + \dbinom{n - 2}{3} = \frac{1}{3}(n - 2)(n^{2} - 4n + 6)$

3 Use the binomial theorem to expand these expressions:

    a $(a + b)^{4}$ **b** $(3x - 2y)^{3}$ **c** $\left(x^{2} - \dfrac{3}{x}\right)^{5}$

4 **a** Write down and simplify the general term in the expansion of $(2 - 3x)^{7}$
Hence, or otherwise, obtain the term in x^{5}.

    b Write down and simplify the general term in the expansion of $\left(x^{2} - \dfrac{3}{x}\right)^{6}$
Hence, or otherwise, obtain the term in x^{3}.

5 Use the binomial expansion to calculate an approximate value for:

    a $1·1^{4}$ **b** $0·97^{3}$

Give your answers correct to 3 significant figures.

- I can express a rational function in partial fractions where there are linear factors in the denominator. ★ Exercise 1A Q2

- I can express a rational function in partial fractions where there is an irreducible quadratic in the denominator. ★ Exercise 1B Q2, Q3

- I can express a rational function in partial fractions where there is a repeated linear factor in the denominator. ★ Exercise 1C Q2

- I can use algebraic division to make proper rational functions. ★ Exercise 1D Q2 ★ Exercise 1E Q1

- I can use the factorial function, $n!$, and the notation and formula for the binomial coefficient, $\binom{n}{r}$ ★ Exercise 1G Q2, Q4, Q5

- I can use the binomial theorem to expand expressions of the form $(a + b)^n$ for $n \in \mathbb{N}$. ★ Exercise 1H Q1

- I can use the general term for a binomial expansion to find a specific term in the expansion. ★ Exercise 1H Q3

- I can use the binomial theorem to obtain an approximation for a number to a given power. ★ Exercise 1I Q1

2 Differentiation

This chapter will show you how to:

- differentiate functions using the chain rule
- differentiate exponential and logarithmic functions
- differentiate functions given in the form of a product and in the form of a quotient
- differentiate to find higher derivatives
- differentiate using standard trigonometric identities
- differentiate inverse trigonometric functions
- find the first and second derivatives of functions defined implicitly
- use logarithmic differentiation
- find the derivatives of functions defined parametrically
- apply differentiation to problems in context.

You should already know how to:

- differentiate an algebraic function which is, or can be simplified to, an expression in powers of x using $\dfrac{d}{dx}(x^n) = nx^{n-1}$
- use a nature table to determine the nature of a stationary point
- differentiate $k\sin x$ and $k\cos x$ using $\dfrac{d}{dx}(\sin x) = \cos x$ and $\dfrac{d}{dx}(\cos x) = -\sin x$
- differentiate a composite function (a function of a function) using the chain rule:

$$y = f\big(g(x)\big) \Rightarrow \frac{dy}{dx} = f'\big(g(x)\big) \times g'(x)$$

Differentiating from first principles

Look at the diagram.

Given the points $T(x, f(x))$ and $M((x+h), f(x+h))$,

the gradient of TM, $m_{TM} = \dfrac{f(x+h) - f(x)}{h}$

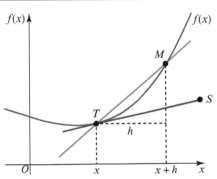

As $h \to 0$, M approaches T along the curve.

As $h \to 0$, the gradient of TM approaches the gradient of TS, the tangent to the curve at T.

As $h \to 0$, the gradient of the tangent at T, $f'(x)$, can be defined as:

$$f'(x) = \lim_{h \to 0}\left[\frac{f(x+h) - f(x)}{h}\right]$$

This is the formula we use when differentiating from **first principles**.

Example 2.1

Find $f'(x)$ of $3x^2$

Let $f(x) = 3x^2$

$$f'(x) = \lim_{h \to 0} \left[\frac{f(x+h) - f(x)}{h} \right]$$

Start with the formula to find $f'(x)$.

$$= \lim_{h \to 0} \left[\frac{3(x+h)^2 - 3x^2}{h} \right]$$

Substitute $f(x) = 3x^2$ and $f(x+h) = 3(x+h)^2$

$$= \lim_{h \to 0} \left[\frac{3x^2 + 6xh + 3h^2 - 3x^2}{h} \right] = \left[\frac{6xh + 3h^2}{h} \right]$$

Multiply out the brackets in the numerator and simplify.

$$= \lim_{h \to 0} \left[\frac{h(6x + 3h)}{h} \right]$$

Cancel out the common factor of h.

$$= \lim_{h \to 0} \left[(6x + 3h) \right] = 6x$$

As h tends to zero, $3h$ also tends to zero and $f'(x)$ tends to $6x$.

Example 2.2

Find $f'(x)$ of $\sin x$

$$f'(x) = \lim_{h \to 0} \left[\frac{f(x+h) - f(x)}{h} \right]$$

$$= \lim_{h \to 0} \left[\frac{\sin(x+h) - \sin x}{h} \right]$$

Substitute $f(x) = \sin x$

$$= \lim_{h \to 0} \left[\frac{\sin x \cos h + \cos x \sin h - \sin x}{h} \right]$$

Expand the numerator using $\sin(A + B) = \sin A \cos B + \cos A \sin B$

$$= \lim_{h \to 0} \left[\frac{\sin x(\cos h - 1) + \cos x \sin h}{h} \right]$$

Factorise the numerator.

$$= \sin x \times \lim_{h \to 0} \left[\frac{(\cos h - 1)}{h} \right] + \cos x \times \lim_{h \to 0} \left[\frac{\sin h}{h} \right]$$

Separate the numerator into two fractions and remove terms in x which are not affected as h tends to zero.

$$\lim_{h \to 0} \left[\frac{(\cos h - 1)}{h} \right] = 0 \text{ and } \lim_{h \to 0} \left[\frac{\sin h}{h} \right] = 1$$

These are two famous limits from calculus which can be proved using the squeeze theorem.

$$f'(x) = \sin x \times 0 + \cos x \times 1 = \cos x$$

Substitute these limits.

Differentiation using the chain rule

The **chain rule** is used to differentiate a function of a function such as $f(g(x))$.

Differentiate the f function first, then differentiate the g function, then multiply both derivatives together.

Applying the chain rule:

$$\frac{dy}{dx} = \frac{d}{dx}f(g(x)) = f'(g(x)) \times g'(x)$$

An alternative method for the chain rule is to let $g(x) = u$ so $y = f(u)$:

$$\frac{dy}{dx} = \frac{dy}{du} \times \frac{du}{dx}$$

Example 2.3

Find $\dfrac{d}{dx}(3x - 2)^4$

Method 1

$$\frac{dy}{dx} = 4(3x - 2)^{4-1} \times \frac{d}{dx}(3x - 2)$$

Differentiate the index power of the bracket, then differentiate the function inside the bracket.

$$= 4(3x - 2)^{4-1} \times 3$$

Multiply both derivatives together.

$$= 12(3x - 2)^3$$

Simplify completely.

Method 2

Let $u = (3x - 2)$ and so $y = u^4$

$$\frac{du}{dx} = 3 \qquad\qquad \frac{dy}{du} = 4u^3$$

Differentiate u with respect to x.
Differentiate y with respect to u.

$$\frac{dy}{dx} = \frac{dy}{du} \times \frac{du}{dx}$$

State the chain rule.

$$= 4u^3 \times 3 = 12u^3$$

Multiply both derivatives together.

$$= 12(3x - 2)^3$$

Substitute for u.

Example 2.4

Find $\dfrac{d}{dx}\cos(\sin x)$

Method 1

$\dfrac{dy}{dx} = -\sin(\sin x) \times \dfrac{d}{dx}(\sin x)$ ●————

> Differentiate the cosine function, then differentiate the function inside the bracket, sin x.

$\qquad = -\sin(\sin x) \times \cos x$

$\qquad = -\sin(\sin x)\cos x$

Method 2

Let $u = \sin x$ and so $y = \cos u$

$\dfrac{du}{dx} = \cos x \qquad \dfrac{dy}{du} = -\sin u$

$\dfrac{dy}{dx} = \dfrac{dy}{du} \times \dfrac{du}{dx}$

$\qquad = -\sin u \, \cos x$

$\qquad = -\sin(\sin x)\cos x$

Example 2.5

Calculate the derivative of $y = \sin^2(x^2)$.

$\dfrac{dy}{dx} = \dfrac{d}{dx}\left(\sin(x^2)\right)^2$

$\qquad = 2\sin(x^2) \times \dfrac{d}{dx}\sin(x^2)$ ●———— Apply the chain rule once.

$\qquad = 2\sin(x^2) \times \cos(x^2) \times 2x$ ●———— Apply the chain rule again.

$\qquad = 4x\sin(x^2)\cos(x^2)$

Example 2.6

Calculate the derivative of $y = (3x - 2\cos\sqrt{x})^3$.

$\dfrac{dy}{dx} = \dfrac{d}{dx}\left(3x - 2\cos x^{\frac{1}{2}}\right)^3$

$\qquad = 3\left(3x - 2\cos x^{\frac{1}{2}}\right)^2 \times \dfrac{d}{dx}\left(3x - 2\cos x^{\frac{1}{2}}\right)$ ●———— Apply the chain rule once.

$$= 3\left(3x - 2\cos x^{\frac{1}{2}}\right)^2 \times \left(3 + 2\sin x^{\frac{1}{2}} \times \frac{1}{2}x^{-\frac{1}{2}}\right)$$

> Apply the chain rule again when differentiating $2\cos x^{\frac{1}{2}}$.

$$= 3\left(3x - 2\cos\sqrt{x}\right)^2 \times \left(3 + \frac{\sin\sqrt{x}}{\sqrt{x}}\right)$$

Exercise 2A

In Questions 1–7, differentiate each function $f(x)$ with respect to x.

1 a $x^4 + \cos x$

   b $\sin 3x + x^5$

   c $\dfrac{2}{x^4} + 5x^2$

   d $\dfrac{3}{\pi^2} + \cos\left(2x + \dfrac{\pi}{3}\right)$

   e $\sin 4x + \cos 4x$

   f $\dfrac{2x^2 + 3x}{x^5}$

   g $\dfrac{1}{2x^4} + 5x^2$

   h $\cos\dfrac{1}{2}x - \dfrac{3}{2x^3}$

   i $(5x + 1)^3$

2 a $(4x^2 - 7x)^3$

   b $\sin(x^2)$

   ★ c $\cos\sqrt[3]{x}$

   d $\cos(x^2 + 3x)$

   e $\sin^3 x$

   f $\dfrac{3}{(1 - 4x)^2}$

   g $\dfrac{\cos^2 3x}{3}$

   h $\dfrac{1}{\sin x}$

   i $\dfrac{1}{\cos x}$

3 a $\sin(\cos x)$

   b $(x^2 + 3x + 1)^4$

   ★ c $\cos(\cos x)$

   d $\sin(\sin x)$

   e $\sin^2 3x$

   f $\cos^2(\sin x)$

   g $(x + \sin 3x)^2$

   ★ h $\cos\left(\dfrac{1}{x^2 + 2x + 1}\right)$

   i $\dfrac{1}{\sin^2(3x + 1)}$

4 $\dfrac{1}{\cos\left(x^2 + x\right)}$

5 $\dfrac{1}{\sin(\cos x)}$

6 $\dfrac{1}{\sqrt{\sin(3x + 2)}}$

7 $\sqrt{\sin^3 2x}$

Differentiating exponential and logarithmic functions

The **natural exponential function**, e^x, and its inverse function, $\ln x$, (where $\ln x = \log_e x$, the natural logarithmic function), are important functions in applications of calculus in the real world. They are used in areas such as probability, radioactive decay and traffic control studies.

$f(x) = e^x$ where $e = e^1 = 2.718\,281\,828$

The Swiss mathematician Leonard Euler was first to use this symbol in his studies of the e function in the eighteenth century.

Differentiating exponential functions

The exponential function e^x is unique. The derivative of e^x is the same as the function itself:

$$\frac{d}{dx}e^x = e^x$$

This is included in the exam formulae list.

Example 2.7

Differentiate $3e^{x^2}$

$$\frac{dy}{dx} = \frac{d}{dx}\left(3e^{x^2}\right)$$

Consider $3e^{(x^2)}$ as $3e^{f(x)}$, and use the chain rule to differentiate as a function of a function.

$$= 3e^{x^2} \times \frac{d}{dx}x^2$$

Differentiate the exponential function first.

$$= 3e^{x^2} \times 2x$$

Differentiate the power x^2

$$= 6xe^{x^2}$$

Differentiating logarithmic functions

The **natural logarithmic function** is the inverse of the exponential function:

$$y = \log_e x \Leftrightarrow y = e^x \text{ so } e^{\log_e x} = x$$

or

$$y = \ln x \Leftrightarrow y = e^x \text{ so } e^{\ln x} = x$$

We can find the standard result for differentiating $\ln x$:

$$e^{\ln x} = x$$

$$\frac{d}{dx}e^{\ln x} = \frac{d}{dx}x$$

Differentiate both sides with respect to x.

$$e^{\ln x}\frac{d}{dx}\ln x = 1$$

Use the chain rule to differentiate the LHS.

$$\frac{d}{dx}\ln x = \frac{1}{e^{\ln x}} = \frac{1}{x}$$

Simplify the LHS.

$$\therefore \frac{d}{dx}\ln x = \frac{1}{x}$$

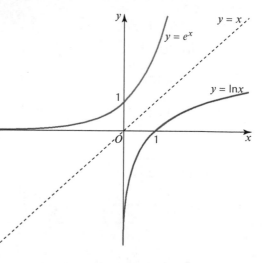

$$\frac{d}{dx}\ln x = \frac{1}{x} \quad \text{and} \quad \frac{d}{dx}e^x = e^x$$

Example 2.8

Differentiate $\ln(2x+3)^3$

Method 1

$$\frac{dy}{dx} = \frac{d}{dx}\ln(2x+3)^3$$

> Apply the chain rule.

$$= 3 \times \frac{1}{2x+3} \times \frac{d}{dx}(2x+3)$$

> Use the laws of logs to reposition the bracket's power. Differentiate the ln function first, and then differentiate the bracket.

$$= 3 \times \frac{1}{2x+3} \times 2$$

$$= \frac{6}{2x+3}$$

Method 2

Let $u = (2x+3)^3$ and so $y = \ln u$

$$\frac{du}{dx} = 3(2x+3)^2 \times 2 \qquad \frac{dy}{du} = \frac{1}{u}$$

$$\frac{dy}{dx} = \frac{dy}{du} \times \frac{du}{dx}$$

$$= \frac{1}{u}3(2x+3)^2 \times 2 = \frac{6(2x+3)^2}{u}$$

$$= \frac{6(2x+3)^2}{(2x+3)^3} = \frac{6}{2x+3}$$

Exercise 2B

1   Evaluate:

a $\dfrac{d}{dx}e^{3x+1}$

b $\dfrac{d}{dx}\left(\ln x - 4x^3\right)$

c $\dfrac{d}{dx}\left(e^{-x} + 4x\right)$

★ d $\dfrac{d}{dx}\left(\ln(x+1) + e^{2x+1}\right)$

e $\dfrac{d}{dx}\left(\dfrac{2}{e^x} - \dfrac{x^3}{3}\right)$

f $\dfrac{d}{dx}\ln(x+2)^2$

g $\dfrac{d}{dx}3e^{x^2+2x}$

h $\dfrac{d}{dx}2e^{\frac{x}{2}}$

i $\dfrac{d}{dx}\ln(\ln x)$

j $\dfrac{d}{dx}\ln\left(x^3+2\right)$

2   For each function y, find $\dfrac{dy}{dx}$

a $y = \sqrt{e^{x^2} - 3}$

b $y = \ln(\cos 2x)$

c $y = e^{\sin 2x}$

Differentiation using the product rule

The **product rule** is used to differentiate two functions which are multiplied together.

$$\frac{d}{dx}(uv) = u\frac{dv}{dx} + v\frac{du}{dx} \quad \text{or} \quad \frac{d}{dx}(uv) = uv' + vu'$$

Example 2.9

Differentiate $y = 6x^2 \cos 3x$

$y = 6x^2 \cos 3x$

$u = 6x^2 \qquad\qquad v = \cos 3x$ — Separate and define the two functions.

$\frac{du}{dx} = 12x \quad \frac{dv}{dx} = -3\sin 3x$ — Differentiate u and v.

$\frac{d}{dx}(uv) = u\frac{dv}{dx} + v\frac{du}{dx}$

$\qquad = 6x^2 \times (-3\sin 3x) + 12x \times \cos 3x$ — Substitute u, v, $\frac{du}{dx}$ and $\frac{dv}{dx}$ to find the derivative of y.

$\qquad = 12x\cos 3x - 18x^2 \sin 3x$ — Write the positive term first and simplify.

$\qquad = 6x(2\cos 3x - 3x\sin 3x)$ — Factorise to simplify further.

Example 2.10

Differentiate $y = (x+4)^5(x-2)^4$

$y = (x+4)^5(x-2)^4$

$u = (x+4)^5 \qquad\qquad v = (x-2)^4$

$\frac{du}{dx} = \frac{d}{dx}(x+4)^5 = 5(x+4)^4 \qquad \frac{dv}{dx} = \frac{d}{dx}(x-2)^4 = 4(x-2)^3$

$\frac{d}{dx}(uv) = u\frac{dv}{dx} + v\frac{du}{dx}$

$\qquad = (x+4)^5 \times 4(x-2)^3 + (x-2)^4 \times 5(x+4)^4$ — Substitute.

$\qquad = 4(x+4)^5(x-2)^3 + 5(x+4)^4(x-2)^4$ — Factorise by taking out $(x+4)^4$ and $(x-2)^3$

$\qquad = (x+4)^4(x-2)^3(4(x+4) + 5(x-2))$

$\qquad = (x+4)^4(x+2)^3(4x+16+5x-10)$

$\qquad = (x+4)^4(x-2)^3(9x+6)$

$\qquad = 3(x+4)^4(x-2)^3(3x+2)$

Exercise 2C

Differentiate the functions in Questions 1–2 with respect to x.

1 a $4x \sin 5x$
   b $5x^4(x-2)^6$
   c $\sqrt{x^3} \cos^2 x$

   d $\sqrt{(x-4)} \cos 3x$
   ★ e $\sqrt{x}(2x+7)^{\frac{2}{3}}$
   f $(x-1)^4(3x-2)^5$

   g $x^5 \ln x$
   h $(x^3+1)\cos 2x$
   i $e^{2x}\sin x$

2 a $e^{-x}\ln x$
   ★ b $e^{\frac{1}{x}}\cos^2 x$
   ★ c $\ln x \sqrt{(x-3)}$

   d $5x^2 e^{\sin x}$
   e $6x^2 \cos 3x$
   f $x^2(x^3+5x)^5$

   g $y = \ln(3+\cos 2x)$
   h $y = e^x \sin x^2$

3 Given that $y = e^{\sin x} \ln(3x+2)$, find $\dfrac{dy}{dx}$ when $x = 0$.

 4 Calculate the gradient when $x = 1$ for the function $f(x) = 3x^3 \ln x$

5 A function is defined as $f(t) = e^t \sin t$. Show that the rate of change when $t = \dfrac{\pi}{6}$ can be written as $e^{\frac{\pi}{6}}\left(\dfrac{\sqrt{3}+1}{2}\right)$

 6 If $f(x) = 4x \cos x$, find the exact value of $f'\left(\dfrac{\pi}{2}\right)$.

Differentiation using the quotient rule

The **quotient rule** is used to differentiate one function being divided by another function:

$$\frac{d}{dx}\left(\frac{u}{v}\right) = \frac{v\dfrac{du}{dx} - u\dfrac{dv}{dx}}{v^2} \qquad \text{or} \qquad \frac{d}{dx}\left(\frac{u}{v}\right) = \frac{vu' - uv'}{v^2}$$

Example 2.11

Find $\dfrac{dy}{dx}$ when $y = \dfrac{x^2+3}{\sin 2x}$

$$\frac{dy}{dx} = \frac{d}{dx}\left(\frac{x^2+3}{\sin 2x}\right)$$

$$u = x^2+3 \qquad\qquad v = \sin 2x$$

$$\frac{du}{dx} = 2x \qquad\qquad \frac{dv}{dx} = 2\cos 2x$$

$$\frac{d}{dx}\left(\frac{u}{v}\right) = \frac{v\dfrac{du}{dx} - u\dfrac{dv}{dx}}{v^2}$$

State the quotient rule.

$$= \frac{\sin 2x \times 2x - (x^2 + 3) \times 2\cos 2x}{(\sin 2x)^2}$$

Substitute to find the derivative of y.

$$= \frac{2x\sin 2x - 2\cos 2x(x^2 + 3)}{\sin^2 2x}$$

Example 2.12

Find $\dfrac{dy}{dx}$ when $y = \dfrac{(3x - 1)^5}{\sqrt{x}}$

$$\frac{dy}{dx} = \frac{d}{dx}\frac{(3x - 1)^5}{x^{\frac{1}{2}}}$$

$$u = (3x - 1)^5 \qquad\qquad v = x^{\frac{1}{2}}$$

$$\frac{du}{dx} = 5(3x - 1)^4 \times 3 \qquad \frac{dv}{dx} = \frac{1}{2}x^{-\frac{1}{2}}$$

$$\frac{d}{dx}\left(\frac{u}{v}\right) = \frac{v\dfrac{du}{dx} - u\dfrac{dv}{dx}}{v^2}$$

$$= \frac{x^{\frac{1}{2}} \times 15(3x - 1)^4 - (3x - 1)^5 \times \frac{1}{2}x^{-\frac{1}{2}}}{\left(x^{\frac{1}{2}}\right)^2}$$

$$= \frac{15x^{\frac{1}{2}}(3x - 1)^4 - \frac{1}{2}x^{-\frac{1}{2}}(3x - 1)^5}{x}$$

$$= \frac{\frac{1}{2}x^{-\frac{1}{2}}(3x - 1)^4(30x - (3x - 1))}{x}$$

$$= \frac{(3x - 1)^4(27x + 1)}{2x^{\frac{3}{2}}}$$

Exercise 2D

Differentiate the functions in Questions 1–3 with respect to x.

1 **a** $\tan x$ **b** $\dfrac{\sin x}{\sqrt{x}}$ **c** $\dfrac{x^2}{\cos x}$

   d $\dfrac{\ln x}{x^3}$ ★**e** $\dfrac{\sqrt{x}}{e^{3x}}$ **f** $\dfrac{e^{2x}}{x^3}$

2 **a** $\dfrac{(x-2)^3}{e^x}$ ★**b** $\dfrac{\cos 2x}{\sin^2 x}$ **c** $\dfrac{\sqrt{x}}{(x-3)(x+1)}$

   ★**d** $y = \dfrac{e^x - e^{-x}}{e^x + e^{-x}}$ **e** $y = \ln\sqrt{x^2 + 1}$ ★**f** $y = \ln\left(\dfrac{1+x}{1-x}\right)$

3 **a** $\dfrac{d}{dx}\left(\dfrac{e^x + 1}{e^x - x}\right)$ **b** $y = \dfrac{x}{\ln 7x}$

   c $y = \dfrac{1 + \ln x}{3x}$

> In Q2 parts **e**, **f** see if you can simplify by using the laws of logs before differentiating.

4 Find $f'(5)$ when $f(x) = \dfrac{x^2}{\sqrt{x-1}}$

5 Find $f'(\pi)$ when $f(x) = \dfrac{\cos^2 x}{x^2}$

★**6** Find $\dfrac{dy}{dx}$ when $y = \dfrac{2x^2 + 3x - 1}{x + 1}$ and $x = 2$

7 Find $\dfrac{d}{dx}\cos\left(\dfrac{\ln x}{e^x}\right)$

Higher derivatives

A function $f(x)$ has a derivative $f'(x)$. If $f'(x)$ is also differentiable then its derivative is denoted by $f''(x)$. (We say 'f double dashed of x'.)

Using Leibniz notation:

Leibniz notation	How notation is read
1st derivative $\dfrac{dy}{dx}$	'd y by d x'
2nd derivative $\dfrac{d^2 y}{dx^2}$	'd 2 y by d x squared'
nth derivative $\dfrac{d^n y}{dx^n}$	'd n y by d x n'

This pattern of derivatives can continue as needed for further higher order derivatives.

Example 2.13

Find the first six derivatives of $y = \sin 2x + e^{3x}$

$y = \sin 2x + e^{3x}$

$\dfrac{dy}{dx} = \dfrac{d}{dx}(y) = 3e^{3x} + 2\cos 2x$ — 1st derivative

$\dfrac{d^2y}{dx^2} = \dfrac{d}{dx}\left(\dfrac{dy}{dx}\right) = 9e^{3x} - 4\sin 2x$ — 2nd derivative

$\dfrac{d^3y}{dx^3} = \dfrac{d}{dx}\left(\dfrac{d^2y}{dx^2}\right) = 27e^{3x} - 8\cos 2x$ — 3rd derivative

$\dfrac{d^4y}{dx^4} = \dfrac{d}{dx}\left(\dfrac{d^3y}{dx^3}\right) = 81e^{3x} + 16\sin 2x$ — 4th derivative

$\dfrac{d^5y}{dx^5} = \dfrac{d}{dx}\left(\dfrac{d^4y}{dx^4}\right) = 243e^{3x} + 32\cos 2x$ — 5th derivative

$\dfrac{d^6y}{dx^6} = \dfrac{d}{dx}\left(\dfrac{d^5y}{dx^5}\right) = 729e^{3x} - 64\sin 2x$ — 6th derivative

The function $y = \sin 2x + e^{3x}$ has endless higher derivatives to be found.

Two important uses of the second derivative

Second derivatives can be used to determine the nature of stationary points, and in displacement, velocity and acceleration problems.

Higher derivatives are used in the Maclaurin series (see page 247).

Stationary points

A **stationary point** can be:

- a maximum

- a minimum

- an inflection (which can be a rising inflection or a falling inflection).

The nature of a stationary point is determined using a table of the gradient or by finding the second derivative.

When $\dfrac{d^2y}{dx^2} > 0$, the stationary point is a **minimum** stationary point (point A).

When $\dfrac{d^2y}{dx^2} < 0$, the stationary point is a **maximum** stationary point (point B).

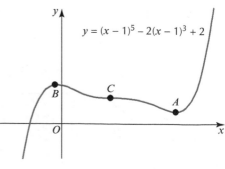

When $\dfrac{d^2y}{dx^2} = 0$, the second derivative may indicate

a **point of inflection** and so the nature table should be used to confirm this (point C).

Example 2.14

Find the stationary points on the curve $y = x^3 - 3x^2 - 9x + 11$ and determine their nature.

$y = x^3 - 3x^2 - 9x + 11$

$\dfrac{dy}{dx} = \dfrac{d}{dx}\left(x^3 - 3x^2 - 9x + 11\right)$ — Find the first derivative.

$= 3x^2 - 6x - 9$

$3x^2 - 6x - 9 = 0$ — Stationary points (SPs) when $\dfrac{dy}{dx} = 0$

$3(x^2 - 2x - 3) = 0$ — Factorise.

$3(x - 3)(x + 1) = 0$ — Solve for x.

$x = 3$ and $x = -1$

$\dfrac{d^2y}{dx^2} = \dfrac{d}{dx}\left(\dfrac{dy}{dx}\right) = \dfrac{d}{dx}\left(3x^2 - 6x - 9\right)$ — Use the second derivative to determine the nature of the SPs.

$= 6x - 6$

When $x = 3$: $\dfrac{dy}{dx} = 12 > 0 \Rightarrow$ minimum SP — Substitute $x = 3$ and $x = -1$ into $6x - 6$. Determine the nature of each SP.

When $x = -1$: $\dfrac{dy}{dx} = -12 < 0 \Rightarrow$ minimum SP

When $x = 3$: $y = -16$: — Substitute $x = 3$ and $x = -1$ into $y = x^3 - 3x^2 - 9x + 11$ to obtain corresponding values for y.

When $x = -1$: $y = 16$

The curve $y = x^3 - 3x^2 + 9x + 11$ has a maximum SP at $(-1, 16)$ and a minimum SP at $(3, -16)$.

Rate of change: displacement, velocity and acceleration problems

A **rate of change** is expressed as a ratio of two quantities with the change in the vertical values being divided by a corresponding change in the horizontal values. Graphically, the gradient or slope of the line represents the rate of change.

Rate of change and gradient are mathematically similar quantities and are calculated from the derived function.

Differentiating a function $f(t)$ gives the rate of change of that function at any point t. The value of the rate of change is calculated by substituting a value for t into the derived function. Two of the most common rates of change are velocity and acceleration.

> **Velocity**, v, is the rate of change of **displacement**, s, with respect to time, t.
>
> $$v(t) = \frac{ds}{dt}$$
>
> Velocity is measured in metres per second, shown as m/s or $\mathrm{ms^{-1}}$
>
> **Acceleration**, a, is the rate of change of velocity, v, with respect to time, t.
>
> $$a(t) = \frac{dv}{dt} = \frac{d^2s}{dt^2}$$
>
> Acceleration is measured in metres per second squared, shown as m/s² or $\mathrm{ms^{-2}}$

Example 2.15

The displacement s of a particle is given by $s = t^4 - 2t^2$

Displacement s is measured in metres and time t is measured in seconds.

Calculate **i** the velocity $v(t)$ and **ii** the acceleration $a(t)$ of this particle after 2 seconds.

i $v(t) = \frac{d}{dt}s(t) = \frac{d}{dt}(t^4 - 2t^2) = 4t^3 - 4t$ — The first derivative gives the equation for velocity.

$v(2) = 4 \times 2^3 - 4 \times 2 = 24\,\mathrm{ms^{-1}}$ — Substitute $t = 2$ into $4t^3 - 4t$

ii $a(t) = \frac{d}{dt}v(t) = \frac{d}{dt}(4t^3 - 4t) = 12t^2 - 4$ — The second derivative gives the equation for acceleration.

$a(2) = 12 \times 4 - 4 = 44\,\mathrm{ms^{-2}}$ — Substitute $t = 2$ into $12t^2 - 4$

Exercise 2E

1 Find the fourth derivative for each function.

　a $y = 2x^4 + x^2 - 1$ 　★ **b** $y = 3x(x^3 + x^2 - 4)$ 　**c** $y = 3e^{2x} + 3\sin 2x$

　d $y = \ln x + 4x^5$ 　**e** $y = (x - 4)\ln(x - 4)$ 　★ **f** $3e^{2x} + \dfrac{1}{x^2}$

2 Sketch the graph of the function $f(x) = (x + 2)(x - 1)^2$

3 Find the coordinates and nature of the stationary points on these curves:

　a $f(x) = x^3 - 81\ln x$ 　**b** $f(x) = e^x - 10x$

4 Use the second derivative to determine the nature of the stationary points of these functions:

　a $y = x^4 - \frac{16}{3}x^3 + 8x^2$ 　★ **b** $y = \sin x + \cos x,\ 0 \leqslant x \leqslant 2\pi$

　c $y = \frac{1}{2}\sin\theta + \sin 2\theta,\ 0 \leq \theta \leq 2\pi$

> You might need to use a trigonometric identity and the quadratic formula here.

5   A particle moves in the xy plane according to the formula
$s(t) = 2t^3 - 27t^2 + 120t - 11$, where the displacement, s, is measured in metres as time, t, passes in seconds.

Calculate the acceleration of this particle after 3 seconds.

★ 6   A vehicle moves according to the formula $s(t) = t^3 + 15t^2 - 30t + 7$, where the displacement, s, is measured in miles as time, t, passes in hours.

Calculate the acceleration of this vehicle after 5 hours.

7   A body is moving in a straight line so that after t seconds its displacement s metres from a fixed point O is given by $s = 9t + 3t^2 - t^3$

   a   Find the initial displacement, velocity and acceleration of the body.

   b   Find the time at which the body is instantaneously at rest.

Derivatives of tan x, cot x, sec x and cosec x

The standard trigonometric functions $\sin x$, $\cos x$ and $\tan x$ have reciprocal functions.

Function	Identity	Graph
cosec$x$ (or cosecant$x$)	$\operatorname{cosec} x = \dfrac{1}{\sin x}$ where $\sin x \neq 0$ $\ldots -2\pi, -\pi, 0 \neq x \neq 0, \pi, 2\pi \ldots$	
sec$x$ (or secant$x$)	$\sec x = \dfrac{1}{\cos x}$ where $\cos x \neq 0$ $\ldots -\dfrac{3\pi}{2}, -\dfrac{\pi}{2} \neq x \neq \dfrac{\pi}{2}, \dfrac{3\pi}{2} \ldots$	
cot$x$ (or cotangent$x$)	$\cot x = \dfrac{1}{\tan x}$ where $\tan x \neq 0$ $\ldots -2\pi, -\pi, 0 \neq x \neq 0, \pi, 2\pi \ldots$	

The derivatives of $\tan x$, $\cot x$, $\sec x$ and $\operatorname{cosec} x$ are shown below, and are provided in the exam formulae list.

f(x)	f′(x)
$\sec x$	$\sec x \tan x$
$\csc x$	$-\csc x \cot x$
$\cot x$	$-\csc^2 x$
$\tan x$	$\sec^2 x$

Examples 2.17 and 2.18 demonstrate how the derivatives of $\sec x$ and $\csc x$ are produced.

Example 2.16

Find $\dfrac{dy}{dx}$ when $y = \sec x$

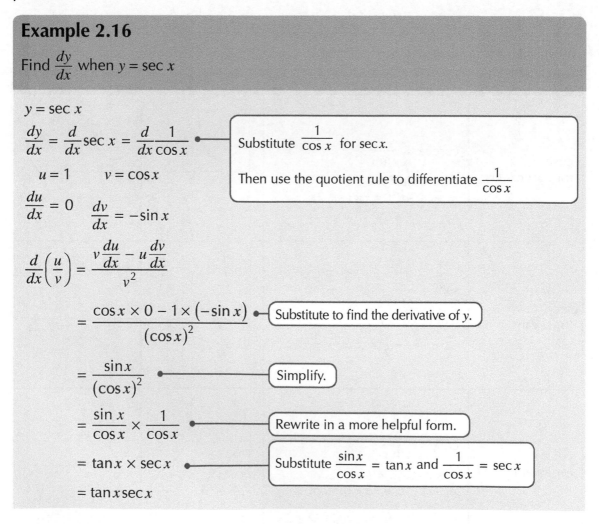

$y = \sec x$

$\dfrac{dy}{dx} = \dfrac{d}{dx}\sec x = \dfrac{d}{dx}\dfrac{1}{\cos x}$

Substitute $\dfrac{1}{\cos x}$ for $\sec x$.

Then use the quotient rule to differentiate $\dfrac{1}{\cos x}$

$u = 1 \qquad v = \cos x$

$\dfrac{du}{dx} = 0 \qquad \dfrac{dv}{dx} = -\sin x$

$\dfrac{d}{dx}\left(\dfrac{u}{v}\right) = \dfrac{v\dfrac{du}{dx} - u\dfrac{dv}{dx}}{v^2}$

$= \dfrac{\cos x \times 0 - 1 \times (-\sin x)}{(\cos x)^2}$

Substitute to find the derivative of y.

$= \dfrac{\sin x}{(\cos x)^2}$

Simplify.

$= \dfrac{\sin x}{\cos x} \times \dfrac{1}{\cos x}$

Rewrite in a more helpful form.

$= \tan x \times \sec x$

Substitute $\dfrac{\sin x}{\cos x} = \tan x$ and $\dfrac{1}{\cos x} = \sec x$

$= \tan x \sec x$

Example 2.17

Find $\dfrac{dy}{dx}$ when $y = \operatorname{cosec} x$

$y = \operatorname{cosec} x$

$\dfrac{dy}{dx} = \dfrac{d}{dx}\operatorname{cosec} x = \dfrac{d}{dx}\dfrac{1}{\sin x}$ — Use the quotient rule to differentiate $\dfrac{1}{\sin x}$

$u = 1 \qquad v = \sin x$

$\dfrac{du}{dx} = 0 \qquad \dfrac{dv}{dx} = \cos x$

$\dfrac{d}{dx}\left(\dfrac{u}{v}\right) = \dfrac{v\dfrac{du}{dx} - u\dfrac{dv}{dx}}{v^2}$

$= \dfrac{\sin x \times 0 - 1 \times \cos x}{(\sin x)^2}$

$= \dfrac{-\cos x}{(\sin x)^2}$ — Simplify.

$= -\dfrac{\cos x}{\sin x} \times \dfrac{1}{\sin x}$

$= -\cot x \times \operatorname{cosec} x$ — Substitute $\dfrac{\cos x}{\sin x} = \dfrac{1}{\tan x} = \cot x$ and $\dfrac{1}{\sin x} = \operatorname{cosec} x$

$= -\cot x \operatorname{cosec} x$

Examples 2.18 and 2.19 make use of the standard derivatives that are provided in the exam formulae list.

Example 2.18

Calculate the derivative of $y = 5x\sec x$

$\dfrac{dy}{dx} = 5\sec x + 5x\left(\dfrac{d}{dx}\sec x\right)$ — Apply the product rule.

$= 5\sec x + 5x\sec x\tan x$ — Use the derivative of sec x given in the exam formulae list.

Example 2.19

Find $\dfrac{dy}{dx}$ for $y = \csc^2(3x)$.

$$y = \csc^2(3x) = (\csc 3x)^2$$

$$\frac{dy}{dx} = 2\csc 3x \times \frac{d}{dx}\csc 3x \quad \bullet\!\!-\!\!\boxed{\text{Apply the chain rule.}}$$

$$= 2\csc 3x \times (-3\csc^2 3x)\bullet\!\!-\!\!\boxed{\begin{array}{l}\text{Use the derivative of } \cot x \text{ given in}\\\text{the exam formulae list along with the chain}\\\text{rule to differentiate } \csc 3x.\end{array}}$$

$$= -6\csc 3x \csc^2 3x$$

Exercise 2F

In Questions 1–3, differentiate with respect to x.

1   a $y = \sec 5x$ b $y = \tan 3x$ c $y = \cot 2x$

    d $y = \csc 4x$ ★e $y = \cot x^2$ ★f $y = \csc\sqrt{2x+1}$

2 ★a $y = \dfrac{\sec x}{\ln x}$ b $y = \cot(\tan x)$ c $y = \ln\left(\sec x^2\right)$

    d $y = \csc^2 3x - 2$ e $\dfrac{e^{\cot x}}{\ln x}$ f $y = \tan^2 4x$

3   a $y = \ln(\csc 2x)$ b $y = \cot(\cot(x+7))$ ★c $y = \dfrac{\ln x}{\tan^2 x}$

4   Find $f'(x)$ when $f(x) = \dfrac{\sec x + \cot x}{x^2 + 2x + 1}$

5   Find $\dfrac{dy}{dx}$ when $y = \dfrac{\cot x + \sec x}{\cot x - \sec x}$

Differentiating inverse trigonometric functions

It is also possible to differentiate the inverse trigonometric functions $\sin^{-1} x$, $\cos^{-1} x$ and $\tan^{-1} x$. The derivatives shown in the table below are provided in the exam formulae list.

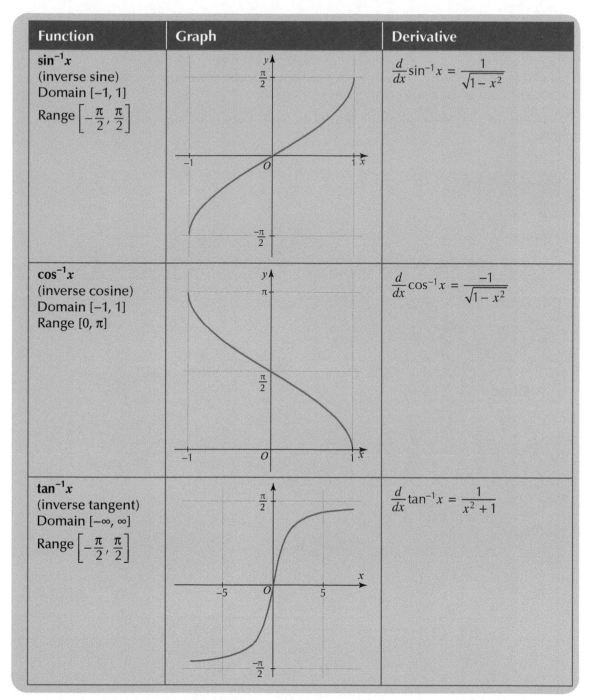

Function	Graph	Derivative
sin⁻¹x (inverse sine) Domain [−1, 1] Range $\left[-\dfrac{\pi}{2}, \dfrac{\pi}{2}\right]$		$\dfrac{d}{dx}\sin^{-1}x = \dfrac{1}{\sqrt{1-x^2}}$
cos⁻¹x (inverse cosine) Domain [−1, 1] Range [0, π]		$\dfrac{d}{dx}\cos^{-1}x = \dfrac{-1}{\sqrt{1-x^2}}$
tan⁻¹x (inverse tangent) Domain [−∞, ∞] Range $\left[-\dfrac{\pi}{2}, \dfrac{\pi}{2}\right]$		$\dfrac{d}{dx}\tan^{-1}x = \dfrac{1}{x^2+1}$

To produce the derivatives of inverse trigonometric functions, we can make use of the following rule, which applies to inverse functions in general.

Let the function f have an inverse f^{-1}:

then $f\left(f^{-1}(x)\right) = x$

$\dfrac{d}{dx} f\left(f^{-1}(x)\right) = \dfrac{d}{dx} x$ ● ⎯⎯⎯⎯ (Differentiate both sides.)

$f'\left(f^{-1}(x)\right) \times \dfrac{d}{dx}\left(f^{-1}(x)\right) = 1$ ● ⎯⎯ (Differentiate the LHS using the chain rule.)

$\dfrac{d}{dx}\left(f^{-1}(x)\right) = \dfrac{1}{f'\left(f^{-1}(x)\right)}$ ● ⎯⎯ (Rearrange algebraically.)

Using Leibniz notation: $y = f^{-1}(x) \Rightarrow \dfrac{dy}{dx} = \dfrac{1}{\dfrac{dx}{dy}}$

This shows that the rate of change of y with respect to x is equal to the reciprocal of the rate of change of x with respect to y.

Therefore, the derivative of a function is equal to the reciprocal of the derivative of its inverse.

Example 2.20

Given $f(x) = x^2$:

a find $f'(x)$

b state the derivative of $f^{-1}(x)$

$f(x) = x^2$

a $\dfrac{dy}{dx} = f'(x) = \dfrac{d}{dx} f(x)$

$\qquad = \dfrac{d}{dx} x^2 = 2x$

b **Method 1**

$f^{-1}(x) = \sqrt{x}$

$\dfrac{dx}{dy} = \dfrac{d}{dx} f^{-1}(x) = \dfrac{d}{dx} x^{\frac{1}{2}}$

$\qquad = \dfrac{1}{2} x^{-\frac{1}{2}} = \dfrac{1}{2\sqrt{x}}$

Method 2

$\dfrac{d}{dx} f^{-1}(x) = \dfrac{1}{f'\left(f^{-1}(x)\right)}$

$\qquad = \dfrac{1}{2\left(f^{-1}(x)\right)}$ ⟵ Substitute $f'(x) = 2x$

$\qquad = \dfrac{1}{2\sqrt{x}}$ ⟵ Substitute $f^{-1}(x) = \sqrt{x}$

Examples 2.21 - 2.23 show how the derivatives of $\sin^{-1} x$, $\cos^{-1} x$ and $\tan^{-1} x$ are produced.

Example 2.21

Given $f(x) = \sin^{-1} x$, $-\dfrac{\pi}{2} \leqslant x \leqslant \dfrac{\pi}{2}$, find $\dfrac{dy}{dx}$

If $y = \sin^{-1} x$
then $x = \sin y$ ① — Change the subject to x.

$\dfrac{dx}{dy} = \dfrac{d}{dy} \sin y = \cos y$ — Differentiate both sides.

$\dfrac{dy}{dx} = \dfrac{1}{\dfrac{dx}{dy}} = \dfrac{1}{\cos y}$ — $\dfrac{dy}{dx}$ is the reciprocal of $\dfrac{dx}{dy}$

$\dfrac{1}{\cos y}$ is the reciprocal of $\cos y$

$= \dfrac{1}{\sqrt{1 - \sin^2 y}}$ — $\sin^2 y + \cos^2 y = 1 \Rightarrow \cos y = \sqrt{1 - \sin^2 y}$

$= \dfrac{1}{\sqrt{1 - x^2}}$ — From ①, $x = \sin y \Rightarrow x^2 = \sin^2 y$. Replace $\sin^2 y$ with x^2

Example 2.22

Find $\dfrac{dy}{dx}$ when $f(x) = \cos^{-1} x$

If $y = \cos^{-1} x$
then $x = \cos y$ ①

$\dfrac{dx}{dy} = \dfrac{d}{dy} \cos y = -\sin y$

$\dfrac{dy}{dx} = \dfrac{1}{\dfrac{dx}{dy}} = -\dfrac{1}{\sin y}$

$= \dfrac{-1}{\sqrt{1 - \cos^2 y}}$ — $\sin^2 y + \cos^2 y = 1 \Rightarrow \sin y = \sqrt{1 - \cos^2 y}$

$= \dfrac{-1}{\sqrt{1 - x^2}}$ — From ①, $x = \cos y \Rightarrow x^2 = \cos^2 y$. Replace $\cos^2 y$ with x^2

Example 2.23

Find $\dfrac{dy}{dx}$ when $f(x) = \tan^{-1} x$

If $y = \tan^{-1} x$

then $x = \tan y$

$\dfrac{dx}{dy} = \dfrac{d}{dy} \tan y = \sec^2 y$

$\dfrac{dy}{dx} = \dfrac{1}{\dfrac{dx}{dy}} = \dfrac{1}{\sec^2 y}$

$= \dfrac{1}{\tan^2 y + 1}$ — Substitute $\sec^2 y = \tan^2 y + 1$

$= \dfrac{1}{x^2 + 1}$ — Replace $\tan^2 y$ with x^2

$\dfrac{1}{\cos^2 y} = \dfrac{\cos^2 y}{\cos^2 y} + \dfrac{\sin^2 y}{\cos^2 y}$

$\sec^2 y = 1 + \tan^2 y$

In the following two examples we make use of the standard derivatives given in the exam formulae list.

Example 2.24

Differentiate $y = \cos^{-1}(5x)$ with respect to x

$\dfrac{dy}{dx} = 5 \times \left(-\dfrac{1}{\sqrt{1 - (5x)^2}} \right)$ — Apply the chain rule and use the standard derivative for $\cos^{-1} x$ from the exam formulae list.

$= \dfrac{-5}{\sqrt{1 - 25x^2}}$

Example 2.25

Find $\dfrac{dy}{dx}$ when $f(x) = \ln\left(\tan^{-1}(1-x)\right)$

$f(x) = \ln\left(\tan^{-1}(1-x)\right)$

$\dfrac{dy}{dx} = \dfrac{d}{dx}f\left(g(x)\right) = f'\left(g(x)\right) \times g'(x)$ — Use the chain rule to differentiate a function of a function.

$= \dfrac{d}{dx}\left(\ln\left(\tan^{-1}(1-x)\right)\right)$ — Differentiate the ln function to get $\dfrac{1}{\tan^{-1}(1-x)}$

$= \dfrac{1}{\tan^{-1}(1-x)} \times \dfrac{d}{dx}\left(\tan^{-1}(1-x)\right)$ — Multiply by the derivative of the bracket, $\tan^{-1}(1-x)$

$= \dfrac{1}{\tan^{-1}(1-x)} \times \dfrac{1}{1+(1-x)^2} \times \dfrac{d}{dx}(1-x)$ — $\dfrac{d}{dx}\tan^{-1}x = \dfrac{1}{1+x^2}$
Replace x^2 with $(1-x)^2$
Multiply by the derivative of the bracket, $\dfrac{d}{dx}(1-x) = (-1)$

$= \dfrac{1}{\tan^{-1}(1-x)} \times \dfrac{1}{1+1-2x+x^2} \times (-1)$ — Multiply out the final bracket.

$= -\dfrac{1}{\left(2-2x+x^2\right)\tan^{-1}(1-x)}$ — Simplify.

Exercise 2G

1 Differentiate with respect to x:

a $y = \sin^{-1}3x$

b $y = \cos^{-1}\sqrt{x}$

★ c $y = \tan^{-1}x^2$

★ d $y = \cos^{-1}\dfrac{1}{x-1}$

★ e $y = \sin^{-1}e^{3x}$

f $y = \ln\left(\cos^{-1}\sqrt[3]{x}\right)$

g $y = e^{3x}\tan^{-1}x$

h $y = \dfrac{x^3}{\tan^{-1}2x}$

2 Calculate the gradient when $y = \tan^{-1}(\sin x)$ and $x = \dfrac{\pi}{3}$

3 Calculate the rate of change of the function $y = \cot\left(\cos^{-1}5x\right)$ when $x = \dfrac{1}{10}$

Differentiating functions defined implicitly

The equation of a line can be expressed in two ways:

$2y - x = 6$ y is expressed **implicitly** (y is not the subject of the formula)

$y = \frac{1}{2}x + 3$ y is expressed **explicitly** (y is the subject of the formula)

A function is said to be **implicit** when:

- the dependent variable y is not the subject of the formula
- there is no distinct letter that is the subject
- the variables are mixed together.

A function is said to be **explicit** when:

- the dependent variable y is expressed in terms of the independent variable x.

Some implicit functions can be written as explicit functions by rearranging, e.g. $x^2 + y^2 = 25$ can be rearranged to $y = \sqrt{25 - x^2}$

Such cases can then be differentiated as usual.

In some cases the function may be too complex or impossible to rearrange (e.g. $4x^2 - 3xy + y^2 = 0$). In such cases we need to use **implicit differentiation**.

Example 2.26

Use implicit differentiation to differentiate $3x^3 + y^3 = 6$, then rearrange to express $\frac{dy}{dx}$ explicitly in terms of x.

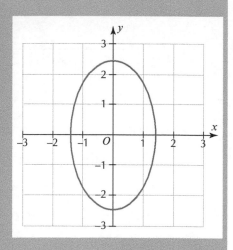

$3x^2 + y^2 = 6$

$\dfrac{d}{dx}3x^2 + \dfrac{d}{dx}y^2 = \dfrac{d}{dx}6$ — Differentiate each term of the function with respect to x.

$6x + \left(\dfrac{d}{dy}y^2\,\dfrac{dy}{dx}\right) = 0$ — $\dfrac{d}{dx} \equiv \dfrac{d}{dy}\dfrac{dy}{dx}$

$6x + 2y\dfrac{dy}{dx} = 0$ — Differentiate y^2 with respect to y to get $2y$.

$\dfrac{dy}{dx} = -\dfrac{6x}{2y}$ — Make $\dfrac{dy}{dx}$ the subject of the formula and simplify.

$= -\dfrac{3x}{y}$ — Derivative expressed **implicitly**.

$3x^2 + y^2 = 6 \;\Rightarrow\; y = \sqrt{6 - 3x^2}$ — Rearrange the original formula to find y.

$\dfrac{dy}{dx} = \dfrac{-3x}{y} = -\dfrac{3x}{\sqrt{6 - 3x^2}}$ — Substitute $y = \sqrt{6 - 3x^2}$

$\dfrac{dy}{dx}$ expressed **explicitly** in terms of x.

Example 2.27

Differentiate $e^x + 3xy + y^3 = 2$ giving an implicit solution.

$e^x + 3xy + y^3 = 2$

$\dfrac{d}{dx}e^x + \dfrac{d}{dx}3xy + \dfrac{d}{dx}y^3 = \dfrac{d}{dx}2$

$e^x + \dfrac{d}{dx}3xy + \dfrac{d}{dy}y^3\,\dfrac{dy}{dx} = 0$

$e^x + \dfrac{d}{dx}3xy + \dfrac{d}{dy}y^3\,\dfrac{dy}{dx} = 0$ — Differentiate $3xy$ using the product rule.

$u = 3x \qquad\qquad v = y$

$\dfrac{du}{dx} = 3 \qquad \dfrac{dv}{dx} = \dfrac{d}{dy}y\,\dfrac{dy}{dx} = 1\dfrac{dy}{dx}$

$\dfrac{dy}{dx} = u\dfrac{dv}{dx} + v\dfrac{du}{dx}$

$e^x + \left[3x\dfrac{dy}{dx} + 3y\right] + 3y^2\dfrac{dy}{dx} = 0$ — $\dfrac{d}{dx}3xy = 3x\dfrac{dy}{dx} + 3y$ and $\dfrac{d}{dx}y^3 = 3y^2\dfrac{dy}{dx}$

$e^x + \dfrac{dy}{dx}\left(3x + 3y^2\right) + 3y = 0$ — Factorise by taking out $\dfrac{dy}{dx}$

$\dfrac{dy}{dx} = \dfrac{-\left(3y + e^x\right)}{3\left(x + y^2\right)}$ — Make $\dfrac{dy}{dx}$ the subject of the formula and factorise. Derivative expressed **implicitly**.

Example 2.28

Find the equation of the tangent to the curve $2x\sin y + x^2 = 1$ at the point $(1, \pi)$.

Gradient of tangent $= \dfrac{dy}{dx}$

$2x\sin y + x^2 = 1$

$\dfrac{d}{dx}(2x\sin y) + \dfrac{d}{dx}x^2 = \dfrac{d}{dx}1$

$\dfrac{d}{dx}2x\sin y + 2x = 0$ ●————— Differentiate $(2x\sin y)$ using the product rule.

$u = 2x \qquad\qquad v = \sin y$ ●————— Separate the two functions.

$\dfrac{du}{dx} = 2 \qquad \dfrac{dv}{dx} = \dfrac{d}{dy}\sin y\dfrac{dy}{dx} = \cos y\dfrac{dy}{dx}$

$2x\cos y\dfrac{dy}{dx} + 2\sin y + 2x = 0$ ●————— Using the product rule,

$\dfrac{d}{dx}(2x\sin y) = 2x\cos y\dfrac{dy}{dx} + 2\sin y$

$\dfrac{dy}{dx} = \dfrac{-(2\sin y + 2x)}{2x\cos y}$ ●————— Make $\dfrac{dy}{dx}$ the subject of the formula to express the derivative implicitly.

$= \dfrac{-(2\sin \pi + 2(1))}{2(1)\cos \pi} = \dfrac{-2}{-2} = 1$ ●————— Substitute $x = 1$ and $y = \pi$ to calculate the gradient at the given point.

$y - b = m(x - a)$

$y - \pi = 1\,(x - 1)$

$y = x - 1 + \pi$ ●————— Substitute $a = 1$, $b = \pi$ and $m = 1$ to find the equation of the tangent.

The equation of the tangent to the curve $2x\sin y + x^2 = 1$ at the point $(1, \pi)$ is
$y = x - 1 + \pi$

⚠

Terms involving only x terms differentiate in the usual way.

Terms involving y differentiate in the usual way but gain $\dfrac{dy}{dx}$

Terms with a combination of x and y should be differentiated using the product rule as well as the above.

Exercise 2H

1 Find $\dfrac{dy}{dx}$ for these implicit functions:

    a $x^2 + y^2 = 9$ ★ b $\ln y = x + y$ ★ c $x^2 + xy = 3$

    d $y^2 - xy = x$ e $x^2 y^2 = x^4 + y^4$ ★ f $x \tan y = e^x$

    g $x^2 + y^2 = \dfrac{y}{x}$ h $5x^2 - 4xy + 3y^2 = 2$

★ 2   The equation $x^4 + y^4 + 9x - 6y = 14$ defines a curve passing through the point $P(1, 2)$.

      Find the equation of the tangent to this curve at P.

3   A curve is defined by the equation $x^2 + y^2 + 2x - 4y = 15$

      Find the equation of the tangent at the point $(3, 4)$ on this curve.

4   Find the equation of the tangent to the curve $2x^2 - 3xy - y^2 = 1$ at the point $(2, 1)$.

5   The curve $x^2 - 4x + y^2 - 6y = 12$ passes through the point $T(5, 7)$.

      Find the equation of the tangent to this curve at the point T.

Second derivatives of implicit functions

Higher derivatives can also be found for implicit functions.

Example 2.29

Find $\dfrac{dy}{dx}$ and $\dfrac{d^2 y}{dx^2}$ for the function $\sqrt{x} + y^3 = 1$

$\sqrt{x} + y^3 = 1$

$\dfrac{d}{dx} x^{\frac{1}{2}} + \dfrac{d}{dx} y^3 = \dfrac{d}{dx} 1$ •——— Differentiate each term of the function with respect to x.

$\dfrac{1}{2} x^{-\frac{1}{2}} + 3y^2 \dfrac{dy}{dx} = 0$

$\dfrac{dy}{dx} = \dfrac{-\frac{1}{2} x^{-\frac{1}{2}}}{3y^2}$

$\qquad = \dfrac{-1}{6y^2 \sqrt{x}}$

$-\dfrac{1}{4} x^{-\frac{3}{2}} + 6y \left(\dfrac{dy}{dx} \right)^2 + 3y^2 \dfrac{d^2 y}{dx^2} = 0$ •——— Differentiate again to find $\dfrac{d^2 y}{dx^2}$

$3y^2 \dfrac{d^2 y}{dx^2} = \dfrac{1}{4x^{\frac{3}{2}}} - 6y \left(\dfrac{dy}{dx} \right)^2$ •——— Rearrange.

$\qquad = \dfrac{1}{4x^{\frac{3}{2}}} - 6y \left(\dfrac{-1}{6y^2 \sqrt{x}} \right)^2$ •——— Substitute for $\dfrac{dy}{dx}$

(continued)

$$= \frac{1}{4x^{\frac{3}{2}}} - 6y\left(\frac{1}{36y^4 x}\right)$$

Simplify.

$$\frac{d^2y}{dx^2} = \frac{1}{3y^2}\left(\frac{1}{4x^{\frac{3}{2}}} - 6y\left(\frac{1}{36y^4 x}\right)\right)$$

$$= \frac{1}{12y^{\frac{7}{2}}} - \frac{1}{18y^5 x}$$

Exercise 2I

1 Find $\dfrac{dy}{dx}$ and $\dfrac{d^2y}{dx^2}$ for these implicit functions:

★ **a** $x^2 + y^2 = 4$ **b** $x^2 + xy = 3$

★ **c** $xy = e^x$ **d** $xy + y^2 = 1$

2 y is defined implicitly in terms of x by the equation $xy - y^2 = 1$. Show that:

    a $\dfrac{dy}{dx} = \dfrac{y}{2y - x}$ **b** $\dfrac{d^2y}{dx^2} = \dfrac{2y(y - x)}{(2y - x)^3}$

3 Show that $(1, 2)$ is a stationary point on the curve $x^2 - xy + y^3 = 7$ and find the nature of the stationary point.

Differentiating logarithmic functions

It is appropriate to use **logarithmic differentiation** in functions with extended products, quotients, and in functions where the variable occurs in an index.

When using logarithmic differentiation (for example with a function where x appears in a power), we take natural logs of both sides first (ln), and then differentiate implicitly.

The only exception is when the e function appears with a power of x. Differentiate such functions as normal – there is no need to take logs first.

Example 2.30

Differentiate the function $y = 6^{2x-1}$

$y = 6^{2x-1}$

Take ln of both sides.

$\ln y = \ln 6^{2x-1}$

Use laws of logs to bring the power to the front. Remember that $\ln 6$ is a constant.

$\quad = (2x - 1) \times \ln 6$

Write in differentiable form and differentiate each term with respect to x.

$\dfrac{d}{dx}\ln y = \dfrac{d}{dx}(2x\ln 6 - \ln 6)$

Use the laws of logs.

$\dfrac{1}{y}\dfrac{dy}{dx} = 2\ln 6 = \ln 6^2$

$\dfrac{dy}{dx} = y\ln 36$

Substitute $y = 6^{2x-1}$
The derivative is expressed explicitly.

$\quad = 6^{2x-1}\ln 36$

Example 2.31

Differentiate the function $y = x^{3x}$

$y = x^{3x}$

$\ln y = \ln x^{3x}$

$\quad = 3x \times \ln x$

$\dfrac{d}{dx} \ln y = \dfrac{d}{dx} 3x \ln x$ ● ──── Differentiate $3x \ln x$ using the product rule.

$u = 3x \qquad\qquad v = \ln x$

$\dfrac{du}{dx} = 3 \qquad \dfrac{dv}{dx} = \dfrac{d}{dx} \ln x = \dfrac{1}{x}$

$\dfrac{1}{y}\dfrac{dy}{dx} = \dfrac{3x}{x} + 3\ln x$ ● ──── Substitute for the product rule on the RHS.

LHS: $\dfrac{d}{dx} \equiv \dfrac{d}{dy}\dfrac{dy}{dx} \Rightarrow \dfrac{d}{dy}\ln y \dfrac{dy}{dx} = \dfrac{1}{y}\dfrac{dy}{dx}$

$\dfrac{dy}{dx} = y(3 + 3\ln x)$

$\dfrac{dy}{dx} = 3x^{3x}(1 + \ln x)$ ● ──── Substitute $y = x^{3x}$

The derivative is expressed explicitly.

Exercise 2J

1 Find $\dfrac{dy}{dx}$ for these functions:

★ **a** $y = 4^{2x}$ **b** $y = \pi^{3x}$ **c** $y = 2^x$ ★**d** $y = \dfrac{5^x}{x}$ **e** $y = x^{\sin x}$

2 Given $y = (\sin x)^x$, show that $\dfrac{dy}{dx} = (\sin x)^x (x \cot x + \ln \sin x)$

3 Given $y = \dfrac{x^2}{\sqrt{x^4 + 1}}$, use logarithmic differentiation to find $\dfrac{dy}{dx}$ in terms of x.

★4 A curve has equation $y = \dfrac{x(2x + 1)^{\frac{3}{2}}}{(3x - 4)^{\frac{1}{3}}}$

    Find the equation of the tangent to the curve where $x = 4$.

5 Find the value of x for which the graph of $y = \dfrac{x^{\frac{1}{2}}(3 - x)^{\frac{1}{6}}}{(2x + 1)^{\frac{2}{3}}}$ is stationary.

Differentiating functions defined parametrically

Many functions are written as $y = f(x)$, so y is expressed in terms of x. However, it is sometimes more convenient to express a function in terms of a third variable such as t.

$x = x(t)$ and $y = y(t)$ are called **parametric equations** where t is the **parameter**.

Consider a point moving in the xy plane:

    Let (x, y) be the coordinates of this point at time t.

    Both x and y must be functions of t.

    Let the parametric equations of x and y be defined as $x = t^2$ and $y = 3t$

    When $t = 2$: $x = 2^2 = 4$ and $y = 3 \times 2 = 6$

    Therefore, the coordinates of the point at time $t = 2$ are $(4, 6)$.

Despite involving another parameter (t), parametric equations can still be differentiated with respect to x to find $\dfrac{dy}{dx}$

We can use the chain rule to work out how to carry out the differentiation:

$$\frac{dy}{dx} = \frac{dy}{dt} \times \frac{dt}{dx} = \frac{dy}{dt} \div \frac{dx}{dt} \Rightarrow \frac{dy}{dx} = \frac{\frac{dy}{dt}}{\frac{dx}{dt}}$$

Similarly, the second derivative can also be established for parametric equations:

$$\frac{d^2y}{dx^2} = \frac{d}{dx}\left(\frac{dy}{dx}\right) = \frac{d}{dt}\left(\frac{dy}{dx}\right)\frac{dt}{dx} = \frac{d}{dt}\left(\frac{\frac{dy}{dt}}{\frac{dx}{dt}}\right) \times \frac{1}{\frac{dx}{dt}}$$

$$\frac{dy}{dx} = \frac{\frac{dy}{dt}}{\frac{dx}{dt}} \quad \text{and} \quad \frac{d^2y}{dx^2} = \frac{d}{dt}\left(\frac{dy}{dx}\right) \times \frac{1}{\frac{dx}{dt}}$$

Example 2.32

Find $\dfrac{dy}{dx}$ for this pair of parametric equations.

$$x = \sin 4\theta \qquad y = -3\cos 2\theta$$

$y = -3\cos 2\theta$

$\dfrac{dy}{d\theta} = \dfrac{d}{d\theta}(-3\cos 2\theta) = 6\sin 2\theta$ — Differentiate the y-equation with respect to θ.

$x = \sin 4\theta$

$\dfrac{dx}{d\theta} = \dfrac{d}{d\theta}(\sin 4\theta) = 4\cos 4\theta$ — Differentiate the x-equation with respect to θ.

$\dfrac{dy}{dx} = \dfrac{\frac{dy}{d\theta}}{\frac{dx}{d\theta}} = \dfrac{6\sin 2\theta}{4\cos 4\theta}$ — Substitute to find $\dfrac{dy}{dx}$

$\qquad = \dfrac{3\sin 2\theta}{2\cos 4\theta}$

Example 2.33

A curve is defined by the parametric equations $x = t^2 - 3$ and $y = 2t^3$
Find the equation of the tangent to this curve when $t = 1$.

Gradient $= \dfrac{dy}{dx}$

$y = 2t^3$

$\dfrac{dy}{dt} = \dfrac{d}{dt}(2t^3) = 6t^2$ — Differentiate the y-equation with respect to t.

$x = t^2 - 3$

$\dfrac{dx}{dt} = \dfrac{d}{dt}\left(t^2 - 3\right) = 2t$ ●————— Differentiate the x-equation with respect to t.

$\dfrac{dy}{dx} = \dfrac{\frac{dy}{dt}}{\frac{dx}{dt}} = \dfrac{6t^2}{2t} = 3t$ ●————— Substitute to find $\dfrac{dy}{dx}$

$\dfrac{dy}{dx} = 3t = 3 \times 1 = 3$ ●————— Substitute $t = 1$ into the gradient equation to find m.

$y = 2t^3 = 2 \times 1^3 = 2$ ●————— Substitute $t = 1$ into the y-equation to find b.

$x = t^2 - 3 = 1^2 - 3 = (-2)$ ●————— Substitute $t = 1$ into the x-equation to find a.

$y - b = m(x - a)$

$y - 2 = 3(x - (-2))$ ●————— Substitute $m = 3$, $b = 2$ and $a = (-2)$ to find the equation of the tangent.

$y = 3x + 8$

The equation of the tangent to this curve when $t = 1$ is $y = 3x + 8$

Exercise 2K

1 A curve is defined parametrically as shown.

   Find an expression for $\dfrac{dy}{dx}$ in terms of the parameter t and simplify where possible.

   a $x = \dfrac{1}{t}$ and $y = \sqrt{t^2 + 1}$ ★ b $x = e^t \cos t$ and $y = e^t \sin t$

   c $x = \dfrac{t - 1}{t + 1}$ and $y = \dfrac{2t - 1}{t - 2}$ ★ d $x = \dfrac{t}{1 + t^2}$ and $y = \dfrac{t}{1 - t^2}$

2 Find the coordinates of the stationary points on the curve defined parametrically as $x = t^2 + t$ and $y = t^3 - 12t$

3 A curve is defined by the parametric equations $x = t - \cos t$ and $y = 1 + \sin t$

   Calculate the gradient to the curve when $t = \pi$

4 A curve is defined by the parametric equations $x = \sin\theta + \cos\theta$ and $y = \sin\theta - \cos\theta$

   Show that $\dfrac{dy}{dx} = \dfrac{1 + \tan\theta}{1 - \tan\theta}$

5 A curve is defined by the parametric equations $x = t^2 + t - 1$ and $y = t^2 - 4t + 1$

   Find the equation of the tangent to the curve at the point with parameter $t = 0$.

6 A curve is defined parametrically by the equations $x = t - \dfrac{1}{t}$ and $y = t + \dfrac{1}{t}$

   a Find the coordinates of the stationary point(s) on the curve.

   b Use the second derivative to determine the nature of the stationary point(s).

★ 7 A curve is defined parametrically by the equations $x = (t^2 + 1)^2$ and $y = t(t - 2)$

   a Find the coordinates of the stationary point(s) on the curve.

   b Use the second derivative to determine the nature of the stationary point(s).

8 Find the stationary points on the curve defined by the parametric equations
   $x = \sin\theta$ and $y = \theta + \cos 2\theta$, $0 < \theta < 2\pi$

Applying differentiation to problems in context

Differentiation is used in a wide range of real-life applications. Engineers differentiate functions to calculate safe gradients in the design of new structures such as theme park rides, bridges, buildings, structures and roads. Aerospace engineers use advanced mathematical equations to build equipment that can safely travel through space.

Most calculus formulae are programmed into computer systems, but engineers will still perform calculations when conducting experiments. Engineers must be able to solve calculus–based equations by hand. Computer designers use calculus to design new games, which involve curves, lines, areas under curves, gradients and stationary points. (Other mathematical skills such as handling matrices and 3-dimensional vectors may also be used in the design of new computer games.)

Scientists differentiate functions when analysing rates of change of functions at different values for a given variable. This process can give vital information about how a function behaves using particular values for the x-variable. Scientists record a series of results to compare and report on their findings.

Modern professionals still need to understand complex mathematical formulae with multiple variables in:

- rectilinear motion
- motion in a plane
- related rates of change
- optimisation.

Example 2.34

The height, h (in pixels), of an object within a computer game's graphic is dependent on the time, t (in seconds). The equation, $h = 5e^{t^2} + (3x + 1)^4$ measures the height of the object above the horizontal level (the ground).

Calculate the rate of change of the height with respect to time, after 1 second.

$h = 5e^{t^2} + (3x + 1)^4$

$\dfrac{dh}{dt} = \dfrac{d}{dt}\left(5e^{t^2} + (3x + 1)^4\right)$ — Use the chain rule to differentiate a function of a function.

$= 5e^{t^2} \times 2t + 4(3x + 1)^3 \times 3$ — Differentiate each term.

$= 10t\, e^{t^2} + 12(3x + 1)^3$ — Simplify the expression.

$= 10 \times 1\, e^{t^2} + 12(3 \times 1 + 1)^3$ — Substitute $t = 1$ and evaluate.

$= 795.1828183 \approx 795$

After 10 seconds, the rate of change is 795 pixels per second.

Example 2.35

An aeroplane performs a turn while waiting to land at an airport. It maintains constant altitude, and relative to suitable axes its coordinates are given parametrically by

$$x = 130t - 0.1t^2 \qquad \text{and} \qquad y = 60\,t\sin\left(\frac{\pi t}{60}\right)$$

where t is the time in seconds, and x and y are measured in metres.

Calculate the speed of the aeroplane after 20 seconds.

$$v_x = \frac{dx}{dt} = 130 - 0.2t$$

$$v_y = \frac{dy}{dt} = 60\sin\left(\frac{\pi t}{60}\right) + \pi t\cos\left(\frac{\pi t}{60}\right)$$ ⟵ Differentiate to find the velocity components.

$$v_x(20) = 130 - 0.2 \times 20 = 126$$

$$v_y(20) = 60\sin\left(\frac{20\pi}{60}\right) + 20\pi\cos\left(\frac{20\pi}{60}\right) = 30\sqrt{3} + 10\pi$$ ⟵ Evaluate when $t = 20$

$$\text{speed} = \sqrt{126^2 + (30\sqrt{3} + 10\pi)^2} = 151.09$$ ⟵ Calculate the magnitude of the velocity vector

After 20 seconds, the speed of the aeroplane is 151.09 m/s (to 2 decimal places).

Exercise 2L

★ 1 Find the equation of the tangent to the curve $y = \tan\left(\sin\frac{x}{3}\right)$ when $x = \pi$.

★ 2 Calculate the rate of change of the function $P(t) = -\sin\frac{t}{6}\cos\frac{t}{6}$ when $t = \pi$.

3 Use the theorems of calculus to show that the function $y = \sin^2\theta\cos^2\theta$ has a stationary point when $\theta = \frac{\pi}{2}$ and use the second derivative to determine its nature.

4 Find the equation of the tangent to the curve defined as $y = 4^x + 5\ln x$ when $x = 1$.

5 A function is defined parametrically by $x = 5t^3 + 3$ and $y = t^5$

Calculate the gradient of the tangent to this curve when $t = 3$.

★ 6 A curve is defined parametrically, for all t, by equations

$$x = 2t + \tfrac{1}{2}t^2 \quad \text{and} \quad y = \tfrac{1}{3}t^3 - 4t$$

a Obtain $\frac{dy}{dx}$ and $\frac{d^2y}{dx^2}$ as functions of t.

b Find the value of t at which the curve has a stationary point and determine its nature.

★ 7 The displacement of a particle is found using the function $s(t) = \cot 3t$ where t is the time in seconds and the displacement s is measured in metres.

Calculate the exact value of the acceleration of this particle when $t = \frac{\pi}{9}$

★ 8 A function is defined explicitly as $y = \frac{e^x}{\sin 2x}$

Find the exact value for the gradient of the tangent to this function when $x = \frac{\pi}{12}$

9   A particle moves in the xy plane according to the formula $s(t) = \sin 3t \tan 3t + 1$ where the displacement s is measured in metres as time t passes in seconds. Calculate the acceleration of this particle after 10 seconds.

10 The fuel efficiency, F, in kilometres per litres, of a vehicle varies with the velocity, v km/h. A particular vehicle has the relationship

$$F = 15 + e^x(\sin x - \cos x - \sqrt{2}), \text{ where } x = \frac{\pi(v - 40)}{80}$$

The velocity range for which this function is valid is $40\,\text{km/h} \leqslant v \leqslant 120\,\text{km/h}$.

a   What are the greatest and least fuel efficiencies over the range for v?

b   At what speeds do these efficiencies occur?

11 The voltage, V volts, decaying across a capacitor in time, t seconds, is given by

$$V = 200\,e^{-0.3t} \tan 2t, \ 1 \leqslant t \leqslant 20$$

Find the rate of change of voltage when $t = 1$ second and $t = 10$ seconds.

12 When cornering during a race, the coordinates (relative to suitable axes) of a Formula 1 car are given by

$$x = 60\cos\left(\frac{2t}{3}\right) \text{ and } y = 40t - t^2$$

where t is the time in seconds, and x and y are measured in metres.

Calculate the speed of the Formula 1 car after 2 seconds.

Chapter review

1   Find:

a $\dfrac{d}{dx}(\ln \sin 2x)$

b $\dfrac{d}{dx}(e^{x+2} \cos 3x)$

c $\dfrac{d}{dx}(\sin 4x \cot 4x)$

d $\dfrac{d}{dx}\left(\dfrac{e^{x^2}}{(x + 2)^4}\right)$

e $\dfrac{d}{dx}(\tan^{-1}(x + 3))$

f $\dfrac{d}{dx}(e^{5x}\operatorname{cosec} x)$

g $\dfrac{d}{dx}(\ln x \cos^{-1}4x)$

h $\dfrac{d}{dx}(2^{x+1} e^{3x})$

i $\dfrac{d}{dx}\left(\dfrac{\cos(\cos x)}{\ln x}\right)$

j $\dfrac{d}{dx}(\sec 4x\, e^{2x+3})$

k $\dfrac{d}{dx}\left(\dfrac{\sec^2 x\, (x + 3)^5}{\ln x}\right)$

l $\dfrac{d}{dx}\left(\dfrac{\cot^3 x(2x - 1)}{\tan x}\right)$

Use logarithmic differentiation for parts k and l.

2   Calculate the fourth derivative of the function $f(x) = \dfrac{1}{\sqrt{(2x - 1)}}$

3   Find the coordinates of the stationary points of the function $y = x^3 - 4x^2 - 11x + 30$

Use the second derivative to determine the nature of the stationary points.

4   The displacement of a particle is determined by the function
    $s = t^3 - 4t^2 - 37t + 40$, where s is the displacement in metres and t is the time in seconds.

    a   Calculate the velocity of this particle after 6 seconds.

    b   Calculate when the acceleration of this particle is zero.

5   Find $\dfrac{dy}{dx}$ and $\dfrac{d^2y}{dx^2}$ in their simplest form for the implicit function $x^2 - xy + 3y^2 = 10$

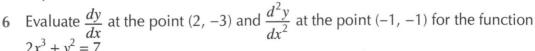

Write each answer as a single fraction in its simplest form.

6   Evaluate $\dfrac{dy}{dx}$ at the point $(2, -3)$ and $\dfrac{d^2y}{dx^2}$ at the point $(-1, -1)$ for the function $2x^3 + y^2 = 7$

7   Show that for the function $3x^2 = 2y^2 \ln y$, $\dfrac{dy}{dx} = \dfrac{3xy}{y^2 + 3x^2}$

8   A curve has equation $x^2 + 4xy + y^2 + 10 = 0$

    Find the values of $\dfrac{dy}{dx}$ and $\dfrac{d^2y}{dx^2}$ at the point $(-2, 3)$.

9   A function is defined by the parametric equations $x = \dfrac{3}{t}$ and $y = 2t^3 + 5$

    Show that $\dfrac{d^2y}{dx^2} = \dfrac{648}{x^5}$

10 A function is defined by the parametric equations $x = 4 + 2\cos\theta$ and $y = 3 + 5\sin\theta$

    Find an expression for $\dfrac{dy}{dx}$ and for $\dfrac{d^2y}{dx^2}$

- • I can differentiate functions using the chain rule. ★ Exercise 2A Q2c, Q3c, Q3h
- • I can differentiate exponential and logarithmic functions. ★ Exercise 2B Q1d
- • I can differentiate functions given in the form of a product. ★ Exercise 2C Q1e, Q2b, Q2c
- • I can differentiate functions given in the form of a quotient. ★ Exercise 2D Q1e, Q2b, Q2d, Q2f, Q6
- • I can differentiate to find higher derivatives. ★ Exercise 2E Q1b, Q1f, Q4b, Q6
- • I can differentiate using standard trigonometric identities. ★ Exercise 2F Q1e, Q1f, Q2a, Q2f
- • I can differentiate inverse trigonometric functions. ★ Exercise 2G Q1c–Q1e
- • I can find the derivative of functions defined implicitly. ★ Exercise 2H Q1b, Q1c, Q1f, Q2
- • I can find the second derivative of functions defined implicitly. ★ Exercise 2I Q1a, Q1c
- • I can use logarithmic differentiation. ★ Exercise 2J Q1a, Q1d, Q4
- • I can find the derivative of functions defined parametrically. ★ Exercise 2K Q1b, Q1d, Q7
- • I can apply differentiation to problems in context. ★ Exercise 2L Q1, Q2, Q6–Q8

3 Integration

This chapter will show you how to:

- integrate expressions using standard results
- integrate by substitution
- integrate by parts
- apply integration to problems in context.

You should already know how to:

- integrate an algebraic function which is, or can be, simplified to an expression in powers of x
- integrate functions of the form $f(x) = (x + q)^n$ and $f(x) = (px + q)^n$
- integrate functions of the form $f(x) = p \sin x$ and $f(x) = p \cos x$;
  $f(x) = p \sin(qx + r)$ and $f(x) = p \cos(qx + r)$
- solve differential equations of the form $\dfrac{dy}{dx} = f(x)$

Uses of integration

Integration is used to find the anti-derivative of functions where each integral needs a constant of integration or to evaluate definite integrals using the limits. Integrating complex functions requires knowing and using standard results, integrating by substitution and using integration by parts.

Integration is used to find the area between a function and the x-axis, to find the area between a function and the y-axis, and to find the area between two curves. Graphs can be rotated 360° around either the x-axis or the y-axis, and integration is then used to find the volume of the resultant solid formed. We can use the limits of integration to calculate the exact value of the area or volume.

There are a number of basic rules you should already know and be able to use:

- $\displaystyle\int x^n \, dx = \frac{1}{n+1} x^{n+1} + c$ and $\displaystyle\int (ax + b)^n \, dx = \frac{1}{a(n+1)}(ax + b)^{n+1} + c$
- $\displaystyle\int a \, f(x) \, dx = \int f(x) \, dx$ where a is a constant
- $\displaystyle\int (f(x) + g(x)) \, dx = \int f(x) \, dx + \int g(x) \, dx$
- $\displaystyle\int_a^c f(x) \, dx = \int_a^b f(x) \, dx + \int_b^c f(x) \, dx$ where $a < b < c$
- $\displaystyle\int_a^b f(x) \, dx = -\int_b^a f(x) \, dx$
- $\displaystyle\int \sin x \, dx = -\cos x + c$ and $\displaystyle\int \sin(ax + b) \, dx = -\frac{1}{a}\cos(ax + b) + c$
- $\displaystyle\int \cos x \, dx = \sin x + c$ and $\displaystyle\int \cos(ax + b) \, dx = \frac{1}{a}\sin(ax + b) + c$

The table below summarises integrations of functions you should already know.

Function	Intermediate stage	Integral
$\dfrac{dy}{dx} = 4x^3$	$\displaystyle\int \dfrac{dy}{dx}\,dx = y = \int 4x^3\,dx = \dfrac{4x^4}{4} + c$	$x^4 + c$
$f(x) = 3x^4 - \dfrac{6}{x^2} - 2 - \sqrt{x}$	$\displaystyle\int f(x)dx$ $= \dfrac{3x^5}{5} - \dfrac{6x^{-1}}{-1} - 2x - \dfrac{x^{\frac{3}{2}}}{\frac{3}{2}} + c$	$\dfrac{3x^5}{5} + \dfrac{6}{x} - 2x - \dfrac{2x\sqrt{x}}{3} + c$
$y = (2x - 1)^4$	$\displaystyle\int y\,dx = \dfrac{(2x-1)^5}{5 \times 2} + c$	$\dfrac{1}{10}(2x-1)^5 + c$
$h(\alpha) = \cos 3\alpha$	$\displaystyle\int \cos 3\alpha\,d\alpha = \sin 3\alpha \times \dfrac{1}{3} + c$	$\dfrac{1}{3}\sin 3\alpha + c$
$g(\theta) = 4\sin(2\theta - \pi)$	$\displaystyle\int 4\sin(2\theta - \pi)\,d\theta$ $= -4\cos(2\theta - \pi) \times \dfrac{1}{2} + c$	$-2\cos(2\theta - \pi) + c$

Exercise 3A

1   Find the following integrals:

   a $\displaystyle\int 4x^6\,dx$ ★b $\displaystyle\int (3x^2 - 7x)\,dx$ c $\displaystyle\int (2x - 3)^2\,dx$

   d $\displaystyle\int \dfrac{x^3 - 7}{x^2}\,dx$ ★e $\displaystyle\int \dfrac{x - 7}{\sqrt{x}}\,dx$ f $\displaystyle\int \cos 2x\,dx$

   g $\displaystyle\int 6\sin 3x\,dx$ h $\displaystyle\int (8t + 3)^3\,dt$ i $\displaystyle\int (5 - 4x)^5\,dx$

   j $\displaystyle\int \dfrac{1}{(4x - 5)^5}\,dx$ k $\displaystyle\int \sin(3\theta - 1)\,d\theta$ l $\displaystyle\int \cos(2 - 4x)\,dx$

★2   a Determine $F(x) = \displaystyle\int \cos(\pi - x)\,dx$, given that $F(0) = 1$.

   ★b Determine $H(x) = \displaystyle\int \dfrac{2}{(3x + 1)^2}\,dx$, given that $H(0) = 2$.

3   a $\displaystyle\int \dfrac{2x - x^3}{\sqrt{x^3}}\,dx$ b $\displaystyle\int (2x + 3)^5\,dx$

4   Determine $F(x) = \text{int}\,(3x^2 - 1)^2\,dx$, given that $F(0) = 1$.

5   Determine the following definite integrals:

   a $\displaystyle\int_{-1}^{5} (1 + 3x)\,dx$ b $\displaystyle\int_{1}^{5} (2 + 3x - x^2)\,dx$ ★c $\displaystyle\int_{\frac{\pi}{8}}^{\frac{\pi}{6}} \cos 2\theta\,d\theta$

   d $\displaystyle\int_{1}^{2} \dfrac{x^2 + 1}{\sqrt{x}}\,dx$ e $\displaystyle\int_{1}^{8} \dfrac{x - 1}{\sqrt[3]{x^2}}\,dx$

6   Calculate the following:

   a $\displaystyle\int_{0}^{1} \left(\sqrt[4]{x^5}\right)\left(\sqrt[4]{x^5} + \sqrt[5]{x^4}\right)dx$ b $\displaystyle\int_{1}^{4} \sqrt{\dfrac{5}{x}}\,dx$

Using trigonometric identities

The following trigonometric identities will help with Exercise 3B.

Useful trigonometric identities		
$\sin^2 A + \cos^2 A = 1$	$\sin 2A = 2\sin A\cos A$	
$\cos 2A = \cos^2 A - \sin^2 A$	$\cos 2A = 2\cos^2 A - 1$	$\cos 2A = 1 - 2\sin^2 A$
$\tan A = \dfrac{\sin A}{\cos A}$	$\tan nA = \dfrac{\sin n A}{\sin n A}$	$\tan^2 A = \dfrac{\sin^2 A}{\cos^2 A}$
$\sec A = \dfrac{1}{\cos A}$	$\operatorname{cosec} A = \dfrac{1}{\sin A}$	$\cot A = \dfrac{1}{\tan A}$
$\sec^2 A = \tan^2 A + 1$	$\sec^2 A = \dfrac{1}{\cos^2 A}$	

Exercise 3B

1 Use an appropriate trigonometric identity to help find these integrals.

    a $\int(\sin^2 x + \cos^2 x)\,dx$ **b** $\int\cos^2 x\,dx$ ★**c** $\int 3\sin^2 x\,dx$

    d $\int(\cos 2\theta - 2\sin^2\theta)\,d\theta$ **e** $\int 2\sin x\cos x\,dx$ ★**f** $\int\cos\theta\tan\theta\,d\theta$

2 Evaluate each integral.

    a $\displaystyle\int_0^{\frac{\pi}{2}} \frac{\sin\phi}{\tan\phi}\,d\phi$ **b** $\displaystyle\int_{\pi}^{\frac{3\pi}{2}} \frac{\sin 2x}{\sin x}\,dx$

    c $\displaystyle\int_0^{\pi}(\cos\theta + \sin\theta)(\cos\theta - \sin\theta)\,d\theta$

    d $\displaystyle\int_{\frac{\pi}{3}}^{\frac{\pi}{2}} \sqrt{\cos^2\psi - \cos 2\psi}\,d\psi$

3 Find $\int\cos^2\left(\tfrac{1}{2}x\right)dx$

Integrating expressions using standard results

You need to know and be able to use standard results to integrate complex functions. The standard results shown here will make the integration of particular complex functions possible.

Standard integral results for exponential functions and $\frac{1}{x}$

The standard results shown here are used in the integration of exponential functions and the reciprocal of x, $\frac{1}{x}$.

It follows that since $\dfrac{d}{dx}(e^x) = e^x$ then $\int e^x\,dx = e^x + c$

This gives new standard results for integrating exponential functions:

$$f(x) = e^x \quad\Rightarrow \int e^x\,dx \quad= e^x + c$$

$$f(x) = e^{ax+b} \Rightarrow \int e^{ax+b}\,dx = \frac{e^{ax+b}}{a} + c$$

> The straight lines in the term $\ln|x|$ are the **modulus** signs, used to indicate the **absolute value** of the expression inside. The absolute value is the value of the expression disregarding any negative sign, e.g. $|-7| = 7$ This is used as we cannot find the logarithm of a negative number.

Note that the power is linear, $ax + b$.

Similarly since we know $\dfrac{d}{dx}\ln x = \dfrac{1}{x}$, then:

$$f(x) = \frac{1}{x} \quad\Rightarrow \int \frac{1}{x}\,dx = \ln|x| + c$$

and:

$$f(x) = \frac{1}{ax + b} \Rightarrow \int \frac{1}{ax + b}\,dx = \frac{1}{a}\ln|ax + b| + c$$

> The modulus signs are needed. The denominator is linear, $ax + b$.

Integrating expressions using standard results for exponential functions and $\frac{1}{x}$

Example 3.1

Find $\displaystyle\int\left(\frac{1}{e^{\frac{1}{2}x}} + 3\sqrt{x}\right)dx$

$$\int\left(\frac{1}{e^{\frac{1}{2}x}} + 3\sqrt{x}\right)dx = \int\left(e^{-\frac{1}{2}x} + 3x^{\frac{1}{2}}\right)dx$$

> Rewrite as a list of terms using index form to replace the fraction and the square root sign.

$$= \frac{e^{-\frac{1}{2}x}}{-\frac{1}{2}} + \frac{3x^{\frac{3}{2}}}{\frac{3}{2}} + c$$

> Use $\displaystyle\int e^{ax+b}\,dx = \frac{e^{ax+b}}{a} + c$

$$= \frac{-2}{e^{\frac{x}{2}}} + 2x^{\frac{3}{2}} + c \text{ or } \frac{-2}{\sqrt{e^x}} + 2\sqrt{x^3} + c$$

$$= 2\sqrt{x^3} - \frac{2}{\sqrt{e^x}} + c$$

Example 3.2

Evaluate $\displaystyle\int_{e}^{2e} \frac{5}{3x}\,dx$ where $x \neq 0$

$$\int_{e}^{2e} \frac{5}{3x}\,dx = \int_{e}^{2e} \frac{5}{3}\frac{1}{x}\,dx$$

> Use the standard integral for $\displaystyle\int \frac{1}{x}\,dx = \ln|x| + c$

$$= \left[\frac{5}{3}\ln x\right]_{e}^{2e}\,dx$$

> There is no constant of integration because you are given the limits of integration. Include the limits $2e$ and e.

$$= \frac{5}{3}(\ln 2e - \ln e)$$ ← Take out the fraction multiplier to ease the calculation.

$$= \frac{5}{3}\left(\ln \frac{2e}{e}\right)$$ ← Use the law of logs, $\ln a - \ln b = \ln \frac{a}{b}$ to simplify.

$$= \frac{5}{3}\ln 2 \text{ or } = \frac{5\ln 2}{3}$$ ← The exact value answer.

$$\approx 1{\cdot}1552$$

Example 3.3

Find $\int \frac{1}{2x - 7} dx$ where $x \neq \frac{7}{2}$

$$\int \frac{1}{2x - 7} dx = \frac{\ln(2x - 7)}{2} + c$$ ←

Use the standard integral
$$\int \frac{1}{ax + b} dx = \frac{\ln(ax + b)}{a} + c$$
Divide by the coefficient of x.

Exercise 3C

Find these integrals.

1   a $\int 4e^x \, dx$

★ b $\int (6e^{3x} - 10e^{-5x}) \, dx$

c $\int \left(5\sqrt{x^3} + \frac{3}{e^{3x}}\right) dx$

d $\int -\frac{7}{3x} \, dx$

★ e $\int \frac{2}{3x + 4} \, dx$

f $\int \frac{8}{7 - 2x} \, dx$

g $\int (e^x + e^{-x})^2 \, dx$

h $\int \frac{e^x + 1}{e^x} \, dx$

i $\int \frac{5}{5x - 6} \, dx$

j $\int e^{-2x} \, dx$

k $\int \frac{1}{1 - x} \, dx$

Evaluate these definite integrals.

2   a $\int_{\frac{1}{2}}^{\frac{3}{2}} \left(8e^{-2x} + 3e^{-x}\right) dx$

★ b $\int_1^2 \left(e^{-4x} + \frac{5}{x}\right) dx$

c $\int_{-2e}^{-e} \frac{1}{x + 1} \, dx$

d $\int_0^2 \frac{3}{5 - 2x} \, dx$

Standard integral results for trigonometric functions

We can integrate complex trigonometric functions involving cosecant (cosec), secant (sec) and tangent (tan), using the following standard results. Refer back to the table on page 78 for useful trigonometric identities.

$$f(x) = \sec^2 x \qquad \Rightarrow \int \sec^2 x \, dx \qquad = \tan x + c$$

$$f(x) = \sec^2 ax \qquad \Rightarrow \int \sec^2 ax \, dx \qquad = \frac{1}{a} \tan ax + c$$

$$f(x) = \sec^2 (ax + b) \Rightarrow \int \sec^2 (ax + b) \, dx = \frac{1}{a} \tan (ax + b) + c$$

Example 3.4

Find $\int 5 \sec^2 \frac{1}{2} x \, dx$

$$\int 5 \sec^2 \frac{1}{2} x \, dx = \frac{5 \tan \frac{1}{2} x}{0.5} + c$$

$$= 5 \tan \frac{1}{2} x \times 2 + c \quad \bullet \!\!-\!\!-\!\!-\!\!- \boxed{\text{Integrate a function of a function using the chain rule.}}$$

$$= 10 \tan \frac{1}{2} x + c \quad \bullet \!\!-\!\!-\!\!-\!\!- \boxed{\text{Simplify.}}$$

Example 3.5

Find $\int \left(\tan^2 x + \frac{1}{2x} \right) dx$

$$\int \left(\tan^2 x + \frac{1}{2x} \right) dx = \int \left(\sec^2 x - 1 + \frac{1}{2} \frac{1}{x} \right) dx \quad \bullet \!\!-\!\!-\!\!- \boxed{\begin{array}{l}\text{Use the identity } \sec^2 x = 1 + \tan^2 x \\ \text{Rearrange to give } \tan^2 x = \sec^2 x - 1 \\ \text{Separate the numerical and algebraic} \\ \text{fractions.}\end{array}}$$

$$= \tan x - x + \frac{1}{2} \ln |x| + c \quad \bullet \!\!-\!\!-\!\!- \boxed{\text{Integrate each term.}}$$

Exercise 3D

1   Find these integrals.

   a $\int 6 \sec^2 3x \, dx$ b $\int \frac{1}{3} \sec^2 2x \, dx$ ★c $\int 4 \sec^2 \left(\frac{x}{2} \right) dx$

   d $\int (\tan^2 x + 1) \, dx$ e $\int \frac{3}{\cos^2 x} \, dx$ f $\int \left(3 \tan^2 x + \frac{6}{\cos^2 x} + 3 \right) dx$

   g $\int \cosec x \tan x \sec x \, dx$ h $\int -\cosec x \cot x \, dx$ i $\int \tan^2 4x \, dx$

2   Evaluate these definite integrals.

   a $\int_{-10}^{10} 4 \sec^2 2x \, dx$ ★b $\int_0^{\frac{\pi}{6}} \left(\cos x \ \sec^3 x \right) dx$

   c $\int_0^{\frac{\pi}{3}} \left(\sin x + \sec^2 x \right) dx$ ★d $\int_{-\frac{\pi}{4}}^{\frac{\pi}{4}} \frac{1 - \cos^2 x}{\cos^2 x} \, dx$

Standard integral results for inverse trigonometric functions

Use these standard integral results for inverse trigonometric functions.
Note that the tan function is the only inverse to bring a fraction to the front.

Function	Intermediate stage	Integral
$f(x) = \dfrac{1}{\sqrt{1 - x^2}}$	$\displaystyle\int \dfrac{1}{\sqrt{1 - x^2}}\, dx$	$\sin^{-1} x + c$
$f(x) = \dfrac{1}{\sqrt{a^2 - x^2}}$	$\displaystyle\int \dfrac{1}{\sqrt{a^2 - x^2}}\, dx$	$\sin^{-1} \dfrac{x}{a} + c$ •
$f(x) = \dfrac{1}{1 + x^2}$	$\displaystyle\int \dfrac{1}{1 + x^2}\, dx$	$\tan^{-1} x + c$
$f(x) = \dfrac{1}{a^2 + x^2}$	$\displaystyle\int \dfrac{1}{a^2 + x^2}\, dx$	$\dfrac{1}{a}\tan^{-1} \dfrac{x}{a} + c$ •

These integrals are provided in the exam formulae list.

Example 3.5

Evaluate $\displaystyle\int_{\frac{3}{2}}^{3} \dfrac{4}{\sqrt{9 - x^2}}\, dx$ where $x \neq 3$ or -3

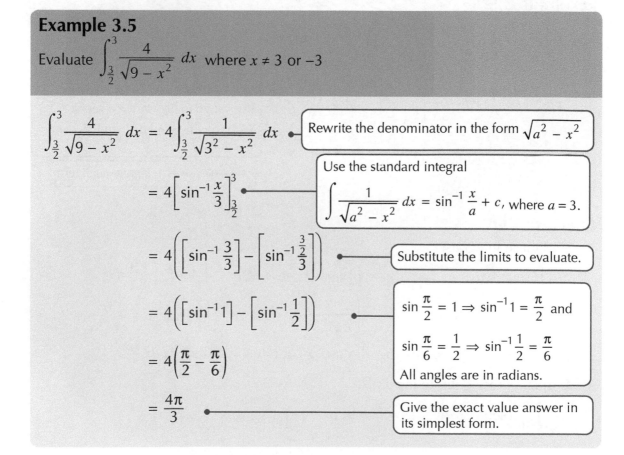

$$\int_{\frac{3}{2}}^{3} \dfrac{4}{\sqrt{9 - x^2}}\, dx = 4\int_{\frac{3}{2}}^{3} \dfrac{1}{\sqrt{3^2 - x^2}}\, dx$$

Rewrite the denominator in the form $\sqrt{a^2 - x^2}$

$$= 4\left[\sin^{-1}\dfrac{x}{3}\right]_{\frac{3}{2}}^{3}$$

Use the standard integral
$$\int \dfrac{1}{\sqrt{a^2 - x^2}}\, dx = \sin^{-1}\dfrac{x}{a} + c,\ \text{where } a = 3.$$

$$= 4\left(\left[\sin^{-1}\dfrac{3}{3}\right] - \left[\sin^{-1}\dfrac{\frac{3}{2}}{3}\right]\right)$$

Substitute the limits to evaluate.

$$= 4\left(\left[\sin^{-1}1\right] - \left[\sin^{-1}\dfrac{1}{2}\right]\right)$$

$\sin\dfrac{\pi}{2} = 1 \Rightarrow \sin^{-1}1 = \dfrac{\pi}{2}$ and

$\sin\dfrac{\pi}{6} = \dfrac{1}{2} \Rightarrow \sin^{-1}\dfrac{1}{2} = \dfrac{\pi}{6}$

All angles are in radians.

$$= 4\left(\dfrac{\pi}{2} - \dfrac{\pi}{6}\right)$$

$$= \dfrac{4\pi}{3}$$

Give the exact value answer in its simplest form.

Example 3.6

Find $\int \dfrac{dx}{1+4x^2}$

$\int \dfrac{dx}{1+4x^2}$

$1 + 4x^2 = 4\left(\dfrac{1}{4} + x^2\right)$ ●————— Rewrite the denominator in the form $a^2 + x^2$

$\therefore a^2 = \dfrac{1}{4} \Rightarrow a = \dfrac{1}{2}$

$\int \dfrac{dx}{1+4x^2} = \int \dfrac{dx}{4\left(\frac{1}{4} + x^2\right)}$

$\qquad = \dfrac{1}{4}\int \dfrac{dx}{4\frac{1}{4} + x^2}$

$\qquad = \dfrac{1}{4} \times \dfrac{1}{\frac{1}{2}}\tan^{-1}\dfrac{x}{\frac{1}{2}} + c$ ●—————

Use the standard integral

$\int \dfrac{1}{a^2 + x^2}\, dx = \dfrac{1}{a}\tan^{-1}\dfrac{x}{a} + c$

$\qquad = \dfrac{1}{2}\tan^{-1}2x + c$

Example 3.7

Find $\int \dfrac{1}{\sqrt{49 - 9x^2}}\, dx$

$\int \dfrac{1}{\sqrt{49 - 9x^2}}\, dx = \int \dfrac{1}{\sqrt{9\left(\frac{49}{9} - x^2\right)}}\, dx$

$\qquad = \int \dfrac{1}{\sqrt{9}\sqrt{\left(\frac{49}{9} - x^2\right)}}\, dx$ ●———— Rewrite the denominator with a unitary x^2

Use the standard integral

$\int \dfrac{1}{\sqrt{a^2 - x^2}}\, dx = \sin^{-1}\left(\dfrac{x}{a}\right) + c$ where $a = \dfrac{7}{3}$

$\qquad = \int \dfrac{1}{3\sqrt{\left(\frac{7}{3}\right)^2 - x^2}}\, dx$ ●————

$\qquad = \dfrac{1}{3}\int \dfrac{1}{\sqrt{\left(\frac{7}{3}\right)^2 - x^2}}\, dx$

$\qquad = \dfrac{1}{3}\sin^{-1}\dfrac{x}{\frac{7}{3}} + c \quad = \dfrac{1}{3}\sin^{-1}\dfrac{3x}{7} + c$

Exercise 3E

1 Find these integrals.

a $\displaystyle\int \frac{dx}{\sqrt{25-x^2}}$

★ b $\displaystyle\int \frac{dx}{\sqrt{2-x^2}}$

c $\displaystyle\int \frac{1}{\sqrt{36-9x^2}}\,dx$

d $\displaystyle\int -\frac{3}{1+9x^2}\,dx$

★ e $\displaystyle\int \frac{dx}{64x^2+4}$

2 Evaluate these integrals.

a $\displaystyle\int_0^3 \frac{dx}{\sqrt{45-5x^2}}$

★ b $\displaystyle\int_0^1 \frac{dx}{\sqrt{3-2x^2}}$

c $\displaystyle\int_{-1}^1 \frac{dx}{5+2x^2}$

★ d $\displaystyle\int_5^{15} \frac{dx}{3x^2+12}$

Integrating expressions using standard results for rational functions

We can integrate rational functions using the technique of integration by partial fractions. See Chapter 1 for more on partial fractions.

When the numerator is of the same or higher order as the denominator, we carry out algebraic division before rewriting in partial fraction form and then integrating.

Example 3.8

Find $\displaystyle\int \frac{3x^3-2x-4}{x^2-x-2}\,dx$

$$
\begin{array}{r}
3x + 3 \\
x^2-x-2\,\overline{\smash{\big)}\,3x^3 \qquad -2x \;-\; 4} \\
-\,(3x^3-3x^2-6x) \\
\hline
3x^2+4x\;-\;4 \\
-\,(3x^2-3x\;-\;6) \\
\hline
7x\;+\;2
\end{array}
$$

First perform algebraic division.

$$\int \frac{3x^3-2x-4}{x^2-x-2}\,dx = \int\left(3x+3+\frac{7x+2}{x^2-x-2}\right)dx$$

Rewrite the given integral.

$$= \int\left(3x+3+\frac{7x+2}{(x+1)(x-2)}\right)dx$$

Factorise the denominator.

Let $\dfrac{7x + 2}{(x + 1)(x - 2)} = \dfrac{A}{x + 1} + \dfrac{B}{x - 2}$ — Use partial fractions to rewrite the algebraic fraction.

$\therefore \dfrac{7x + 2}{(x + 1)(x - 2)} = \dfrac{A(x - 2) + B(x + 1)}{(x + 1)(x - 2)}$ — Write as a single fraction.

$A(x - 2) + B(x + 1) = 7x + 2$ — Equate numerators.

Let $x = 2$

$A(2 - 2) + B(2 + 1) = (7 \times 2) + 2$ — Solve for A and B.

$$3B = 16 \Rightarrow B = \dfrac{16}{3}$$

Let $x = -1$

$A(-1 - 2) + B(-1 + 1) = (7 \times -1) + 2$

$$-3A = -5$$

$$A = \dfrac{5}{3}$$

$$\int \left(3x + 3 + \dfrac{7x + 2}{x^2 - x - 2} \right) dx = \int \left(3x + 3 + \dfrac{5}{3(x + 1)} + \dfrac{16}{3(x - 2)} \right) dx$$

$$= \dfrac{3x^2}{2} + 3x + \dfrac{5}{3} \ln|x + 1| + \dfrac{16}{3} \ln|x - 2| + c$$

Exercise 3F

1 Find these integrals:

★ a $\displaystyle\int \dfrac{x\, dx}{(x - 3)(x - 1)(x + 2)}$

b $\displaystyle\int \dfrac{x^2 + 4x - 1}{x^3 + 2x^2 + x}\, dx$

c $\displaystyle\int \dfrac{x^2}{x^2 - 4}\, dx$

d $\displaystyle\int \dfrac{x^3 + 2}{x(x + 1)}\, dx$

e $\displaystyle\int_{1}^{\frac{3}{2}} \dfrac{4}{x(x - 2)^2}\, dx$

2 Evaluate these integrals:

★ a $\displaystyle\int_{-10}^{-5} \dfrac{x^2 + 2}{x^2 + 5x + 6}\, dx$

b $\displaystyle\int_{\frac{1}{5}}^{\frac{4}{5}} \dfrac{x^2 - 4}{x(x - 1)}\, dx$

Integrating by substitution to evaluate integrals given the substitution function

Integrating by substitution is used to make integration easier. It can be applied to composite functions. Making the correct substitution can make difficult integrals much more straightforward.

For straightforward integrals where the substitution is obvious, the function used to substitute will not be given. In some situations the function to use in the substitution process is not obvious. A suitable function will be suggested in such questions. When integrating using substitution, all terms in the original variable x, must be replaced, leaving the new integral in terms only of the new given variable, often expressed as u. This can make the integration easier to carry out. Once the substitution is complete, resubstitute for u so that the answer is expressed in terms of the original variable.

Example 3.9

Find $\int \dfrac{2\sqrt{x} + 3}{\sqrt{x}} dx$ using the substitution $u = 2\sqrt{x} + 3$

$u = 2\sqrt{x} + 3 \quad \dfrac{du}{dx} = \dfrac{d}{dx}\left(2x^{\frac{1}{2}} + 3\right) = x^{-\frac{1}{2}} = \dfrac{1}{\sqrt{x}}$ ● — Differentiate the function u with respect to x.

$du = \dfrac{1}{\sqrt{x}} dx$ ● — Rearrange to get du on the LHS.

$\int \dfrac{2\sqrt{x} + 3}{\sqrt{x}} dx = \int u\, du$ ● — Rewrite the integral using the substitutions for u and du.

$= \dfrac{u^2}{2} + c$

$= \dfrac{(2\sqrt{x} + 3)^2 + c}{2}$ ● — Substitute $2\sqrt{x} + 3$ for u to obtain the answer in terms of the original variable x.

Example 3.10

Find $\int 4x\sqrt{x^2 - 1}\, dx$ using the substitution $u^2 = x^2 - 1$

$u^2 = x^2 - 1 \Rightarrow u = \sqrt{x^2 - 1} = \left(x^2 - 1\right)^{\frac{1}{2}}$

$\dfrac{du}{dx} = \dfrac{d}{dx}\left(x^2 - 1\right)^{\frac{1}{2}}$ ● — Differentiate both sides of $u^2 = x^2 - 1$ and simplify.

$2u\dfrac{du}{dx} = 2x$

$u\dfrac{du}{dx} = x$

$\int 4x\sqrt{x^2 - 1}\, dx = \int 4u^2\, du$ ● — Rewrite the integral using the substitutions for u and du.

$= \dfrac{4u^3}{3} + c$

$= \dfrac{4\left(\sqrt{x^2 - 1}\right)^3}{3} + c$ ● — Use the substitution $u = \sqrt{x^2 - 1}$ to rewrite the answer in terms of the original variable x.

Example 3.11

Use the substitution $x = 2\sin u$ to find $\displaystyle\int \frac{8}{\sqrt{4 - x^2}}\, dx$

$x = 2\sin u$

$\dfrac{dx}{du} = 2\cos u$ —————————————————— $\boxed{\text{Differentiate and simplify.}}$

$dx = 2\cos u\, du$

$\displaystyle\int \frac{8}{\sqrt{4 - x^2}}\, dx = \int \frac{8}{\sqrt{4 - (2\sin u)^2}} 2\cos u\, du$ —— $\boxed{\begin{array}{l}\text{Substitute using } x = 2\sin u \\ \text{and } dx = 2\cos u\, du\end{array}}$

$\displaystyle = \int \frac{16\cos u}{\sqrt{4 - 4\sin^2 u}}\, du$

$\displaystyle = \int \frac{16\cos u}{\sqrt{4(1 - \sin^2 u)}}\, du$ ———— $\boxed{\text{Use } 1 - \sin^2 u = \cos^2 u}$

$\displaystyle = \int \frac{16\cos u}{\sqrt{4\cos^2 u}}\, du$

$\displaystyle = \int \frac{16\cos u}{2\cos u}\, du$

$\displaystyle = \int 8\, du$

$= 8u + c$

$= 8\sin^{-1}\dfrac{x}{2} + c$ ————— $\boxed{\begin{array}{l}\text{Present the answer in terms of} \\ \text{the original variable } x \text{ by} \\ \text{substituting } \sin^{-1}\dfrac{x}{2} \text{ for } u\end{array}}$

Exercise 3G

1 Find these integrals using the substitution given in square brackets.

   a $\displaystyle\int \cos x\, e^{\sin x}\, dx$ $[u = \sin x]$ ★ b $\displaystyle\int e^{x^2 + 4x}(x + 2)\, dx$ $[u = x^2 + 4x]$

   c $\displaystyle\int 2e^x(e^x + 1)^4\, dx$ $[u = e^x + 1]$ d $\displaystyle\int \cos^3 x \sin x\, dx$ $[u = \cos x]$

   e $\displaystyle\int \frac{e^{-2x}}{1 - e^{-2x}}\, dx$ $[u = 1 - e^{-2x}]$ ★ f $\displaystyle\int \frac{\cos x}{\sin^3 x}\, dx$ $[u = \sin x]$

2 $\displaystyle\int \frac{\cos x - \sin x}{\cos x + \sin x}\, dx$ $[u = \cos x + \sin x]$

3 $\displaystyle\int \frac{3}{\sqrt{9 - x^2}}\, dx$ $[x = 3\sin u]$ $\boxed{\text{Simplify first, then integrate.}}$ ⚠

4 $\int \sqrt{3 - x^2}\, dx$ $[x = \sqrt{3} \sin u]$

> Use $\cos 2u = 2\cos^2 u - 1$, then simplify, then integrate.

5 $\int 2x\sqrt{3 + x^2}\, dx$ $[u^2 = 3 + x^2]$

★6 $\int e^x \sqrt{e^x - 1}\, dx$ $[u^2 = e^x - 1]$

★7 $\int \dfrac{x}{\sqrt{(4 - x^2)}}\, dx$ $[u^2 = 4 - x^2]$

8 Find $\int \dfrac{\operatorname{cosec}^2 x}{\cot^3 x}\, dx$

9 Find $\int -\operatorname{cosec}^2 x\, dx$

> Use trig identities to rewrite the integral, then let $u = \tan x$

Integrating by substitution using special cases

There are some special substitutions that are so commonly used that they can be treated and used as standard integrals.

1 $\int \dfrac{f'(x)}{f(x)}\, dx$

Let $u = f(x)$ then $\dfrac{du}{dx} = f'(x) \Rightarrow du = f'(x)dx$

$\int \dfrac{f'(x)}{f(x)}\, dx = \int \dfrac{du}{u} = \ln|u| + c$ •————————— Substituting.

$\int \dfrac{f'(x)}{f(x)}\, dx = \ln|f(x)| + c$ •————————— Substituting back.

This can be used as a standard integral, for example:

$\int \dfrac{4x - 3}{2x^2 - 3x}\, dx = \ln\left|2x^2 - 3\right| + c$ $\int \cot\theta\, d\theta = \int \dfrac{\cos\theta}{\sin\theta}\, d\theta = \ln|\sin\theta| + c$

2 $\int f'(x)f(x)dx$

Let $u = f(x)$ then $\dfrac{du}{dx} = f'(x) \Rightarrow du = f'(x)dx$

$\int f'(x)f(x)dx = \int u\, du = \tfrac{1}{2}u^2 + c$ •————————— Substituting.

$\int f'(x)f(x)\, dx = \tfrac{1}{2}(f(x))^2 + c$ •————————— Substituting back.

This can be used as a standard integral, for example:

$\int (4x - 3)(2x^2 - 3x)\, dx = \tfrac{1}{2}\left(2x^2 - 3x\right)^2 + c$

$\int 3x\left(x^2 + 2\right) dx = \tfrac{3}{2}\int 2x\left(x^2 + 2\right) dx = \tfrac{3}{2} \times \tfrac{1}{2}\left(x^2 + 2\right)^2 + c = \tfrac{3}{4}\left(x^2 + 2\right)^2 + c$

To integrate a product of functions in the form $\int g(f(x))f'(x)\,dx$:

- multiply out the brackets and integrate term by term
- use the given substitution.

Examples 3.12–3.14 illustrate the method, replacing x terms with u.

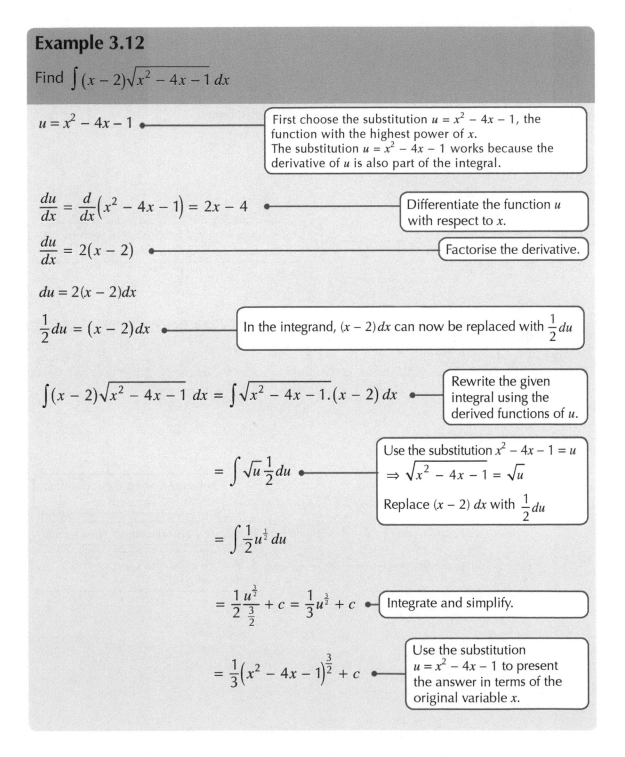

Example 3.12

Find $\int (x-2)\sqrt{x^2-4x-1}\,dx$

$u = x^2 - 4x - 1$

> First choose the substitution $u = x^2 - 4x - 1$, the function with the highest power of x.
> The substitution $u = x^2 - 4x - 1$ works because the derivative of u is also part of the integral.

$\dfrac{du}{dx} = \dfrac{d}{dx}\left(x^2 - 4x - 1\right) = 2x - 4$

> Differentiate the function u with respect to x.

$\dfrac{du}{dx} = 2(x - 2)$

> Factorise the derivative.

$du = 2(x - 2)dx$

$\dfrac{1}{2}du = (x - 2)dx$

> In the integrand, $(x - 2)\,dx$ can now be replaced with $\dfrac{1}{2}du$

$\int (x-2)\sqrt{x^2-4x-1}\,dx = \int \sqrt{x^2-4x-1}\cdot(x-2)\,dx$

> Rewrite the given integral using the derived functions of u.

$= \int \sqrt{u}\,\dfrac{1}{2}du$

> Use the substitution $x^2 - 4x - 1 = u$
> $\Rightarrow \sqrt{x^2 - 4x - 1} = \sqrt{u}$
> Replace $(x - 2)\,dx$ with $\dfrac{1}{2}du$

$= \int \dfrac{1}{2}u^{\frac{1}{2}}\,du$

$= \dfrac{1}{2}\dfrac{u^{\frac{3}{2}}}{\frac{3}{2}} + c = \dfrac{1}{3}u^{\frac{3}{2}} + c$

> Integrate and simplify.

$= \dfrac{1}{3}\left(x^2 - 4x - 1\right)^{\frac{3}{2}} + c$

> Use the substitution $u = x^2 - 4x - 1$ to present the answer in terms of the original variable x.

Example 3.13

Find $\displaystyle\int \frac{3(\ln x)^2}{x}\,dx$

$u = \ln x$

$\dfrac{du}{dx} = \dfrac{d}{dx}(\ln x) = \dfrac{1}{x}$

$du = \dfrac{1}{x}dx$

$\displaystyle\int \frac{3(\ln x)^2}{x}\,dx = \int 3(\ln x)^2 \cdot \frac{1}{x}\,dx$ ●────── Rewrite the integral using the substitutions.

$\qquad = \displaystyle\int 3u^2\,du$

$\qquad = \dfrac{3u^3}{3} + c$

$\qquad = u^3 + c$ ●────── Integrate and simplify. Replace u with $\ln x$ to obtain your answer in terms of x.

$\qquad = (\ln x)^3 + c$

Example 3.14

Find $\displaystyle\int \sin^3 x \sin 2x\,dx$

$u = \sin x$

$\dfrac{du}{dx} = \dfrac{d}{dx}(\sin x) = \cos x$

$du = \cos x\,dx$

$\displaystyle\int \sin^3 x \sin 2x\,dx = \int \sin^3 x\, 2\sin x \cos x\,dx$ ●────── Substitute using the indentify $\sin 2x = 2\sin x \cos x$

$\qquad = \displaystyle\int 2\sin^4 x \cos x\,dx$ ●────── Simplify by combining all $\sin x$ terms.

$\qquad = \displaystyle\int 2u^4\,du$ ●────── Substitute u for $\sin x \cos x\,dx$ and replace with du.

$\qquad = \dfrac{2u^5}{5} + c$

$\qquad = \dfrac{2}{5}\sin^5 x + c$

Exercise 3H

Find these indefinite integrals. Choose a suitable function to substitute or use the suggested function.

★ 1 a $\int \dfrac{6x+5}{3x^2+5x}\,dx$ b $\int \tan x\,dx$ c $\int \dfrac{e^{2x}}{e^{2x}+1}\,dx$ d $\int x(x^2-3)\,dx$

★ 2 a $\int (2x+3)(x^2+3x-6)^5\,dx$ $[u=x^2+3x-6]$

   b $\int x^2(x^3-5)^4\,dx$ $[u=x^3-5]$ c $\int x^3(x^4+2)^7\,dx$ $[u=x^4+2]$

3 a $\int 2x\sqrt{(x^2-3)}\,dx$ b $\int 3x^2\sqrt{(x^3-8)}\,dx$ c $\int x^3\sqrt{(5+x^4)}\,dx$

4 a $\int \dfrac{4\ln x}{x}\,dx$ b $\int \dfrac{3}{x\ln x}\,dx$ ★ c $\int \dfrac{\cos(\ln x)}{x}\,dx$

5 a $\int \dfrac{1}{3x}\ln x\,dx$ b $\int 3x^2(1-x^3)^3\,dx$ c $\int 2x(x^2+3)^{-4}\,dx$

   d $\int 8x^3\sqrt{x^4+1}\,dx$ e $\int 3x^2\sqrt[3]{(x^3-3)}\,dx$

★ 6 $\int \dfrac{2x^3-x-1}{(x-3)(x^2+1)}\,dx$

★ 7 $\int \dfrac{x+2}{(3x^2+12x-7)^3}\,dx$ 8 $\int \dfrac{4x^3-3}{\sqrt{x^4-3x}}\,dx$ 9 $\int \dfrac{\sec^2 x}{\tan x}\,dx$

Integrating by substitution to evaluate definite integrals

Assuming that the function to be integrated is continuous over the interval of integration, then, exchanging the limits for x with the corresponding limits for u will save you having to substitute back after the integration process.

Example 3.15

Evaluate $\int_3^5 4x\sqrt{x^2-9}\,dx$ using the substitution $u=x^2-9$

$u=x^2-9$

$\dfrac{du}{dx}=\dfrac{d}{dx}(x^2-9)=2x$

$du=2x\,dx$

$\int_3^5 4x\sqrt{x^2-9}\,dx = \int_a^b 2\sqrt{u}\,du$ ◄— Substitute u and du. The new variable requires different limits.

When $x = 3$, $u = 3^2 - 9 = 0$

When $x = 5$, $u = 5^2 - 9 = 16$ ●————— Calculate the limits for u

$\int_0^{16} 2\sqrt{u}\ du = \int_0^{16} 2u^{\frac{1}{2}}\ du$ ●————— Rewrite in index form to replace the square root sign.

$$= \left[\frac{2u^{\frac{3}{2}}}{\frac{3}{2}} \right]_0^{16} = \left[\frac{4}{3} u^{\frac{3}{2}} \right]_0^{16}$$

$$= \frac{4}{3}(16)^{\frac{3}{2}} - \frac{4}{3}(0)^{\frac{3}{2}} = \frac{4}{3} \times 4^3 - 0 = 85\frac{1}{3}$$

Example 3.16

Evaluate $\displaystyle\int_0^1 \frac{3}{\sqrt{\left(4 - x^2\right)^3}}\ dx$ using the substitution $x = 2\sin q$

$x = 2\sin q$

$\dfrac{dx}{dq} = \dfrac{d}{dq}\left(2\sin q\right) = 2\cos q$

$dx = 2\cos q\ dq$

Replace x with $2 \sin q$ and replace dx with $2\cos q\, dq$. The new variable requires different limits.

$$\int_0^1 \frac{3}{\sqrt{\left(4 - x^2\right)^3}}\ dx = \int_a^b \frac{3 \times 2\cos q}{\sqrt{\left(4 - \left(2\sin q\right)^2\right)^3}}\ dq$$

Evaluate the squared terms and simplify the numerator.

$$= \int_a^b \frac{6\cos q}{\sqrt{\left(4 - 4\sin^2 q\right)^3}}\ dq$$

Factorise the denominator.

$$= \int_a^b \frac{6\cos q}{\sqrt{\left(4\left(1 - \sin^2 q\right)\right)^3}}\ dq$$

$$= \int_a^b \frac{6\cos q}{\sqrt{\left(4\cos^2 q\right)^3}}\ dq$$

Begin to evaluate $\sqrt{\left(4\cos^2 q\right)^3}$

Substitute $1 - \sin^2 q = \cos^2 q$

For trig substitutions it is highly likely that the denominator will become one of:

$\sqrt{1 - \cos^2 x} = \sin x$

$\sqrt{1 - \sin^2 x} = \cos x$

$\sqrt{1 + \tan^2 x} = \sec x$

$$= \int_a^b \frac{6\cos q}{\sqrt{4^3}\sqrt{(\cos^2 q)^3}}\, dq$$

> Separate the denominator to evaluate.

$$= \int_a^b \frac{6\cos q}{8\cos^3 q}\, dq = \int_a^b \frac{3}{4\cos^2 q}\, dq$$

> Evaluate the square roots in the denominator. Simplify by cancelling.

$$= \int_a^b \frac{3}{4}\frac{1}{\cos^2 q}\, dq$$

> Separate the numeric and trigonometric terms.

$$= \int_a^b \frac{3}{4}\sec^2 q\, dq$$

$$= \left[\frac{3}{4}\tan q\right]_a^b$$

> Use the standard integral
> $$\int \sec^2 x\, dx = \tan x + c$$

When $x = 0$, $2\sin q = 0 \Rightarrow q = \sin^{-1} 0 \Rightarrow q = 0$

When $x = 1$, $2\sin q = 1 \Rightarrow \sin^{-1}\frac{1}{2} \Rightarrow q\frac{\pi}{6}$

> Calculate the new limits using $x = \sin q$.

$$\left[\frac{3}{4}\tan q\right]_a^b = \left[\frac{3}{4}\tan q\right]_0^{\frac{\pi}{6}}$$

> Replace a and b with the new limits.

$$= \frac{3}{4}\left(\tan\frac{\pi}{6} - \tan 0\right)$$

> Substitute the new limits to find the exact value answer in its simplest form.

$$= \frac{3}{4}\left(\frac{1}{\sqrt{3}} - 0\right) = \frac{3}{4\sqrt{3}}$$

$$= \frac{3}{4\sqrt{3}} \times \frac{\sqrt{3}}{\sqrt{3}} = \frac{\sqrt{3}}{4}$$

> Rationalise the denominator.

Exercise 3I

Evaluate these definite integrals using the substitution in square brackets:

1 $\int_{-1}^{0} 6x(x^2 - 3)^{-5}\, dx$ $[u = x^2 - 3]$

2 $\int_{3}^{4} (2 - x^2)(5 + 6x - x^3)^3\, dx$ $[u = 5 + 6x - x^3]$

★ 3 $\int_{5}^{6} x^3\sqrt{x^4 - 2}\, dx$ $[u = x^4 - 2]$

4 $\int_{\sqrt{2}}^{\sqrt{6}} \frac{4x}{\sqrt{x^2 - 2}}\, dx$ $[u = x^2 - 2]$

★ 5 $\displaystyle\int_3^5 \frac{x}{\sqrt{x^2 - 5}}\,dx$ $\qquad [u^2 = x^2 - 5]$

6 $\displaystyle\int_0^2 \sqrt{16 - x^2}\,dx$ $\qquad [x = 4\sin\theta]$

★ 7 $\displaystyle\int_{-10}^{10} \frac{\sec^2 3x}{1 + \tan 3x}\,dx$ $\qquad [u = 1 + \tan 3x]$

✖ ★ 8 $\displaystyle\int_{-\sqrt{3}}^{\sqrt{3}} \frac{9}{x^2 + 9}\,dx$ $\qquad [x = 3\tan u]$

✖ 9 $\displaystyle\int_{-\frac{3}{2}}^{\frac{3}{2}} \frac{1}{\sqrt{3 - x^2}}\,dx$ $\qquad [x = \sqrt{3}\sin u]$

10 $\displaystyle\int_e^{e^2} \frac{dx}{x\ln x}\,dx$ $\qquad [u = \ln x]$

Integrating higher powers of trigonometric functions

By using trigonometric identities and an appropriate substitution it is possible to integrate higher powers of trigonometric functions such as $\int \sin^n x\,dx$ and $\int \cos^n x\,dx$
Example 3.17 shows how to deal with:

* **odd** powers in part **a**

* **even** powers in part **b**.

Example 3.17

Find these indefinite integrals:

a $\int \sin^3 x\,dx$ $\qquad$ b $\int \cos^4 x\,dx$

a $\int \sin^3 x\,dx = \int \sin^2 x\,\sin x\,dx = \int(1 - \cos^2 x)\sin x\,dx$ ⟵ Substitute $\sin^2 x = 1 - \cos^2 x$

Let $u = \cos x$

$\dfrac{du}{dx} = -\sin x$

$du = -\sin x\,dx$

$\int(1 - \cos^2 x)\sin x\,dx = \int -(1 - u^2)\,du$

$\qquad\qquad\qquad\qquad = \int\left(-1 + u^2\right)du$

$\qquad\qquad\qquad\qquad = -u + \dfrac{u^3}{3} + c$

$\qquad\qquad\qquad\qquad = -\cos x + \dfrac{\cos^3 x}{3} + c$ ⟵ Change back to x using $u = \cos x$

b $\displaystyle\int \cos^4 x\, dx = \int \left(\cos^2 x\right)^2 dx = \int \left(\frac{1}{2}(1+\cos 2x)\right)^2 dx$

> Use
> $\cos^2 x = \frac{1}{2}(1 + \cos 2x)$

$\displaystyle = \frac{1}{4}\int \left(1 + 2\cos 2x + \cos^2 2x\right) dx$

> Use
> $\cos^2 x = \frac{1}{2}(1 + \cos 2x)$
> again.

$\displaystyle = \frac{1}{4}\int \left(1 + 2\cos 2x + \frac{1}{2} + \frac{1}{2}\cos 4x\right) dx$

$\displaystyle = \frac{1}{4}\left(x + \sin 2x + \frac{1}{2}x + \frac{1}{8}\sin 4x\right) + c$

$\displaystyle = \frac{3}{8}x + \frac{1}{4}\sin 2x + \frac{1}{32}\sin 4x + c$

Exercise 3J

★ 1 Find:

   a $\displaystyle\int \cos^3 x\, dx$ **b** $\displaystyle\int \sin^5 x\, dx$

2 Find:

   a $\displaystyle\int \sin^4 x\, dx$ **b** $\displaystyle\int \cos^2 2x\, dx$

3 Find these expressions using the identity $\tan^2 x = \sec^2 x - 1$

   a $\displaystyle\int \tan^3 x\, dx$ **b** $\displaystyle\int \tan^5 x\, dx$

Integrating by parts

Integration by parts is used to evaluate a product of functions written as u and $\dfrac{dv}{dx}$

The formula tells us that one function, $\dfrac{dv}{dx}$, is being integrated and the other function, u, is being differentiated.

A general rule for deciding which function to differentiate and which to integrate is to choose the functions so that the next integral is reduced to a single function that can be evaluated. Examples 3.18 and 3.19 show how this works.

Using integration by parts once

Using integration by parts once works for some products of functions. The integral of a product of two functions does not equal the product of the integrals of the two functions, that is:

$$\int f(x)\, g(x)\, dx \neq \int f(x)\, dx \times \int g(x)\, dx$$

Integration by parts is developed from the product rule for differentiating, as shown below.

$$\frac{d}{dx}(uv) = v\frac{du}{dx} + u\frac{dv}{dx}$$

$$\int \frac{d}{dx}(uv)\,dx = \int \left(v\frac{du}{dx} + u\frac{dv}{dx}\right)dx$$ ● ——— Integrate both sides.

$$uv = \int v\frac{du}{dx}\,dx + \int u\frac{dv}{dx}\,dx$$ ● Simplify LHS by cancelling and separate the RHS.

$$\int u\frac{dv}{dx}\,dx = uv - \int v\frac{du}{dx}\,dx$$ ● Rearrange to give the formula for integration by parts.

Remember $\int u\dfrac{dv}{dx} = uv - \int v\dfrac{du}{dx}$ is the same as $\int uv' = uv - \int vu'$

Example 3.18

Find the integral $\int 4x\cos 2x\,dx$

$$\int u\frac{dv}{dx} = uv - \int v\frac{du}{dx}$$ ● ——— State the formula for integration by parts.

Let $\int 4x\cos 2x\,dx = \int u\dfrac{dv}{dx}$

$$= uv - \int v\frac{du}{dx}$$

Let $u = 4x$

$\dfrac{du}{dx} = 4$

Let $\dfrac{dv}{dx} = \cos 2x$

$v = \int \cos 2x\,dx = \dfrac{1}{2}\sin 2x$

$$\int 4x\cos 2x\,dx = 4x \times \frac{1}{2}\sin 2x - \int 4 \times \frac{1}{2}\sin 2x\,dx$$ ● Substitue for u, v and $\dfrac{du}{dx}$

$$= 2x\sin 2x - \int 2\sin 2x\,dx$$ ● Simplify the coefficient of each term on RHS.

$$= 2x\sin 2x - 2\left(-\frac{1}{2}\cos 2x\right) + c$$ The resultant integral can be evaluated in one step.

$$= 2x\sin 2x + \cos 2x + c$$

Example 3.19

Evaluate $\displaystyle\int_1^2 \frac{\ln x}{x^3}\, dx$

$$\int u \frac{dv}{dx} = uv - \int v \frac{du}{dx}$$

Let $\displaystyle\int_1^2 \ln x\, x^{-3}\, dx = \int u \frac{dv}{dx}$

Let $u = \ln x$

$$\frac{du}{dx} = \frac{d}{dx}\ln x = \frac{1}{x}$$

Let $\dfrac{dv}{dx} = x^{-3}$

$$v = \int x^{-3}\, dx = \frac{x^{-2}}{-2}$$

$$\int_1^2 \ln x\, x^{-3}\, dx = uv - \int v \frac{du}{dx}$$

$$uv - \int v \frac{du}{dx} = \left[\ln x \frac{x^{-2}}{-2} - \int_1^2 \frac{1}{x}\frac{x^{-2}}{-2}\, dx \right]$$

> Simplifiy the resultant as far as possible then substitute the limits to evaluate the integral.

$$= -\frac{\ln x}{2x^2} + \int_1^2 \frac{1}{2} x^{-3}\, dx$$

$$= \left[-\frac{\ln x}{2x^2} + \frac{1}{2} \times \frac{x^{-2}}{2} \right]_1^2$$

$$= \left[-\frac{\ln x}{2x^2} - \frac{1}{4x^2} \right]_1^2$$

$$= \left[-\frac{\ln 2}{2 \times 2^2} - \frac{1}{4 \times 2^2} \right] - \left[-\frac{\ln 1}{2 \times 1^2} - \frac{1}{4 \times 1^2} \right]$$

$$= \left[-\frac{\ln 2}{8} - \frac{1}{16} \right] - \left[0 - \frac{1}{4} \right] = \frac{3}{16} - \frac{\ln 2}{8}$$

$$\approx 0{\cdot}1009$$

Integration by parts allows us to integrate some functions, which appear as if they cannot be integrated. Example 3.20 shows that by using a dummy variable (e. g. 1) we can use this technique to find a solution. The same applies to integrating the inverse trig functions.

Example 3.20

Evaluate $\displaystyle\int_1^2 \ln x\, dx$

$$\int u \frac{dv}{dx} = uv - \int v \frac{du}{dx}$$

$$\int_1^2 \ln x \ dx = \int_1^2 \ln x \times 1 dx$$

> Use a dummy variable to ease the integration.

Let $\int_1^2 \ln x \times 1 dx = \int u \dfrac{dv}{dx}$

$$= uv - \int v \dfrac{du}{dx}$$

Let $u = \ln x$

$$\dfrac{du}{dx} = \dfrac{d}{dx}\ln x = \dfrac{1}{x}$$

Let $\dfrac{dv}{dx} = 1$

$$v = \int 1 dx = x$$

$$\int_1^2 \ln x \times 1 dx = uv - \int v \dfrac{du}{dx}$$

$$uv - \int v \dfrac{du}{dx} = \left[x\ln x - \int_1^2 \dfrac{1}{x} x \ dx \right]$$

$$= \left[x\ln x - \int_1^2 1 dx \right]$$

$$= [x\ln x - x]_1^2 = \left[(2\ln 2 - 2) - (1\ln 1 - 1)\right]$$

$$= 2\ln 2 - 2 - 0 + 1 = 2\ln 2 - 1$$

$$\approx 0{\cdot}3863$$

Exercise 3K

Find these integrals using integration by parts.

1 $\int x\sin x dx$

★2 $\int_{-1}^1 e^x(2x + 3)dx$

3 $\int_0^2 (5 - 3x)e^x \ dx$

★4 $\int 5x\sec^2 5x dx$

★5 $\int \dfrac{3x - 1}{\cos^2 x} dx$

6 $\int_0^2 e^x(4x + 1)dx$

7 $\int \sin^{-1} x dx$

Using integration by parts twice

Sometimes you need to use the method of integration by parts twice, as shown in Examples 3.21 and 3.22.

Example 3.21

Find the integral $\int 9x^2 e^{3x} dx$

$$\int u\dfrac{dv}{dx} = uv - \int v \dfrac{du}{dx}$$

Let $\int 9x^2 e^{3x}\, dx = \int u\dfrac{dv}{dx}$ ──────── First use of integration by parts.

$$= uv - \int v\dfrac{du}{dx}$$

Let $u = 9x^2$

$$\dfrac{du}{dx} = 18x$$

Let $\dfrac{dv}{dx} = e^{3x}$

$$v = \int e^{3x}\, dx = \tfrac{1}{3}e^{3x}$$

$\int 9x^2 e^{3x}\, dx = 9x^2 \times \dfrac{1}{3}e^{3x} - \int 18x \times \dfrac{1}{3}e^{3x}\, dx$ ──────── The first substitution process.

$$= 3x^2 e^{3x} - \int 6x e^{3x}\, dx$$ ──────── Simplify throughout. The resultant integral is a product of two functions.

Now let $\int 6x e^{3x}\, dx = \int u\dfrac{dv}{dx}$ ──────── Second use of integration by parts for the new second element.

Let $u = 6x$

$$\dfrac{du}{dx} = 6$$

Let $\dfrac{dv}{dx} = e^{3x}$

$$v = \int e^{3x}\, dx = \dfrac{1}{3}e^{3x}$$

$3x^2 e^{3x} - \int 6x e^{3x}\, dx = 3x^2 e^{3x} - \left(uv - \int v\dfrac{dv}{dx}\right)$

$$= 3x^2 e^{3x} - \left(6x \times \dfrac{1}{3}e^{3x} - \int 6 \times \dfrac{1}{3}e^{3x}\, dx\right)$$ ──────── The second substitution process.

$$= 3x^2 e^{3x} - 2x e^{3x} + \int 2 e^{3x}\, dx$$ ──────── Simplify throughout.

$$= 3x^2 e^{3x} - 2x e^{3x} + \dfrac{2}{3}e^{3x} + c$$ ──────── Integrate the final term. Simplify by taking out the common factor e^{3x}

$$= e^{3x}\left(3x^2 - 2x + \dfrac{2}{3}\right) + c$$

Example 3.22

Evaluate the definite integral of $\int_{-2}^{0} x^2 e^{2x}\, dx$ using integration by parts.

$$\int u\dfrac{dv}{dx} = uv - \int v\dfrac{du}{dx}$$

Let $\int_{-2}^{0} x^2 e^{2x}\, dx = \int u\dfrac{dv}{dx}$ ──────── First use of integration by parts.

$$= uv - \int v\dfrac{du}{dx}$$

Let $u = x^2$

$\dfrac{du}{dx} = 2x$

Let $\dfrac{dv}{dx} = e^{2x} dx$

$v = \int e^{2x} dx$

$\quad = \dfrac{1}{2}e^{2x}$

$\displaystyle\int_{-2}^{0} x^2 e^{2x}\, dx = \left[\dfrac{1}{2}x^2 e^{2x} - \int x e^{2x} dx\right]_{-2}^{0}$

Now let $\displaystyle\int x e^{2x}\, dx = \int u\,\dfrac{dv}{dx}$

> Second use of integration by parts for the new second element.

$\left[\dfrac{1}{2}x^2 e^{2x} - \displaystyle\int x e^{2x} dx\right]_{-2}^{0} = \left[\dfrac{1}{2}x^2 e^{2x} - \left(uv - \displaystyle\int v\,\dfrac{du}{dx}\right)\right]_{-2}^{0}$

Let $u = x$

$\dfrac{du}{dx} = 1$

Let $\dfrac{dv}{dx} = e^{2x}$

$v = \displaystyle\int e^{2x} dx = \dfrac{1}{2}e^{2x}$

$\left[\dfrac{1}{2}x^2 e^{2x} - \displaystyle\int x e^{2x} dx\right] = \left[\dfrac{1}{2}x^2 e^{2x} - \left(\dfrac{1}{2}x e^{2x} - \displaystyle\int \dfrac{1}{2}e^{2x} dx\right)\right]_{-2}^{0}$

$= \left[\dfrac{1}{2}x^2 e^{2x} - \dfrac{1}{2}x e^{2x} + \dfrac{1}{4}e^{2x}\right]_{-2}^{0}$

$= \left[\dfrac{1}{2}x e^{2x}(x - 1) + \dfrac{1}{4}e^{2x}\right]_{-2}^{0}$

$= \left(\dfrac{1}{2}(0)e^{0}(0 - 1) + \dfrac{1}{4}e^{0}\right) - \left(\dfrac{1}{2}(-2)e^{-4}(-2 - 1) + \dfrac{1}{4}e^{-4}\right)$

$= \dfrac{1}{4} - \left(3e^{-4} + \dfrac{1}{4}e^{-4}\right) = \dfrac{1}{4} - 3e^{-4} - \dfrac{1}{4}e^{-4}$

$= \dfrac{1}{4} - \dfrac{13}{4}e^{-4} = \dfrac{1}{4}\left(1 - 13e^{-4}\right)$

$\approx 0{\cdot}1905$

Exercise 3L

Find these integrals using integration by parts.

1 $\int 3x^2 \cos x \, dx$ ★ 2 $\int_{-5}^{0} e^x \left(6x^2 + x\right) dx$ 3 $\int x^2 \cos 4x \, dx$

★ 4 $\int x^2 \cos x \, dx$ ★ 5 $\int \left(8x^2 + x - 5\right) \sin 2x \, dx$ 6 $\int 8x^2 e^{4x} \, dx$

When the integration by parts process reproduces the integration of the original product, there is no need to carry out further integration. This is demonstrated in Example 3.23.

Example 3.23

Find the integral $\int e^x \sin x \, dx$

$$u\frac{dv}{dx} = uv - \int v\frac{du}{dx}$$

Let $\int e^x \sin x \, dx = \int u\frac{dv}{dx}$ ● ——— (First use of integration by parts.)

$$= uv - \int v\frac{du}{dx}$$

Let $u = \sin x$

$$\frac{du}{dx} = \cos x$$

Let $\frac{dv}{dx} = e^x$

$$v = \int e^x \, dx = e^x$$

$\int e^x \sin x \, dx = e^x \sin x - \int e^x \cos x \, dx$

Now let $\int e^x \cos x \, dx = \int u\frac{dv}{dx}$ ● ——— (Second use of integration by parts for the new second element.)

$e^x \sin x - \int e^x \cos x \, dx = e^x \sin x - \left(uv - \int v\frac{du}{dx} \right)$

Let $u = \cos x$

$$\frac{du}{dx} = -\sin x$$

Let $\frac{dv}{dx} = e^x$

$$v = \int e^x \, dx = e^x$$

$\int e^x \sin x - \int e^x \cos x \, dx = e^x \sin x - \left[e^x \cos x - \int e^x (-\sin x) \, dx \right]$

$$= e^x \sin x - \left[e^x \cos x + \int e^x \sin x \, dx \right]$$

$$= e^x \sin x - e^x \cos x - \int e^x \sin x \, dx$$

$\therefore \int e^x \sin x \, dx = e^x \sin x - e^x \cos x - \int e^x \sin x \, dx$

$\int e^x \sin x \, dx + \int e^x \sin x \, dx = e^x \sin x - e^x \cos x$

$$2 \int e^x \sin x \, dx = e^x \sin x - e^x \cos x$$

$$\int e^x \sin x \, dx = \frac{1}{2}\left(e^x \sin x - e^x \cos x\right)$$

Exercise 3M

Find these integrals.

1 $\int \sin 3x \cos 4x \, dx$

2 $\int e^x \sin 2x \, dx$

★3 $\int e^{2x} \cos 3x \, dx$

4 $\int \sqrt{4 - x^2} \, dx$

5 $\int \sin x \sin 4x \, dx$

★6 $\int \cos 2x \cos 3x \, dx$

Applications of integration

Integration has direct practical applications involving:

- areas and volumes: industrial processes can involve the design and manufacture of objects whose shape is defined by rotations about an axis

- displacement, velocity and acceleration in rectilinear motion: movement of objects can be defined in terms of position and time.

Areas and volumes

Areas and volumes of shapes can be modelled as functions.

The area under a curve is given by the formula $\int_a^b f(x)\,dx$, where a and b are the limits bounding the area.

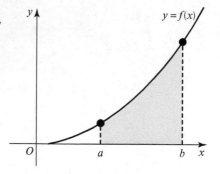

Often a single function is not complex enough to represent the shape or area to be constructed. We find the area enclosed between two functions using this integration:

$$\text{area between two curves} = \int_a^b \big(f(x) - g(x)\big)\,dx$$

where $f(x)$ is the top curve and $g(x)$ is the bottom curve and the limits a and b are the points of intersection of the two curves.

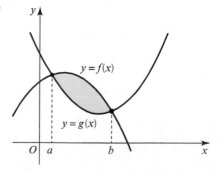

We can rotate the area defined by a single function or from the area between two curves to create a **volume of revolution**.

For example, a rectangle rotated about an axis creates a cylinder, and a right-angled triangle rotated round an axis creates a cone.

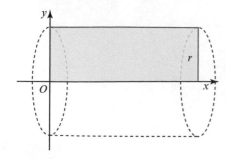

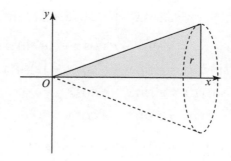

Rotation about the *x*-axis

When a line or curve is rotated about the *x*-axis between two limits *a* and *b*, the volume, *V*, of the solid generated is given by the formula:

$$V = \int_a^b \pi y^2 \, dx$$

Here we need a function in terms of *x* in order to integrate with respect to *x*.

For example, the function $y = x^2 + 2$ is rotated about the *x*-axis and the resultant volume calculated between $x = 1$ and $x = 3$.

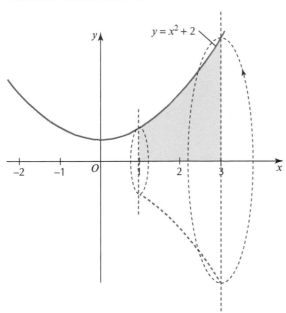

The volume of revolution is: $V = \displaystyle\int_1^3 \pi y^2 \, dx = \int_1^3 \pi \left(x^2 + 2\right)^2 \, dx$

Rotation about the *y*-axis

When a line or curve is rotated about the *y*-axis between two limits *a* and *b*, the volume, *V*, of the solid generated is given by the formula:

$$V = \int_a^b \pi x^2 \, dy$$

Here we need a function in terms of *y* in order to integrate with respect to *y*.

Here, the function $y = x^2 + 2$ is rotated about the *y*-axis and the resultant volume calculated between $y = 3$ and $y = 11$.

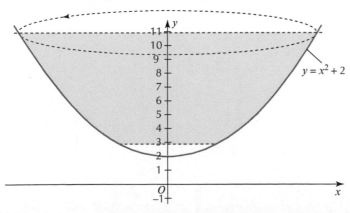

The volume of revolution is: $V = \int_3^{11} \pi x^2 \, dy = \int_3^{11} \pi(y-2) \, dy$ ⟵ Rearranging $y = x^2 + 2$
$\Rightarrow x^2 = y - 2$

Rotation about the x-axis: use dx and formula with $y^2 = \ldots$
Rotation about the y-axis: use dy and formula with $x^2 = \ldots$

Example 3.24

Calculate the area under the curve $y = e^x$ between 0 and 5 on the x-axis.

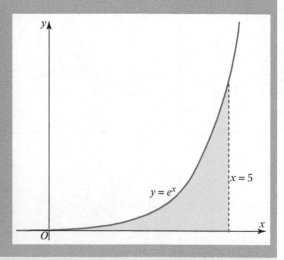

Area $= \int_0^5 e^x \, dx$

$= \left[e^x \right]_0^5$ ⟵ Integrate the function with respect to x.

$= e^5 - e^0$ ⟵ Substitute the limits 5 and 0.

$= e^5 - 1 \text{ units}^2 \approx 147 \cdot 4132 \text{ units}^2$ ⟵ Simplify the answer.

Example 3.25

Find the area enclosed between the curve $y = x^2 + 1$ and the y-axis between $y = 2$ and $y = 5$, $x \geqslant 0$.

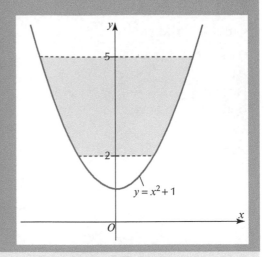

$$\text{Area} = \int_a^b f(y)\, dy$$

a and b are the y-coordinates that form the upper and lower bounds of the area.

$$y = x^2 + 1 \Rightarrow x^2 = y - 1 \Rightarrow x = (y-1)^{\frac{1}{2}}$$

$$\text{Area} = \int_2^5 (y-1)^{\frac{1}{2}}\, dy = \left[\frac{2}{3}(y-1)^{\frac{3}{2}} \right]_2^5$$

Rearrange the function to make y the variable.

$$= \frac{2}{3}(4)^{\frac{3}{2}} - \frac{2}{3}(1)^{\frac{3}{2}}$$

$$= \frac{16}{3} - \frac{2}{3}$$

$$= \frac{14}{3} \text{ units}^2$$

Example 3.26

Calculate the area between the curves $y = e^{3x}$ and $y = e^{-3x}$, the y-axis and the line $x = 3$.

$$\text{Area} = \int_0^3 \left(e^{3x} - e^{-3x} \right) dx$$

Set up the integral to calculate the area.

$$= \left[\frac{e^{3x}}{3} + \frac{e^{-3x}}{3} \right]_0^3$$

$$= \frac{1}{3}\left[\left(e^9 + e^{-9} \right) - \left(e^0 + e^0 \right) \right]$$

$$= \frac{1}{3}\left[e^9 + e^{-9} - 2 \right] \text{ units}^2 \approx 2700{\cdot}361 \text{ units}^2$$

Example 3.27

Calculate the volume generated when the curve $y = \dfrac{1}{\sqrt{x^2 + 4}}$ is rotated 360° about the x-axis between the y-axis $(x = 0)$ and $x = 20$.

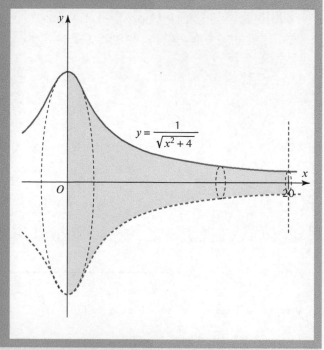

$V = \displaystyle\int_a^b \pi y^2 \, dx$

> The shaded area is rotated about the x-axis, so the function is integrated with respect to x.

$= \displaystyle\int_0^{20} \pi \left(\dfrac{1}{\sqrt{x^2 + 4}} \right)^2 dx$

> Substitute for y.

$= \pi \displaystyle\int_0^{20} \dfrac{1}{x^2 + 4} \, dx$

> Take out the constant, π, and square the bracket.

$= \pi \left[\dfrac{1}{2} \tan^{-1} \dfrac{x}{2} \right]_0^{20}$

> Use the standard integral
> $\displaystyle\int \dfrac{1}{x^2 + 4} \, dx = \dfrac{1}{2} \tan^{-1} \dfrac{x}{2}$

$= \dfrac{\pi}{2} \left(\tan^{-1} \dfrac{20}{2} - \tan^{-1} \dfrac{0}{2} \right)$

$= \dfrac{\pi}{2} \tan^{-1} 10 \text{ units}^3 \approx 2 \cdot 311 \text{ units}^3$

Example 3.28

Calculate the volume generated when the curve $y = x^2 - 25$, is rotated 360° about the y-axis between the x-axis ($y = 0$) and $y = 10$.

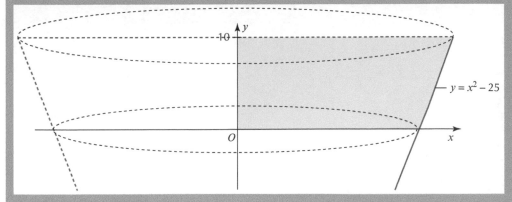

$y = x^2 - 25 \Rightarrow x^2 = y + 25$

> The shaded area is rotated about the y-axis, so the function is integrated with respect to y. Use algebraic manipulation to get the function in terms of y.

$V = \int_a^b \pi x^2 \, dy$

$= \int_0^{10} \pi (y + 25) \, dy$

> Substitute $x^2 = y + 25$ and the limits 0 and 10.

$= \pi \left[\dfrac{y^2}{2} + 25y \right]_0^{10}$

$= \pi \left[\left(\dfrac{10^2}{2} + 25(10) \right) - (0) \right]$

$= 300\pi \text{ units}^3 \approx 942 \cdot 478 \text{ units}^3$

Displacement, velocity and acceleration in rectilinear motion

The velocity v of a body is the rate of change of the displacement s of a body from some fixed origin, with respect to time t. The formula is:

$v = \dfrac{ds}{dt}$

The acceleration a of a body is the rate of change of the velocity v of a body from some fixed origin, with respect to time t. The formula is:

$a = \dfrac{dv}{dt} = \dfrac{d^2s}{dt^2}$

It follows that if $v = \dfrac{ds}{dt} \Rightarrow s = \int v \, dt$

and $a = \dfrac{dv}{dt} \Rightarrow v = \int a \, dt$

The diagram shows these relationships.

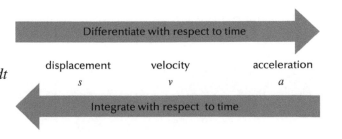

Example 3.29

The velocity of an object at time t seconds is $2t^2 - \sqrt{1+t} + 2$ metres per second. At $t = 0$ the position of the object is 0. What is the position of the object 3 seconds later?

$v(t) = 2t^2 - \sqrt{1+t} + 2$

$s(t) = \int \left(2t^2 - \sqrt{1+t} + 2\right) dt$

$\quad = \int \left(2t^2 - (1+t)^{\frac{1}{2}} + 2\right) dt$

$\quad = \frac{2}{3}t^3 - \frac{2}{3}(1+t)^{\frac{3}{2}} + 2t + c$

When $t = 0$, $s(t) = 0$

$s(0) = \frac{2}{3}0^3 - \frac{2}{3}(1+0)^{\frac{3}{2}} + 2(0) + c \Rightarrow c = \frac{2}{3}$ •————— Substitute initial conditions to evaluate c.

$s(t) = \frac{2}{3}t^3 - \frac{2}{3}(1+t)^{\frac{3}{2}} + 2t + \frac{2}{3}$

At $t = 3$: $s(3) = \frac{2}{3}3^3 - \frac{2}{3}(1+3)^{\frac{3}{2}} + 2(3) + \frac{2}{3}$

$\qquad\qquad = 18 - \frac{16}{3} + 6 + \frac{2}{3} = 19\frac{1}{3}\text{m}$

Exercise 3N

1 Calculate the area enclosed by the curve $y = \dfrac{4x}{x^2 + 5}$, the x-axis, the y-axis and the line $x = 10$.

★ 2 Work out the total area enclosed by the curve $y = \dfrac{3x}{x + 2}$, the x-axis, the y-axis and the line $x = 6$.

★ 3 Work out the total area between the curves $y = 100 - x^2$, $y = \ln 6x$, and the lines $x = 1$ and $x = 10$.

▥ 4 Find the area enclosed by the curve $y = x^2$ between $y = 1$ and $y = 4$, $x \geqslant 0$.

★ 5 Find the area enclosed by the curve $y = \ln x$ between $y = 2$ and $y = 5$, $x \geqslant 0$.

▥ 6 The graph of $y = x^2 - 16$ is drawn between zero and 4 on the y-axis.

The whole shape is rotated about the y-axis to form a solid vase shape.

Calculate the volume of the solid formed.

7 A drinking goblet is to be modelled by rotating the graph of $y = e^x$ about the y-axis over the interval $1 \leqslant y \leqslant 5$, where 1 unit represents 1 centimetre.

According to the manufacturer's specifications, each goblet needs to hold at least $12 \cdot 5 \, \text{cm}^3$.

Do the specifications of the proposed drinking goblet meet the manufacturer's requirements?

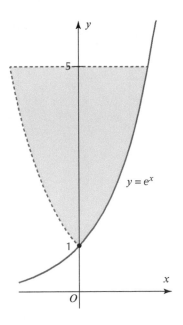

8 Calculate the exact volume obtained by revolving the part of the graph of $y = \sqrt{x + 1}$ between $x = 3$ and $x = 5$ through 360° around the x-axis.

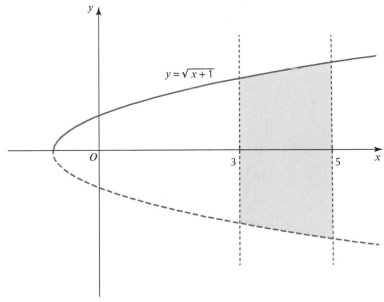

9 The region R is bounded by the x-axis, the y-axis, the line $y = 12$ and the part of the curve with equation $y = x^2 - 4$ which lies between $x = 2$ and $x = 4$.

a Sketch this region, showing clearly where all the boundaries start and stop.

Rotating the region R through 360° about the y-axis forms the inside of a vase.

Each unit of x and y represents 5 centimetres.

b Write down an expression for the volume of revolution of region R about the y-axis.

c Find the capacity of the vase in litres.

d Show that if a vase generated in this way is filled to $\frac{5}{6}$ of its internal height, it is $\frac{3}{4}$ full.

10 Find the volume of the solid of revolution formed when the region in the first quadrant bounded by the function $y = \cos x$ and the y-axis is rotated 360° about the x-axis.

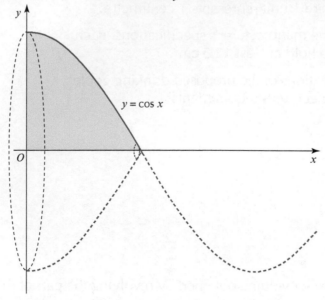

11 The shaded area between two curves, $y = x^2 + 2$, and $y = 8 - 2x^2$, and the y-axis is rotated about the y-axis. By splitting the shaded area into two parts, or otherwise, find the volume of the solid formed.

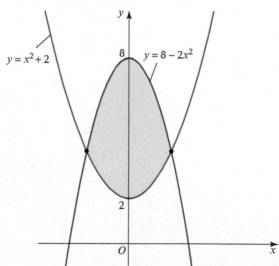

12 A particle is moving in a straight line. Its velocity at time t is given by $2\sin^2 t$.

   a Find an expression for the displacement, $s(t)$, of the particle.

   b When $t = 0$, $s(t) = 2$. Calculate the displacement of the particle after $\dfrac{\pi}{2}$ seconds.

13 A particle starts from rest and proceeds in a straight line with an acceleration of $a(t) = e^{2t} - 3$.

   How far from the starting point is the particle after 4 seconds and in which direction is it moving?

Chapter review

Integrate these expressions:

1 $\int \left(\dfrac{1}{3x} + \sqrt{e^x} \right) dx$

2 $\int \dfrac{4}{4x - 1}\, dx$

3 $\int_1^2 \left(e^{-x} + \dfrac{3}{x} \right) dx$

★ 4 $\int_{-2e}^{-e} \dfrac{4}{x - 4}\, dx$

5 $\int \left(\tan 3x + \dfrac{1}{\cos^2 2x} - 5 \right) dx$

6 $\int \dfrac{1}{\cos^2 (3x + 1)}\, dx$

7 $\int_0^{\frac{\pi}{9}} \tan 4x\, dx$

☒ 8 $\int_0^{\frac{\pi}{3}} \sin^2 3x\, dx$

9 $\int_{-\frac{3}{2}}^{\frac{3}{2}} \dfrac{9}{\sqrt{4 - x^2}}\, dx$

10 $\int_{-2}^{2} \dfrac{3}{1 + 81x^2}\, dx$

11 $\int \dfrac{5x^2 + 2x + 3}{\left(x^2 + 1\right)(x + 1)}\, dx$

12 $\int \dfrac{4x^3 - 5x - 1}{x^2 + 2x + 1}\, dx$

13 Integrate $\int 2x\, e^{x^2 - 5}\, dx$ using the substitution $u = x^2 - 5$

★ 14 Use partial fractions to find the exact value of the integral $\int_{\frac{6}{5}}^{\frac{12}{5}} \dfrac{1}{1 - x^2}\, dx$

Give your answer as a simplified, single logarithm.

15 Integrate $\int_{-1}^{1} \sqrt{3 - x^2}\, dx$ using the substitution $x = \sqrt{3}\sin u$

☒ 16 Using the substitution $2x = \sin \theta$, or otherwise, find the exact value of $\int_0^{\frac{1}{4}} \dfrac{1}{\sqrt{1 - 4x^2}}\, dx$

17 Find the volume generated when the area between the curve

$y = \dfrac{1}{\sqrt{x(5 - x)}}$, the x-axis and the lines $x = 5$ and $x = 10$ is rotated completely

about the x-axis.

18 Find y given $\dfrac{dy}{dx} = \sec^2 2x$ when $x = 0$ and $y = \pi$.

19 A drinking cup can be modelled by rotating the relevant parts of
the line $y = \frac{x}{20} + 4$ and the curve $y = \frac{91}{20} - \frac{x^2}{2}$ through 360°
about the x-axis.

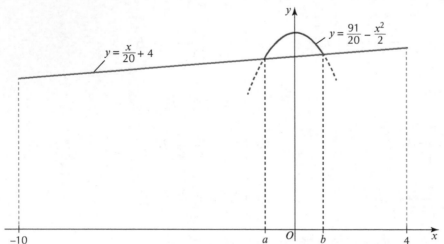

a Calculate the volume generated by
rotating the curve $y = \frac{91}{20} - \frac{x^2}{2}$ about the
x-axis from $x = a$ to $x = b$.

> Form and solve a quadratic equation
> to find a and b.

b Show that the volume, V, of the drinking cup is given by:

$$V = \pi\left[\int_{-10}^{-1}\left(\frac{x}{20} + 4\right)^2 dx + \int_{-1}^{1\cdot1}\left(\frac{91}{20} - \frac{x^2}{2}\right)^2 dx + \int_{1\cdot1}^{4}\left(\frac{x}{20} + 4\right)^2 dx\right]$$

c Calculate the volume of the drinking cup.

- I can integrate expressions using standard results. ★ Exercise 3C Q1b, Q1e,
  Q2b ★ Exercise 3D Q1f, Q2b, Q2d ★ Exercise 3E Q1b, Q1e, Q2b, Q2d
- I can integrate using partial fractions. ★ Exercise 3F Q1a, Q1e, Q2a
- I can use integration by substitution. ★ Exercise 3G Q1b, 1f, Q6, Q7;
  ★ Exercise 3H Q1a, Q2a, Q4c, Q6; ★ Exercise 3I Q3, Q5, Q7, Q8;
  ★ Exercise 3J Q1
- I can use integration by parts. ★ Exercise 3K Q2, Q4, Q5;
  ★ Exercise 3L Q2, Q4; ★ Exercise 3M Q3, Q6
- I can apply integration to problems in context. ★ Exercise 3N Q3, Q5 and Q7

4 Complex numbers

This chapter will show you how to:

- define a complex number
- add, subtract, multiply, divide and take the square root of a complex number
- solve equations involving complex numbers
- use an Argand diagram
- convert a complex number from Cartesian to polar form and vice versa
- multiply and divide numbers in polar form
- use de Moivre's theorem with integer and fractional indices
- apply de Moivre's theorem to multiple-angle trigonometric formulae and to find nth roots of a complex number
- find the roots of a polynomial (including complex roots)
- interpret geometrically equations/inequalities in the complex plane.

You should already know:

- how to solve quadratic equations using the quadratic formula
- how to use synthetic division.

Complex numbers

In 1545, the Italian mathematician Girolamo Cardano solved the problem of finding two numbers that have a sum of 10 and a product of 40. The numbers are $5 + \sqrt{-15}$ and $5 - \sqrt{-15}$:

- $5 + \sqrt{-15} + 5 - \sqrt{-15} = 5 + 5 = 10$ and

- $\left(5 + \sqrt{-15}\right)\left(5 - \sqrt{-15}\right) = 25 - 5\sqrt{-15} + 5\sqrt{-15} - \sqrt{-15}\sqrt{-15}$

$$= 25 - (-15)$$
$$= 40$$

However, this led to the necessity for dealing with the square root of a negative number. Similar problems arose when finding the roots of cubic equations. The term 'fictitious' was used to describe such roots and they were seen as a sign that a mathematical problem had no solutions. In 1572, Rafael Bombelli set down the rules for multiplying **complex numbers** and over many years, the idea of 'imaginary' numbers gradually became accepted.

The square root of a negative number

Consider the equation $x^2 + 2x + 5 = 0$.

This equation is solved using the quadratic formula to give $x = \dfrac{-2 \pm \sqrt{-16}}{2}$

To get a solution, we need to know the solution to $\sqrt{-16}$.
Mathematicians define the square root of -1 to be i.

> $i = \sqrt{-1}$ and since $i = \sqrt{-1}$ then $i^2 = -1$ ($i \notin \mathbb{R}$)

The equation $x^2 + 1 = 0$ has two solutions: $\pm\sqrt{-1}$, that is, $\pm i$. The number i, and any multiple of i (such as $3i$, $-8i$, $\frac{i}{2}$, $\sqrt{2}i$) are **imaginary numbers**.

If we consider the equation $x^2 + 2x + 5 = 0$ then given $\sqrt{-16} = \sqrt{16}\sqrt{-1} = 4i$, the solutions of the equation are $\frac{-2 \pm 4i}{2} = -1 \pm 2i$.

The solution is in the form $a + ib$, where $a, b \in \mathbb{R}$.

- The letter z is usually used to denote a complex number, with $z = a + ib$ where $a, b \in \mathbb{R}$.

- Sometimes a complex number is written as $z = x + iy$ as it is easier to remember how to plot on a Argand diagram (see page 120). $\mathbb{C}$ is the set of complex numbers.

- a is called the **real** part of z: $\text{Re}(z) = a$

- b is called the **imaginary** part of z: $\text{Im}(z) = b$

Example 4.1

State the real and imaginary parts of these complex numbers.

a $z = 5 - 2i$

b $z = i - \frac{1}{2}$

c $z = 7$

d $z = -4i$

a $\text{Re}(z) = 5$ $\text{Im}(z) = -2$

b $\text{Re}(z) = -\frac{1}{2}$ $\text{Im}(z) = 1$

c $\text{Re}(z) = 7$ $\text{Im}(z) = 0$

d $\text{Re}(z) = 0$ $\text{Im}(z) = -4$

Arithmetic with complex numbers

Complex numbers can be added to (or subtracted from) each other by adding (or subtracting) the real parts together and then adding (or subtracting) the imaginary parts together:

$z_1 = a + ib$ and $z_2 = c + id$

$z_1 + z_2 = (a + ib) + (c + id)$
$\qquad = (a + c) + (b + d)i$

$z_1 - z_2 = (a + ib) - (c + id)$
$\qquad = (a - c) + (b - d)i$

To multiply complex numbers together, the normal algebraic rules for multiplying two brackets together are applied.

$z_1 = a + ib$ and $z_2 = c + id$,

$$z_1 z_2 = (a + ib)(c + id)$$
$$= ac + adi + bci + i^2 bd$$
$$= ac + adi + bci - bd \qquad \text{(since } i^2 = -1\text{)}$$
$$= (ac - bd) + (ad + bc)i$$

Example 4.2

If $z_1 = 6 + i$ and $z_2 = 3 - 2i$, calculate:

a $z_1 + z_2$ **b** $z_1 - z_2$ **c** $z_1 z_2$

a $z_1 + z_2 = (6 + i) + (3 - 2i)$
$$= 6 + i + 3 - 2i = 9 - i$$

b $z_1 - z_2 = (6 + i) - (3 - 2i)$
$$= 6 + i - 3 + 2i = 3 + 3i$$

c $z_1 z_2 = (6 + i)(3 - 2i)$
$$= 18 - 12i + 3i - 2i^2$$
$$= 18 - 12i + 3i - 2(-1) \qquad \boxed{\text{Since } i^2 = -1}$$
$$= 20 - 9i$$

Condition for two complex numbers to be equal

If two complex numbers are equal then their real and imaginary parts are equal. For two complex numbers $a + ib$ and $c + id$:

if $a + ib = c + id$

then $a = c$ and $b = d$.

Example 4.3

Solve $4 + 5i = z - (3 - 2i)$ for z where $z = a + ib$, $a, b \in \mathbb{R}$.

Let z be a complex number such that $z = a + ib$, $a, b \in \mathbb{R}$.

Then $4 + 5i = (a + ib) - (3 - 2i)$
$$4 + 5i = a + ib - 3 + 2i$$
$$4 + 5i = a - 3 + bi + 2i$$
$$4 + 5i = a - 3 + (b + 2)i \qquad \boxed{\text{Collect real and imaginary parts.}}$$
$$a - 3 = 4 \Rightarrow a = 7 \qquad \boxed{\text{Equate real parts and solve for } a.}$$
$$b + 2 = 5 \Rightarrow b = 3 \qquad \boxed{\text{Equate imaginary parts and solve for } b.}$$

So $z = 7 + 3i$

Example 4.4

Solve $4 + 5i = (3 - 2i)\,z$ for z where $z = a + ib$, $a, b \in \mathbb{R}$.

Let z be a complex number such that $z = a + ib$, $a, b \in \mathbb{R}$.

Then $4 + 5i = (3 - 2i)(a + ib)$

$\quad 4 + 5i = 3a + 3bi - 2ai - 2bi^2$

$\quad 4 + 5i = 3a + 3bi - 2ai - 2b(-1)$ ●——————————— Since $i^2 = -1$

$\quad 4 + 5i = 3a + 3bi - 2ai + 2b$

$\quad 4 + 5i = (3a + 2b) + (3b - 2a)i$

$\quad 3a + 2b = 4$

$\quad 3b - 2a = 5$

$a = \frac{2}{13},\ b = \frac{23}{13}$ ●——————————— Solve simultaneously.

$\qquad$ So $z = \frac{2}{13} + \frac{23}{13}i$

Exercise 4A

1   Express in terms of i:

    a $\sqrt{-8}$ b $\sqrt{-49}$ c $\sqrt{-100}$

    d $4 + \sqrt{-64}$ e $\sqrt{-81} - \sqrt{-16}$ f $\sqrt{-625} - \sqrt{-225}$

2   Given that $z_1 = 5 + 2i$ and $z_2 = 3 - 4i$, calculate:

    a $z_1 + z_2$ b $z_1 - z_2$ c $z_2 - z_1$

    d $5z_2$ e $z_1 z_2$ f $2z_1 + 3z_2$

    g $-z_2$ h $z_1^{\,2} z_2$ i $z_2^{\,3}$

★ 3   Simplify these terms, giving your answer in the form $a + ib$ where $a, b \in \mathbb{R}$.

    a $(4 + 3i) + (5 - 2i)$ b $(2 + 7i) - (3 - 2i)$ c $5(2 + i)$

    d $2i(3i + 1)$ e $(2 + i)(3 - i)$ f $(4 + 3i)(6 - i)$

    g $(5 - 3i)(2 - 5i)$ h $(2 + i)^2$ i $(5 + 3i)^3$

4   Solve these equations.

    a $z^2 + 16 = 0$ b $z^3 + 25z = 0$ c $z^2 - 4z + 5 = 0$

    d $z^2 + 2z + 5 = 0$ e $3z^2 + 2z + 1 = 0$ f $2z^2 + z + 10 = 0$

    g $z^2 = z - 9$ h $3z(z + 2) = -8$

5 **a** Calculate i^2, i^3, i^4, ..., i^8, i^9

    b Hence find the smallest value of n ($n \in \mathbb{N}$, $n \neq 0$) such that:

        i $i^n = 1$ **ii** $i^n = -i$

    c From your answer to part **b**, what can you conclude about all the values of n such that:

        i $i^n = 1$ **ii** $i^n = -i$

6 Solve each equation for z where $z = a + ib$, $a, b \in \mathbb{R}$.

    a $3 + 5i = (1 + i) + z$ **b** $(2 + 4i) - z = 6 - 5i$

    c $z - (3 - i) = 2 + 3i$ **d** $3z + (2 - 3i) = 8$

7 Solve each equation for z where $z = a + ib$, $a, b \in \mathbb{R}$.

    a $\dfrac{z}{2 + 3i} = 3 - i$ **b** $\dfrac{z}{1 - i} = 1 - 4i$

    c $3 + 5i = (1 + i)z$ **d** $(2 + i)z = (3 - i)$

    e $(1 + i)z = 2 + i$ **f** $3 - 2i = (5 - 2i)z$

Division and square roots of complex numbers

If $z = a + ib$, where $a, b \in \mathbb{R}$, then the complex number $\bar{z} = a - ib$ is the **complex conjugate** of z (and vice versa). The complex conjugate of a complex number is denoted as $\bar{z}$ (pronounced 'z bar').

For example:

$z^2 + 2z + 5 = 0$ has solutions $z = -1 + 2i$ and $-1 - 2i$.

One root is the complex conjugate of the other root, and complex roots of equations occur in pairs.

Consider $z\bar{z}$ where $z = a + ib$ then $\bar{z} = a - ib$:

then $z\bar{z} = (a + ib)(a - ib)$

$$= a^2 - iab + iab - i^2b^2$$

$$= a^2 - (-1)b^2$$

$$= a^2 + b^2$$

and $a^2 + b^2 \in \mathbb{R}$ since $a, b \in \mathbb{R}$.

Hence, multiplying a complex number by its conjugate results in a real number. This fact is used when dividing complex numbers.

Division of complex numbers

Example 4.5

Calculate $(6 + i) \div (3 - 2i)$ giving your answer in the form $a + ib$ where $a, b \in \mathbb{R}$.

$(6 + i) \div (3 - 2i) = \dfrac{6 + i}{3 - 2i} \times \dfrac{3 + 2i}{3 + 2i}$

When dividing, always set up the question as a fraction. Then multiply the numerator and denominator by the complex conjugate of $3 - 2i$ to give a real number on the denominator.

$= \dfrac{18 + 12i + 3i + 2i^2}{9 + 6i - 6i - 4i^2}$

Multiply the fractions together.

$= \dfrac{18 + 15i + 2i^2}{9 - 4i^2}$

$= \dfrac{16 + 15i}{13}$

Simplify using $i^2 = -1$

$= \dfrac{16}{13} + \dfrac{15}{13}i$

State answer in required form.

Square root of a complex number

Example 4.6

Calculate $\sqrt{3 + 4i}$ giving your answer in the form $a + ib$ where $a, b \in \mathbb{R}$.

Let $a + ib = \sqrt{3 + 4i}$ where $a, b \in \mathbb{R}$.

This is always the starting line in this type of question. Remember to write down $a, b \in \mathbb{R}$.

$(a + ib)^2 = 3 + 4i$

Square both sides of the equation.

$a^2 + 2abi + i^2b^2 = 3 + 4i$

$a^2 - b^2 + 2abi = 3 + 4i$

Multiply out brackets and simplify.

$a^2 - b^2 = 3$ ①

Compare real and imaginary parts.

$2ab = 4$ ②

$b = \dfrac{4}{2a} = \dfrac{2}{a}$ ③

Rearrange ②

$a^2 - b^2 = 3$

$a^2 - \left(\dfrac{2}{a}\right)^2 = 3$

Substitute ③ into ①

$a^2 - \dfrac{4}{a^2} = 3$

$a^4 - 4 = 3a^2$

Simplify by multiplying through by a^2.

$a^4 - 3a^2 - 4 = 0$

Rearrange to make equation equal to 0.

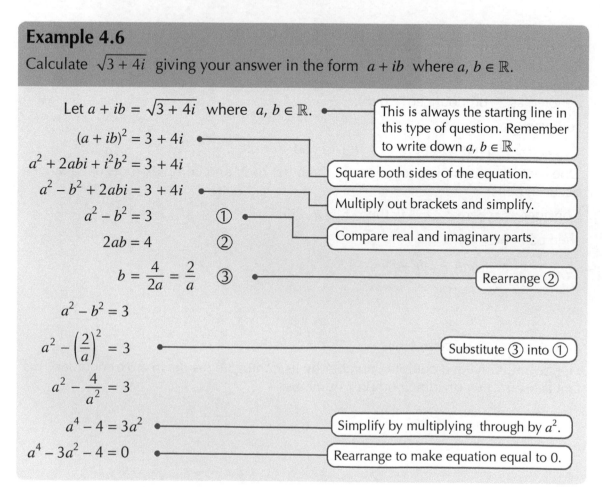

$(a^2 + 1)(a^2 - 4) = 0$ ⸺ Factorise.

So:

$a^2 + 1 = 0$ or $a^2 - 4 = 0$

So:

$a^2 = -1$ ⸺ Impossible since $a \in \mathbb{R}$

$a = \pm 2$

$\Rightarrow a = \pm 2$

When $a = 2$, $b = 1$

When $a = -2$, $b = -1$ ⸺ Substitute the two possible values of a into ③

$\sqrt{3 + 4i}$ is $2 + i$ and $-2 - i$ ⸺ State final answer. There should always be two answers.

Exercise 4B

1 State $\bar{z}$ for each of these complex numbers z.

  a $1 + i$
  b $2 + 3i$
  c $4 - 2i$

  d $-5 - \frac{1}{2}i$
  e i
  f 8

  g $-7i$
  h $-5 + 2i$
  i $-4i + 7$

★ 2 Calculate these fractions, giving your final answer in the form $a + ib$ where $a, b \in \mathbb{R}$.

  a $\dfrac{5}{2 + i}$
  b $\dfrac{2i}{3 + 2i}$
  c $\dfrac{1 + 5i}{2 - i}$

  d $\dfrac{3 + i}{1 - 2i}$
  e $\dfrac{7 + 3i}{1 - i}$
  f $\dfrac{5 + i}{i - 3}$

3 Let $z_1 = 5 + 2i$ and $z_2 = 3 - 4i$.

Express these terms in the form $a + ib$ where $a, b \in \mathbb{R}$.

  a $z_1 z_2$
  b $\dfrac{z_1}{z_2}$
  c $\dfrac{z_2}{z_1}$

  d $z_1 \bar{z}_1$
  e $z_1 \bar{z}_2$

4 Express these fractions in the form $a + ib$ where $a, b \in \mathbb{R}$.

  a $\dfrac{2}{(i - 2)(1 + 2i)}$
  b $\dfrac{i}{(3 + i)(5 - 2i)}$
  c $\dfrac{3 - 2i}{(1 - 2i)(3 + i)}$

5 Given that $z = -2 + i$, express $z + \dfrac{3}{z}$ in the form $a + ib$ where $a, b \in \mathbb{R}$.

6 Express these terms in the form $a + ib$ where $a, b \in \mathbb{R}$.

  a $\dfrac{1}{1 + 3i} + \dfrac{1}{1 - 3i}$
  b $\dfrac{2}{1 - 2i} - \dfrac{1}{1 + 4i}$
  c $3 + 4i + \dfrac{1}{4 - 5i}$

7 Given that $z_1 = a + ib$ and $z_2 = c + id$, show that:

a $\overline{z_1 + z_2} = \overline{z_1} + \overline{z_2}$ **b** $\overline{-z} = -\overline{z}$ **c** $\overline{z_1 - z_2} = \overline{z_1} - \overline{z_2}$

d $\overline{z_1 z_2} = \overline{z_1}\,\overline{z_2}$ **e** $\overline{\left(\dfrac{1}{z}\right)} = \dfrac{1}{\overline{z}}$ **f** $\overline{\left(\dfrac{z_1}{z_2}\right)} = \dfrac{\overline{z_1}}{\overline{z_2}}$

★ 8 Find the complex number z such that:

a $z - 3\overline{z} = -6 + 4i$ **b** $3z + 5\overline{z} = 16 + i$

c $z + 2i\overline{z} = 16 + 23i$ **d** $z - 4i\overline{z} = 10 + 5i$

9 Find all the possible values of z such that $z + \overline{z} = 10$

10 Solve these pairs of simultaneous equations for z and w, where $z, w \in \mathbb{C}$.

$$(1 + i)z + (3 + 2i)w = 21 + 6i$$

$$2z - (2 - i)w = 5 + 14i$$

★ 11 Find the square root of these complex numbers.

a $5 + 12i$ **b** $7 + 24i$

c $15 + 8i$ **d** $9 - 40i$

Argand diagrams

Consider the complex number $3 + 4i$. If we draw a pair of Cartesian axes with the x-axis as the real axis and the y-axis as the imaginary axis, then the complex number $3 + 4i$ can be represented as the point $(3, 4)$.

The **Argand diagram** shown is a geometrical representation of complex numbers. It is named after the Swiss amateur mathematician Jean Robert Argand and was published in 1806.

In general, the complex number $a + ib$ is represented on an Argand diagram with the point $P(a, b)$ and the position vector $\overrightarrow{OP}$.

The real part of the number is the x-coordinate and the imaginary part is the y-coordinate. Both axes **must** be labelled Re and Im as shown.

The sum and difference of two complex numbers can be represented geometrically on an Argand diagram.

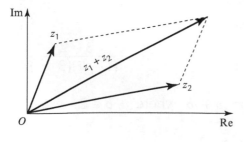 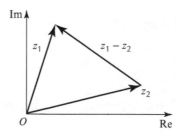

Modulus and argument of a complex number

Consider $z = a + ib$, drawn on an Argand diagram as shown here.

The **modulus** of a complex number $z = a + ib$ is defined as the magnitude of z, that is, the **length** of the corresponding vector on the Argand diagram. It is written as $|z|$ or r and it can be seen that $r^2 = a^2 + b^2$, so $r = |z| = \sqrt{a^2 + b^2}$

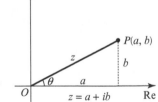

The **argument** of a complex number $z = a + ib$ is defined as the magnitude of the **anticlockwise angle** (in radians) between the positive x-axis and the corresponding vector on the Argand diagram.

In the diagram, the argument is θ where $\tan\theta = \dfrac{b}{a}$. Note that the argument of z could also be $\theta \pm 2\pi$, $\theta \pm 4\pi$, $\theta \pm 6\pi$, etc. The principal argument of z must be between $-\pi < \theta \leqslant \pi$. We write arg z (using a lower case 'a') for the principal argument. To find the principal argument:

- find the first quadrant angle α using $\tan^{-1}\left|\dfrac{b}{a}\right|$

- plot the Argand diagram to decide which quadrant the point (a, b) is in

- use the quadrant rules (see diagram on the right) to find the value of $\arg z$

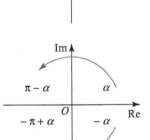

Additionally since the principal angle is between $-\pi < \theta \leqslant \pi$, then when finding θ we need to remember to find the angle:

- **anticlockwise** between the positive real axis and the complex number if it is in quadrants 1 and 2

- **clockwise** between the positive real axis and the complex number if it is in quadrants 3 and 4.

Example 4.7

Calculate the modulus ($|z|$) and argument (arg z or θ) of these complex numbers.

a $1 + i$ **b** $3 - 4i$ **c** $-\sqrt{3} - i$

a $1 + i$

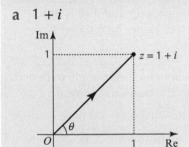

> To find the modulus and argument of a complex number, always sketch it on an Argand diagram before starting the question.

$|z| = \sqrt{1^2 + 1^2} = \sqrt{2}$

$\tan \theta = \frac{1}{1} = 1$

$\Rightarrow \arg z = \theta = \frac{\pi}{4}$

b $3 - 4i$

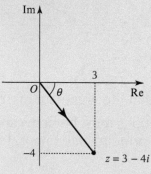

$|z| = \sqrt{3^2 + 4^2} = 5$

$\tan \alpha = \frac{4}{3}$

$\Rightarrow \alpha = 0.927$ radians (3 s.f.)

$\Rightarrow \theta = -0.927$ radians

> α has been used in this line of working as you have calculated an angle in quadrant 1. From the diagram, the complex number is in quadrant 4. So far you have not calculated θ which is the argument of z.

> θ is defined to be the anticlockwise angle between the positive x-axis and the position vector. The negative in front of the 0.927 shows you have moved in a clockwise direction from the x-axis.

c $-\sqrt{3} - i$

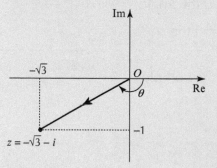

$|z| = \sqrt{\left(\sqrt{3}\right)^2 + 1^2} = 2$

$\tan \alpha = \frac{1}{\sqrt{3}}$

$\Rightarrow \alpha = \frac{\pi}{6}$

$\Rightarrow \theta = -\frac{5\pi}{6}$

> A sketch of an Argand diagram helps get the argument in the correct quadrant.

> From the diagram, the complex number is in quadrant 3. So far you have not calculated θ which is the argument of z.

> θ is defined to be the anticlockwise angle between the positive x-axis and the position vector. The negative in front of the $\frac{5\pi}{6}$ shows you have moved in a clockwise direction from the x-axis. $\frac{5\pi}{6}$ is calculated by subtracting $\frac{\pi}{6}$ from π.

Polar form of a complex number

Consider the complex number $z = a + ib$, represented on an Argand diagram as shown on the right. We will call $|z| = r$ and $\arg z = \theta$.

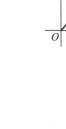

$\cos\theta = \dfrac{a}{r}$, so $a = r\cos\theta$

$\sin\theta = \dfrac{b}{r}$, so $b = r\sin\theta$

So $z = a + ib$

$\qquad = r\cos\theta + i(r\sin\theta)$

$\qquad = r(\cos\theta + i\sin\theta)$.

Thus $z = a + ib$ can be rewritten as

$\qquad z = r(\cos\theta + i\sin\theta)$

$z = a + ib$ is the **Cartesian** form of a complex number

$z = r(\cos\theta + i\sin\theta)$ is the **polar form** of a complex number

---

Example 4.8

Express $z = -\sqrt{3} + i$ in polar form.

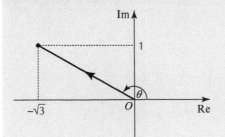

> A sketch of an Argand diagram helps get the argument in the correct quadrant. ⚠

$r = |z| = \sqrt{\left(\sqrt{3}\right)^2 + 1^2} = 2$

$\tan\alpha = \dfrac{1}{\sqrt{3}}$

$\Rightarrow \alpha = \dfrac{\pi}{6}$

> From the diagram, the complex number is in quadrant 2. So far you have not calculated θ which is the argument of z.

$\Rightarrow \theta = \pi - \dfrac{\pi}{6} = \dfrac{5\pi}{6}$

$\Rightarrow z = r(\cos\theta + i\sin\theta) = 2\left(\cos\dfrac{5\pi}{6} + i\sin\dfrac{5\pi}{6}\right)$

> State z in polar form.

Check:

$z = 2\left(\cos\dfrac{5\pi}{6} + i\sin\dfrac{5\pi}{6}\right)$

> As a check, rewrite the final answer (which is in polar form) into Cartesian form. You should have the complex number from the original question.

$= 2\left(-\dfrac{\sqrt{3}}{2} + \dfrac{1}{2}i\right) = -\sqrt{3} + i$

Exercise 4C

★ 1 Represent each of these terms on an Argand diagram.

    a $3 + 2i$ **b** $-3 + 2i$ **c** $4 - i$ **d** $-4 - 3i$

    e $7 + 2i$ **f** $-3i + 5$ **g** $-6 - \frac{1}{2}i$ **h** $-5 + 4i$

2   **a** If $z = 4 + 3i$, plot z and $\bar{z}$ on an Argand diagram.

    b If $z = -2 + i$, plot z and $\bar{z}$ on an Argand diagram.

    c What conclusion can you make about how z and $\bar{z}$ are related geometrically?

3 Find the modulus ($|z|$) and argument (arg z or θ) of these complex numbers. Give your answers as exact values or, if required, to 3 significant figures.

    a $3 + 4i$ **b** $2 - i$ **c** $-5 + 3i$

    d $-2 - 4i$ **e** $7 + 2i$ **f** $\sqrt{3}i + 1$

    g $\frac{2}{\sqrt{3}}i - 2$ **h** $-\frac{1}{2} - \frac{1}{2}i$

4 Find the modulus and argument of these complex numbers.

    a $(2 + i)(3 - 2i)$ **b** $\dfrac{1 + i}{1 - i}$ **c** $\dfrac{-1 - 7i}{4 + 3i}$

★ 5   **i** Write these complex numbers in polar form.

    ii Convert them to Cartesian form.

    a $|z| = 2$ and arg $z = \dfrac{\pi}{3}$ **b** $|z| = 1$ and arg $z = \dfrac{\pi}{2}$

    c $|z| = 3$ and arg $z = -\dfrac{\pi}{6}$ **d** $|z| = 2$ and arg $z = \dfrac{2\pi}{3}$

    e $|z| = 5$ and arg $z = -\dfrac{5\pi}{6}$ **f** $|z| = 4$ and arg $z = -\dfrac{3\pi}{4}$

★ 6 Write these complex numbers in polar form.

    a $1 + i$ **b** $1 + \sqrt{3}i$ **c** $\sqrt{3} - i$ **d** $-\sqrt{2} + \sqrt{2}i$

    e -5 **f** $\dfrac{1}{2} + \dfrac{\sqrt{3}}{2}i$ **g** $2i$ **h** $5\sqrt{3} - 5i$

7 If $z = a + ib$, solve $z^2 = |z|^2 - 8$.

Multiplication and division of complex numbers in polar form

The polar form of a complex number is $z = r(\cos\theta + i\sin\theta)$

Consider $z_1 = r_1(\cos\theta_1 + i\sin\theta_1)$ and $z_2 = r_2(\cos\theta_2 + i\sin\theta_2)$

then $z_1 z_2 = r_1(\cos\theta_1 + i\sin\theta_1) \times r_2(\cos\theta_2 + i\sin\theta_2)$

$$= r_1 r_2(\cos\theta_1 + i\sin\theta_1)(\cos\theta_2 + i\sin\theta_2)$$

$$= r_1 r_2(\cos\theta_1\cos\theta_2 + i\sin\theta_2\cos\theta_1 + i\sin\theta_1\cos\theta_2 + i^2\sin\theta_1\sin\theta_2)$$

> Collect real and imaginary parts.

$$= r_1 r_2((\cos\theta_1\cos\theta_2 - \sin\theta_1\sin\theta_2) + i(\sin\theta_1\cos\theta_2 + \cos\theta_1\sin\theta_2))$$

> Use $\cos(A+B) = \cos A\cos B - \sin A\sin B$ and $\sin(A+B)$

$$= r_1 r_2(\cos(\theta_1 + \theta_2) + i\sin(\theta_1 + \theta_2))$$

> $= \sin A\cos B + \cos A\sin B$

So $|z_1 z_2| = |z_1||z_2| = r_1 \times r_2$ and $\arg z_1 z_2 = \arg z_1 + \arg z_2 = \theta_1 + \theta_2$

Since we would be required to find the principal argument, you must add or subtract 2π, if required, from $\arg(z_1 z_2)$ to ensure that $-\pi < \arg(z_1 z_2) \leqslant \pi$.

Consider $z_1 = r_1(\cos\theta_1 + i\sin\theta_1)$ and $z_2 = r_2(\cos\theta_2 + i\sin\theta_2)$, then:

$$\frac{z_1}{z_2} = \frac{r_1(\cos\theta_1 + i\sin\theta_1)}{r_2(\cos\theta_2 + i\sin\theta_2)}$$

$$= \frac{r_1(\cos\theta_1 + i\sin\theta_1)}{r_2(\cos\theta_2 + i\sin\theta_2)} \times \frac{(\cos\theta_2 - i\sin\theta_2)}{(\cos\theta_2 - i\sin\theta_2)}$$

> Multiply the numerator and denominator by the complex conjugate of $\cos\theta_2 + i\sin\theta_2$

$$= \frac{r_1(\cos\theta_1 + i\sin\theta_1)(\cos\theta_2 - i\sin\theta_2)}{r_2(\cos\theta_2 + i\sin\theta_2)(\cos\theta_2 - i\sin\theta_2)}$$

$$= \frac{r_1(\cos\theta_1\cos\theta_2 - i\cos\theta_1\sin\theta_2 + i\sin\theta_1\cos\theta_2 - i^2\sin\theta_1\sin\theta_2)}{r_2(\cos\theta_2\cos\theta_2 - i\cos\theta_2\sin\theta_2 + i\sin\theta_2\cos\theta_2 - i^2\sin\theta_2\sin\theta_2)}$$

> Multiply out the brackets.

$$= \frac{r_1((\cos\theta_1\cos\theta_2 + \sin\theta_1\sin\theta_2) + i(\sin\theta_1\cos\theta_2 - \cos\theta_1\sin\theta_2))}{r_2(\cos\theta_2\cos\theta_2 + \sin\theta_2\sin\theta_2)}$$

> Collect real and imaginary parts.

$$= \frac{r_1(\cos(\theta_1 - \theta_2) + i\sin(\theta_1 + \theta_2))}{r_2(\cos^2\theta_2 + \sin^2\theta_2)}$$

> Simplify numerator using the trigonometric identities $\cos(A-B) = \cos A\cos B + \sin A\sin B$ and $\sin(A-B) = \sin A\cos B - \cos A\sin B$ and simplify the denominator.

$$= \frac{r_1}{r_2}(\cos(\theta_1 - \theta_2) + i\sin(\theta_1 - \theta_2))$$

> Simplify denominator using $\sin^2\theta + \cos^2\theta = 1$.

So $\left|\dfrac{z_1}{z_2}\right| = \dfrac{|z_1|}{|z_2|} = \dfrac{r_1}{r_2}$ and $\arg\dfrac{z_1}{z_2} = \arg z_1 - \arg z_2 = \theta_1 - \theta_2$

If $z_1 = r_1(\cos\theta_1 + i\sin\theta_1)$ and $z_2 = r_2(\cos\theta_2 + i\sin\theta_2)$:

$|z_1 z_2| = r_1 \times r_2$ and $\arg(z_1 z_2) = \theta_1 + \theta_2$ $\qquad \left|\dfrac{z_1}{z_2}\right| = \dfrac{r_1}{r_2}$ and $\arg\dfrac{z_1}{z_2} = \theta_1 - \theta_2$

Example 4.9

If $z_1 = 2\left(\cos\dfrac{2\pi}{3} + i\sin\dfrac{2\pi}{3}\right)$ and $z_2 = 6\left(\cos\dfrac{3\pi}{4} - i\sin\dfrac{3\pi}{4}\right)$, simplify $z_1 z_2$ and $\dfrac{z_1}{z_2}$ leaving your answer in polar form.

Remember:
$\cos(-\theta) = \cos\theta$
$\sin(-\theta) = -\sin\theta$

$z_1 z_2 = 2\left(\cos\dfrac{2\pi}{3} + i\sin\dfrac{2\pi}{3}\right) \times 6\left(\cos\dfrac{3\pi}{4} - i\sin\dfrac{3\pi}{4}\right)$

z_2 needs to be written in polar form: $\cos\theta + i\sin\theta$

$= 2\left(\cos\dfrac{2\pi}{3} + i\sin\dfrac{2\pi}{3}\right) \times 6\left(\cos\left(-\dfrac{3\pi}{4}\right) + i\sin\left(-\dfrac{3\pi}{4}\right)\right)$

cos is an even function: $\cos\theta = \cos(-\theta)$
sin is an odd function: $\sin(-\theta) = -\sin\theta$

$= 12\left(\cos\left(\dfrac{2\pi}{3} + -\dfrac{3\pi}{4}\right) + i\sin\left(\dfrac{2\pi}{3} + -\dfrac{3\pi}{4}\right)\right)$

Multiply moduli together; add arguments together.

$= 12\left(\cos\left(-\dfrac{\pi}{12}\right) + i\sin\left(-\dfrac{\pi}{12}\right)\right)$ or $12\left(\cos\dfrac{\pi}{12} - i\sin\dfrac{\pi}{12}\right)$

$\dfrac{z_1}{z_2} = \dfrac{2\left(\cos\dfrac{2\pi}{3} + i\sin\dfrac{2\pi}{3}\right)}{6\left(\cos\dfrac{3\pi}{4} - i\sin\dfrac{3\pi}{4}\right)}$

z_2 needs to be written in polar form: $\cos\theta + i\sin\theta$

$= \dfrac{2\left(\cos\dfrac{2\pi}{3} + i\sin\dfrac{2\pi}{3}\right)}{6\left(\cos\left(-\dfrac{3\pi}{4}\right) + i\sin\left(-\dfrac{3\pi}{4}\right)\right)}$

cos is an even function: $\cos\theta = \cos(-\theta)$
sin is an odd function: $\sin(-\theta) = -\sin\theta$

$= \dfrac{1}{3}\left(\cos\left(\dfrac{2\pi}{3} - -\dfrac{3\pi}{4}\right) + i\sin\left(\dfrac{2\pi}{3} - -\dfrac{3\pi}{4}\right)\right)$

Divide moduli; subtract arguments.

$= \dfrac{1}{3}\left(\cos\left(\dfrac{17\pi}{12}\right) + i\sin\left(\dfrac{17\pi}{12}\right)\right)$

$= \dfrac{1}{3}\left(\cos\left(-\dfrac{7\pi}{12}\right) + i\sin\left(-\dfrac{7\pi}{12}\right)\right)$

Ensure principal argument is in final answer.

Exercise 4D

1 Simplify these expressions, leaving your answer in polar form.

a $2\left(\cos\dfrac{\pi}{4} + i\sin\dfrac{\pi}{4}\right) \times 3\left(\cos\dfrac{\pi}{3} + i\sin\dfrac{\pi}{3}\right)$

b $4\left(\cos\dfrac{\pi}{2} + i\sin\dfrac{\pi}{2}\right) \times 2\left(\cos\dfrac{\pi}{6} + i\sin\dfrac{\pi}{6}\right)$

c $\left(\cos\dfrac{3\pi}{4} + i\sin\dfrac{3\pi}{4}\right) \times 5\left(\cos\dfrac{\pi}{3} + i\sin\dfrac{\pi}{3}\right)$

d $2\left(\cos\left(-\dfrac{\pi}{4}\right) + i\sin\left(-\dfrac{\pi}{4}\right)\right) \times 6\left(\cos\left(-\dfrac{2\pi}{3}\right) + i\sin\left(-\dfrac{2\pi}{3}\right)\right)$

e $3\left(\cos\dfrac{\pi}{3} - i\sin\dfrac{\pi}{3}\right) \times 4\left(\cos\dfrac{5\pi}{6} + i\sin\dfrac{5\pi}{6}\right)$

f $6\left(\cos\dfrac{\pi}{3} + i\sin\dfrac{\pi}{3}\right) \div 3\left(\cos\dfrac{\pi}{4} + i\sin\dfrac{\pi}{4}\right)$

g $10\left(\cos\dfrac{7\pi}{15} + i\sin\dfrac{7\pi}{15}\right) \div 2\left(\cos\dfrac{\pi}{3} + i\sin\dfrac{\pi}{3}\right)$

h $3\left(\cos\dfrac{\pi}{6} + i\sin\dfrac{\pi}{6}\right) \div \dfrac{1}{2}\left(\cos\dfrac{2\pi}{3} + i\sin\dfrac{2\pi}{3}\right)$

i $6\left(\cos\dfrac{\pi}{4} + i\sin\dfrac{\pi}{4}\right) \div 2\left(\cos\dfrac{\pi}{2} - i\sin\dfrac{\pi}{2}\right)$

j $5\left(\cos\dfrac{7\pi}{12} - i\sin\dfrac{7\pi}{12}\right) \div 9\left(\cos\dfrac{2\pi}{3} + i\sin\dfrac{2\pi}{3}\right)$

★ 2 Simplify these expressions giving your answer in the form $a + ib$ where $a, b \in \mathbb{R}$. Round answers to 3 significant figures if required.

a $2\left(\cos\dfrac{2\pi}{3} + i\sin\dfrac{2\pi}{3}\right) \times 3\left(\cos\left(-\dfrac{3\pi}{4}\right) + i\sin\left(-\dfrac{3\pi}{4}\right)\right)$

b $10\left(\cos\dfrac{5\pi}{9} + i\sin\dfrac{5\pi}{9}\right) \times 10\left(\cos\dfrac{13\pi}{9} + i\sin\dfrac{13\pi}{9}\right)$

c $2\left(\cos\dfrac{5\pi}{7} + i\sin\dfrac{5\pi}{7}\right) \times 5\left(\cos\dfrac{2\pi}{7} + i\sin\dfrac{2\pi}{7}\right)$

d $\sqrt{2}\left(\cos\dfrac{\pi}{6} - i\sin\dfrac{\pi}{6}\right) \times 3\left(\cos\dfrac{\pi}{4} + i\sin\dfrac{\pi}{4}\right)$

e $6\left(\cos\left(-\dfrac{3\pi}{4}\right) + i\sin\left(-\dfrac{3\pi}{4}\right)\right) \div 2\left(\cos\dfrac{2\pi}{3} + i\sin\dfrac{2\pi}{3}\right)$

f $3\left(\cos\dfrac{3\pi}{4} + i\sin\dfrac{3\pi}{4}\right) \div 2\left(\cos\dfrac{5\pi}{4} + i\sin\dfrac{5\pi}{4}\right)$

g $10\left(\cos\dfrac{\pi}{4} + i\sin\dfrac{\pi}{4}\right) \div 2\left(\cos\dfrac{5\pi}{6} + i\sin\dfrac{5\pi}{6}\right)$

h $2\left(\cos\dfrac{5\pi}{6} + i\sin\dfrac{5\pi}{6}\right) \div 10\left(\cos\dfrac{\pi}{4} + i\sin\dfrac{\pi}{4}\right)$

3   a   Plot these complex numbers on the same Argand diagram.

$$z_1 = 2\left(\cos\frac{\pi}{2} + i\sin\frac{\pi}{2}\right), \; z_2 = 4\left(\cos\frac{\pi}{6} + i\sin\frac{\pi}{6}\right)$$

$$z_3 = 3\left(\cos\left(-\frac{3\pi}{4}\right) + i\sin\left(-\frac{3\pi}{4}\right)\right)$$

   b   Let $z = \cos\frac{\pi}{2} + i\sin\frac{\pi}{2}$

   Calculate $z \times z_1$, $z \times z_2$ and $z \times z_3$.

   c   Plot $z \times z_1$, $z \times z_2$ and $z \times z_3$ on the Argand diagram drawn in part **a**. Compare the position of the complex number you have plotted for z_1 with the position of the complex number you have plotted for $z \times z_1$. Compare the position of the complex number you have plotted for z_2 with the position of the complex number you have plotted for $z \times z_2$. What do you notice?

   d   What do you think will happen if you change the value of the argument in z? Test your hypothesis using z_1, z_2 and z_3 in part **a**.

   e   What do you think will happen if you change the value of the modulus in z? Test your hypothesis using z_1, z_2 and z_3 in part **a**.

4   Given that $z = \left(\cos\frac{\pi}{6} + i\sin\frac{\pi}{6}\right)$, calculate these values, leaving your answer in polar form.

   a z^2 b z^3 c z^4 d z^5

   e   Can you find z^n?

de Moivre's theorem

de Moivre's theorem states that for any complex number:

$(\cos x + i\sin x)^n = \cos nx + i\sin nx$

You will examine a proof of this theorem using proof by induction in Chapter 11.

Consider $z = r(\cos\theta + i\sin\theta)$:

   then $z^2 = r(\cos\theta + i\sin\theta) \times r(\cos\theta + i\sin\theta)$

   $\qquad = r^2(\cos 2\theta + i\sin 2\theta)$ ———— Multiply the modulus and add the arguments.

   and $z^3 = z^2 \times z$

   $\qquad = r^2(\cos 2\theta + i\sin 2\theta) \times r(\cos\theta + i\sin\theta)$

   $\qquad = r^3(\cos 3\theta + i\sin 3\theta)$

   and $z^4 = z^3 \times z$

   $\qquad = r^3(\cos 3\theta + i\sin 3\theta) \times r(\cos\theta + i\sin\theta)$

   $\qquad = r^4(\cos 4\theta + i\sin 4\theta)$

In general:

$[r(\cos\theta + i\sin\theta)]^n - [r^n(\cos n\theta + i\sin n\theta)]$ ———— This is included in the exam formulae list.

Example 4.10

Express $2\sqrt{3} + 2i$ in polar form.

Use de Moivre's theorem to find

a z^2 b z^5 c z^8

giving your answer in both polar form and Cartesian form.

$r = |z| = \sqrt{\left(2\sqrt{3}\right)^2 + 2^2} = 4$

$\tan\theta = \dfrac{2}{2\sqrt{3}}$ so $\theta = \dfrac{\pi}{6}$

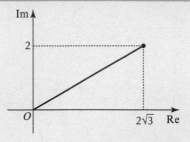

Sketch the Argand diagram.

$\therefore z = 4\left(\cos\dfrac{\pi}{6} + i\sin\dfrac{\pi}{6}\right)$

a $z^2 = \left(4\left(\cos\dfrac{\pi}{6} + i\sin\dfrac{\pi}{6}\right)\right)^2$

$= 4^2\left(\cos\left(2 \times \dfrac{\pi}{6}\right) + i\sin\left(2 \times \dfrac{\pi}{6}\right)\right)$ ← Using de Moivre's theorem.

$= 16\left(\cos\dfrac{\pi}{3} + i\sin\dfrac{\pi}{3}\right)$ ← Answer in polar form.

$= 16\left(\dfrac{1}{2} + \dfrac{\sqrt{3}}{2}i\right) = 8 + 8\sqrt{3}i$ ← Answer in Cartesian form.

b $z^5 = \left(4\left(\cos\dfrac{\pi}{6} + i\sin\dfrac{\pi}{6}\right)\right)^5$

$= 4^5\left(\cos\left(5 \times \dfrac{\pi}{6}\right) + i\sin\left(5 \times \dfrac{\pi}{6}\right)\right)$

$= 1024\left(\cos\dfrac{5\pi}{6} + i\sin\dfrac{5\pi}{6}\right)$

$= 1024\left(-\dfrac{\sqrt{3}}{2} + \dfrac{1}{2}i\right) = -512\sqrt{3} + 512i$

c $z^8 = \left(4\left(\cos\dfrac{\pi}{6} + i\sin\dfrac{\pi}{6}\right)\right)^8$

$= 4^8\left(\cos\left(8 \times \dfrac{\pi}{6}\right) + i\sin\left(8 \times \dfrac{\pi}{6}\right)\right)$

$= 65\,536\left(\cos\dfrac{4\pi}{3} + i\sin\dfrac{4\pi}{3}\right)$

$= 65\,536\left(\cos\left(-\dfrac{2\pi}{3}\right) + i\sin\left(-\dfrac{2\pi}{3}\right)\right)$ ← Answer in polar form. Remember to calculate the principal argument in your final answer by subtracting 2π from the argument: $\dfrac{4\pi}{3} - 2\pi = \dfrac{-2\pi}{3}$

$= 65\,536\left(-\dfrac{1}{2} - \dfrac{\sqrt{3}}{2}i\right) = -32768 - 32768\sqrt{3}i$

Example 4.11

Simplify $\left(\cos\dfrac{\pi}{3} + i\sin\dfrac{\pi}{3}\right)^2\left(\cos\dfrac{\pi}{12} - i\sin\dfrac{\pi}{12}\right)^3$

$$\left(\cos\frac{\pi}{3} + i\sin\frac{\pi}{3}\right)^2 = 1^2\left(\cos\left(2\times\frac{\pi}{3}\right) + i\sin\left(2\times\frac{\pi}{3}\right)\right)$$

> Apply de Moivre's theorem to each bracket separately before multiplying them together.

$$= \cos\frac{2\pi}{3} + i\sin\frac{2\pi}{3}$$

and

$$\left(\cos\frac{\pi}{12} - i\sin\frac{\pi}{12}\right)^3 = \left(\cos\left(-\frac{\pi}{12}\right) + i\sin\left(-\frac{\pi}{12}\right)\right)^3$$

> Rewrite in the form $\cos\theta + i\sin\theta$ so that de Moivre's theorem can be applied.

$$= 1^3\left(\cos\left(3\times-\frac{\pi}{12}\right) + i\sin\left(3\times-\frac{\pi}{12}\right)\right)$$

> Use de Moivre's theorem

$$= \cos\left(-\frac{\pi}{4}\right) + i\sin\left(-\frac{\pi}{4}\right)$$

$$\left(\cos\frac{\pi}{3} + i\sin\frac{\pi}{3}\right)^2\left(\cos\frac{\pi}{12} - i\sin\frac{\pi}{12}\right)^3 = \left(\cos\frac{2\pi}{3} + i\sin\frac{2\pi}{3}\right)\left(\cos\left(-\frac{\pi}{4}\right) + i\sin\left(-\frac{\pi}{4}\right)\right)$$

$$= \cos\frac{5\pi}{12} + i\sin\frac{5\pi}{12}$$

> Add the arguments together.

Exercise 4E

1 Use de Moivre's theorem to simplify:

a $\left(\cos\dfrac{\pi}{7} + i\sin\dfrac{\pi}{7}\right)^5$

b $\left(3\left(\cos\dfrac{\pi}{11} + i\sin\dfrac{\pi}{11}\right)\right)^4$

c $\left(2\left(\cos\dfrac{2\pi}{3} + i\sin\dfrac{2\pi}{3}\right)\right)^{10}$

d $\left(10\left(\cos\dfrac{3\pi}{4} + i\sin\dfrac{3\pi}{4}\right)\right)^3$

2 a If $z = 2\left(\cos\dfrac{\pi}{3} + i\sin\dfrac{\pi}{3}\right)$, find z^6.

b If $z = 3\left(\cos\dfrac{\pi}{4} + i\sin\dfrac{\pi}{4}\right)$, find z^5.

★ 3 Using de Moivre's theorem, calculate these terms giving your answer in the form $a + ib$.

a z^2 where $z = 1 + \sqrt{3}i$

b z^4 where $z = 1 - i$

c z^8 where $z = 2 - 2\sqrt{3}i$

d z^{10} where $z = 1 + i$

4 a Use de Moivre's theorem to find the real and imaginary parts of $\left(\sqrt{3} - i\right)^{10}$

b Find the real and imaginary parts of $\left(i - \sqrt{3}\right)^7$

5 Simplify these expressions leaving your answer in polar form.

a $\left(\cos\dfrac{\pi}{6} + i\sin\dfrac{\pi}{6}\right)^2 \left(\cos\dfrac{\pi}{18} + i\sin\dfrac{\pi}{18}\right)^4$

b $\left(\cos\dfrac{3\pi}{5} + i\sin\dfrac{3\pi}{5}\right)^4 \left(\cos\dfrac{5\pi}{7} + i\sin\dfrac{5\pi}{7}\right)^3$

c $\left(\cos\dfrac{2\pi}{9} - i\sin\dfrac{2\pi}{9}\right)^4 \left(\cos\dfrac{3\pi}{4} + i\sin\dfrac{3\pi}{4}\right)^5$

d $\left(\cos\dfrac{2\pi}{9} + i\sin\dfrac{2\pi}{9}\right)^3 \div \left(\cos\dfrac{\pi}{6} + i\sin\dfrac{\pi}{6}\right)^2$

e $\left(\cos\dfrac{4\pi}{11} + i\sin\dfrac{4\pi}{11}\right)^3 \div \left(\cos\dfrac{\pi}{3} + i\sin\dfrac{\pi}{3}\right)$

f $\left(\cos\dfrac{4\pi}{7} + i\sin\dfrac{4\pi}{7}\right)^3 \div \left(\cos\dfrac{3\pi}{4} - i\sin\dfrac{3\pi}{4}\right)^4$

6 a Given that $z = \cos\theta + i\sin\theta$, show that $z^2 + \dfrac{1}{z^2} = 2\cos 2\theta$

b Given that $z = \cos\theta + i\sin\theta$, show that $z^n + \dfrac{1}{z^n} = 2\cos n\theta$

7 a Use the binomial theorem to expand $(\cos\theta + i\sin\theta)^2$

b Use de Moivre's theorem to expand $(\cos\theta + i\sin\theta)^2$

c By equating real parts, express $\cos 2\theta$ in terms of $\cos\theta$ and $\sin\theta$

d By equating imaginary parts, express $\sin 2\theta$ in terms of $\cos\theta$ and $\sin\theta$

★ 8 a Use the binomial theorem to expand $(\cos\theta + i\sin\theta)^3$

b Use de Moivre's theorem to expand $(\cos\theta + i\sin\theta)^3$

c By equating real parts, express $\cos 3\theta$ in terms of $\cos\theta$ and $\sin\theta$

d By equating imaginary parts, express $\sin 3\theta$ in terms of $\cos\theta$ and $\sin\theta$

e Express $\sin 3\theta$ in terms of $\sin\theta$

9 Express $\cos 5\theta$ in terms of $\cos\theta$

10 Express $\cos 7\theta$ in terms of $\cos\theta$

★ 11 a Expand $(\cos\theta + i\sin\theta)^4$ using

i the binomial theorem   ii de Moivre's theorem.

b Prove that   i $\cos 4\theta = \cos^4\theta - 6\cos^2\theta\sin^2\theta + \sin^4\theta$

ii $\sin 4\theta = 4\cos^3\theta\sin\theta - 4\cos\theta\sin^3\theta$

c Hence express $\tan 4\theta$ in terms of $\tan\theta$

Roots of a complex number

Consider the complex number $z = r(\cos\theta + i\sin\theta)$, where θ is the **principal argument** of z, then z can also be written as:

$$z = r\big(\cos(\theta + 2k\pi) + i\sin(\theta + 2k\pi)\big), k \in \mathbb{Z}$$

because, if 2π (or multiples of) is added to (or subtracted from) θ then we move to the same position on the Argand diagram.

So if:

$$z^2 = r\big(\cos(\theta + 2k\pi) + i\sin(\theta + 2k\pi)\big)$$

then:

$$z = \Big(r\big(\cos(\theta + 2k\pi) + i\sin(\theta + 2k\pi)\big)\Big)^{\frac{1}{2}}$$

$$= r^{\frac{1}{2}}\Big(\cos\tfrac{1}{2}(\theta + 2k\pi) + i\sin\tfrac{1}{2}(\theta + 2k\pi)\Big), k = 0, 1$$

> Taking the square root of a number results in two solutions: one when $k = 0$ and one when $k = 1$.

And if:

$$z^3 = r\big(\cos(\theta + 2k\pi) + i\sin(\theta + 2k\pi)\big)$$

then:

$$z = \Big(r\big(\cos(\theta + 2k\pi) + i\sin(\theta + 2k\pi)\big)\Big)^{\frac{1}{3}}$$

$$= r^{\frac{1}{3}}\Big(\cos\tfrac{1}{3}(\theta + 2k\pi) + i\sin\tfrac{1}{3}(\theta + 2k\pi)\Big), k = 0, 1, 2$$

> Taking the cube root of a number results in three solutions: one when $k = 0$, one when $k = 1$ and one when $k = 2$.

So we can find all the roots by adding $\dfrac{2k\pi}{n}$ to θ where $k = 0, 1, \ldots, n - 1$.

If $z = r(\cos\theta + i\sin\theta)$, then the n solutions of the equation $z_k^n = z$ are given by:

$$z_k = r^{\frac{1}{n}}\left(\cos\left(\frac{\theta + 2k\pi}{n}\right) + i\sin\left(\frac{\theta + 2k\pi}{n}\right)\right) \text{ where } k = 0, 1, 2, \ldots, n - 1$$

When plotted on an Argand diagram, the position vectors of the solutions will divide the circle of radius, r, (centre at O), into n equal sectors.

The diagram shows the three equally spaced solutions (represented by a dot) of the equation $z^3 = 1$. The three solutions all have $\dfrac{2\pi}{3}$ radians between them.

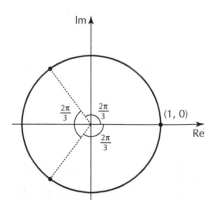

Example 4.12

Solve the equation $z^2 = \cos\dfrac{\pi}{3} + i\sin\dfrac{\pi}{3}$

If $z^2 = \cos\dfrac{\pi}{3} + i\sin\dfrac{\pi}{3}$

then the two solutions are of the form:

$$z_k = \left(\cos\frac{\pi}{3} + i\sin\frac{\pi}{3}\right)^{\frac{1}{2}}$$

$$= \cos\frac{1}{2}\left(\frac{\pi}{3} + 2k\pi\right) + i\sin\frac{1}{2}\left(\frac{\pi}{3} + 2k\pi\right) \text{ where } k = 0, 1$$

When $k = 0$:

$$z_0 = \cos\frac{1}{2}\left(\frac{\pi}{3} + 0\right) + i\sin\frac{1}{2}\left(\frac{\pi}{3} + 0\right)$$

$$= \cos\frac{\pi}{6} + i\sin\frac{\pi}{6}$$

$$= \frac{\sqrt{3}}{2} + \frac{1}{2}i$$

When $k = 1$:

$$z_1 = \cos\frac{1}{2}\left(\frac{\pi}{3} + 2\pi\right) + i\sin\frac{1}{2}\left(\frac{\pi}{3} + 2\pi\right)$$

$$= \cos\frac{7\pi}{6} + i\sin\frac{7\pi}{6}$$

$$= \cos\left(-\frac{5\pi}{6}\right) + i\sin\left(-\frac{5\pi}{6}\right)$$

$$= -\frac{\sqrt{3}}{2} - \frac{1}{2}i$$

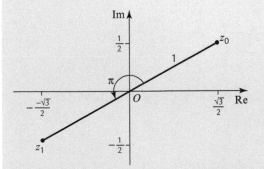

The solutions divide circle of radius 1 (r) into 2 equal sectors. The position vectors of the two solutions are π radians apart.

Example 4.13

Solve the equation $z^3 = 8\left(\cos\dfrac{\pi}{3} + i\sin\dfrac{\pi}{3}\right)$

If $z^3 = 8\left(\cos\dfrac{\pi}{3} + i\sin\dfrac{\pi}{3}\right)$

then the three solutions are of the form:

$$z_k = \left(8\left(\cos\dfrac{\pi}{3} + i\sin\dfrac{\pi}{3}\right)\right)^{\frac{1}{3}}$$

$$= 8^{\frac{1}{3}}\left(\cos\dfrac{1}{3}\left(\dfrac{\pi}{3} + 2k\pi\right) + i\sin\dfrac{1}{3}\left(\dfrac{\pi}{3} + 2k\pi\right)\right)$$

$$= 2\left(\cos\dfrac{1}{3}\left(\dfrac{\pi}{3} + 2k\pi\right) + i\sin\dfrac{1}{3}\left(\dfrac{\pi}{3} + 2k\pi\right)\right) \text{ where } k = 0, 1, 2$$

When $k = 0$:

$$z_0 = 2\left(\cos\dfrac{1}{3}\left(\dfrac{\pi}{3} + 0\right) + i\sin\dfrac{1}{3}\left(\dfrac{\pi}{3} + 0\right)\right)$$

$$= 2\left(\cos\dfrac{\pi}{9} + i\sin\dfrac{\pi}{9}\right)$$

When $k = 1$:

$$z_1 = 2\left(\cos\dfrac{1}{3}\left(\dfrac{\pi}{3} + 2\pi\right) + i\sin\dfrac{1}{3}\left(\dfrac{\pi}{3} + 2\pi\right)\right)$$

$$= 2\left(\cos\dfrac{7\pi}{9} + i\sin\dfrac{7\pi}{9}\right)$$

When $k = 2$:

$$z_2 = 2\left(\cos\dfrac{1}{3}\left(\dfrac{\pi}{3} + 4\pi\right) + i\sin\dfrac{1}{3}\left(\dfrac{\pi}{3} + 4\pi\right)\right)$$

$$= 2\left(\cos\dfrac{13\pi}{9} + i\sin\dfrac{13\pi}{9}\right)$$

$$= 2\left(\cos\left(-\dfrac{5\pi}{9}\right) + i\sin\left(-\dfrac{5\pi}{9}\right)\right)$$

The solutions divide a circle of radius 2 (r) into 3 equal sectors. The position vectors of the three solutions are $\dfrac{2\pi}{3}$ radians apart.

Example 4.14

Find the cube roots of unity.

> This is another way of asking you to 'Solve the equation $z^3 = 1$.'

$z^3 = 1$

> Express the right-hand side of the equation in polar form.

arg z is 0

$\Rightarrow 1 = \cos 0 + i\sin 0$

$\Rightarrow z^3 = 1(\cos 0 + i\sin 0)$,

then solutions are of the form:

> Sketch $z = 1$ on an Argand diagram.

$z_k = \left(1(\cos 0 + i\sin 0)\right)^{\frac{1}{3}}$

$= 1^{\frac{1}{3}}\left(\cos\frac{1}{3}(0 + 2k\pi) + i\sin\frac{1}{3}(0 + 2k\pi)\right)$

> General case

$= \cos\frac{1}{3}(0 + 2k\pi) + i\sin\frac{1}{3}(0 + 2k\pi)$ where $k = 0, 1, 2$

When $k = 0$:

$z_0 = \cos\frac{1}{3}(0 + 0) + i\sin\frac{1}{3}(0 + 0)$

$= \cos 0 + i\sin 0$

$= 1$

When $k = 1$:

$z_1 = \cos\frac{1}{3}(0 + 2\pi) + i\sin\frac{1}{3}(0 + 2\pi)$

$= \cos\frac{2\pi}{3} + i\sin\frac{2\pi}{3}$

$= -\frac{1}{2} + \frac{\sqrt{3}}{2}i$

When $k = 2$:

$z_2 = \cos\frac{1}{3}(0 + 4\pi) + i\sin\frac{1}{3}(0 + 4\pi)$

$= \cos\frac{4\pi}{3} + i\sin\frac{4\pi}{3}$

$= \cos\left(-\frac{2\pi}{3}\right) + i\sin\left(-\frac{2\pi}{3}\right)$

$= -\frac{1}{2} - \frac{\sqrt{3}}{2}i$

Solutions divide circle of radius 1 (r) into 3 equal sectors. The position vectors of the three solutions are $\frac{2\pi}{3}$ radians apart.

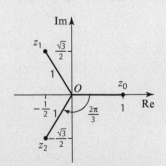

Exercise 4F

1 Solve each of these equations leaving your answer in polar form.

a $z^3 = \cos\dfrac{\pi}{3} + i\sin\dfrac{\pi}{3}$

b $z^3 = 8\left(\cos\dfrac{\pi}{4} + i\sin\dfrac{\pi}{4}\right)$

c $z^5 = 32\left(\cos\dfrac{2\pi}{5} + i\sin\dfrac{2\pi}{5}\right)$

d $z^3 = 64\left(\cos\dfrac{3\pi}{4} + i\sin\dfrac{3\pi}{4}\right)$

e $z^2 = 1 + i$

f $z^5 = \sqrt{3} + i$

g $z^4 = -2 - 2\sqrt{3}i$

h $z^3 = 3 - 3\sqrt{3}i$

2 Solve these equations.

a $z^3 = 1$

b $z^4 = 1$

c $z^5 = 1$

> ⚠ Write the right-hand side of the equation in polar form.

★ 3 a Find the cube roots of unity.

b Find the sixth roots of unity.

★ 4 Solve these equations.

a $z^4 = -8$

b $z^5 = i$

c $z^3 = -64i$

d $z^4 = 625i$

e $z^5 + 1 = 0$

f $z^3 + 64 = 0$

Calculating roots of polynomials

A polynomial of degree n will have n complex roots. The proof of this is beyond the scope of this course but is based on the **fundamental theorem of algebra**. If a polynomial is of degree n, then there are n roots in the set of complex numbers. These roots may be real or complex. If z is a root, by the fundamental theorem of algebra, $\bar{z}$ is also a root.

Polynomials with an odd degree will have at least one real root. Further roots will be pairs of complex roots or real roots.

Using synthetic division to find roots of polynomials

Based on our knowledge of synthetic division, the quadratic formula and complex numbers we can solve problems of the form $z^3 - 4z^2 + 6z - 4 = 0$.

Example 4.15

Solve $z^3 - 4z^2 + 6z - 4 = 0$

$$
\begin{array}{c|cccc}
2 & 1 & -4 & 6 & -4 \\
  &   & + & + & + \\
  &   & 2 & -4 & 4 \\
\hline
  & 1 & -2 & 2 & \boxed{0}
\end{array}
$$

Use synthetic division to show that $z = 2$ is a root and to find the other factor.

Remainder is 0

$\Rightarrow z - 2$ is a factor of $z^3 - 4z^2 + 6z - 4$

and $z^2 - 2z + 2$ is the other factor

So:

$z^3 - 4z^2 + 6z - 4 = 0$

$(z - 2)(z^2 - 2z + 2) = 0$

$z = \dfrac{2 \pm \sqrt{(-2)^2 - 4 \times 1 \times 2}}{2 \times 1}$

Use the quadratic formula to factorise $z^2 - 2z + 2$

$= \dfrac{2 \pm \sqrt{-4}}{2}$

$= \dfrac{2 \pm 2i}{2}$

$= 1 \pm i$

Solutions are $z = 2,\ 1 + i,\ 1 - i$

In Example 4.15, there are three roots:

- one real, $z = 2$
- two complex roots, $z = 1 + i$ and $1 - i$

It should be noted that the roots $1 + i$ and $1 - i$ are complex conjugates.

If a polynomial equation, $f(z) = 0$, **with real coefficients**, has a root $a + ib$ where $a, b \in \mathbb{R}$, then the complex conjugate $a - ib$ is also a root of $f(z) = 0$.

Calculating roots of polynomials

Example 4.16

If $f(z) = az^2 + bz + c$ has a root $z = 2 - 3i$, find the values of a, b and c where $a, b, c \in \mathbb{R}$.

If $2 - 3i$ is a root, then $2 + 3i$ is also a root

> By the fundamental theorem of algebra, a conjugate pair exists.

So the function $f(z) = (z - (2 - 3i))(z - (2 + 3i))$
$$= z^2 - (2 + 3i)z - (2 - 3i)z + (2 - 3i)(2 + 3i)$$
$$= z^2 - 2z + 3iz - 2z - 3iz + 4 - 6i + 6i - 9i^2$$
$$= z^2 - 4z + 4 - 9(-1)$$
$$= z^2 - 4z + 13$$

So $a = 1$, $b = -4$, $c = 13$

Example 4.17

Show that $1 - i$ is a root of
$z^3 - 3z^2 + 4z - 2 = 0$
and solve completely.

> Since the degree of the polynomial is 3, they you would expect to get 3 solutions.

$(1 - i)^2 = 1 - 2i + i^2 = -2i$

$(1 - i)^3 = (1 - i)^2(1 - i) = -2i(1 - i) = -2i + 2i^2 = -2 - 2i$

> It is easier to multiply out and simplify $(1 - i)^2$ and $(1 - i)^3$ before substituting them into the equation to be solved in the question.

so $z^3 - 3z^2 + 4z - 2 = (1 - i)^3 - 3(1 - i)^2 + 4(1 - i) - 2$
$$= (-2 - 2i) - 3(-2i) + 4(1 - i) - 2$$
$$= -2 - 2i + 6i + 4 - 4i - 2$$
$$= 0 \text{ as required so } 1 - i \text{ is a root of } z^3 - 3z^2 + 4z - 2 = 0$$

If $1 - i$ is a root, then $1 + i$ is also a root

So $(z - (1 + i))$ and $(z - (1 - i))$ are factors

Hence $(z - (1 + i)(z - (1 - i))$ is a factor

i.e. $z^2 - 2z + 2$ is a factor

> By the fundamental theorem of algebra, a conjugate pair exists.

> Multiply out the brackets.

$$
\begin{array}{r}
-z - 1 \\
z^2 - 2z + 2 \overline{\smash{)}\ z^3 - 3z^2 + 4z - 2} \\
\underline{z^3 - 2z^2 + 2z} \\
-z^2 + 2z - 2 \\
\underline{-z^2 + 2z - 2} \\
0
\end{array}
$$

> To find the other factors, divide $z^2 - 2z + 2$ into $z^3 - 3z^2 + 4z - 2$

So $z^3 - 3z^2 + 4z - 2 = 0$

can be factorised as $(z^2 - 2z + 2)(z - 1) = 0$

> Factorise.

$\therefore z = 1, 1 + i, 1 - i$

Exercise 4G

1 Solve these equations.

   a $z^2 + 1 = 0$

   b $z^2 - 4z + 5 = 0$

   c $z^2 + 2z + 5 = 0$

   d $z^2 - 2z + 2 = 0$

   e $z^2 - 4z + 13 = 0$

   f $4z^2 - 16z + 17 = 0$

2 For each of these equations, one root is given. Show that it is a root and find the other roots.

   a $z^3 - 3z^2 + z + 5 = 0$ $z = -1$

   b $z^3 + z - 10 = 0$ $z = 2$

   c $z^3 + z^2 - 4z + 6 = 0$ $z = -3$

   d $4z^3 - 20z^2 + 33z - 17 = 0$ $z = 1$

   e $2z^3 + 6z^2 + 5z + 2 = 0$ $z = -2$

   f $z^3 - 9z^2 + 33z - 65 = 0$ $z = 5$

3 For each of these equations, one root is given. Show that it is a root and find the other roots.

   a $z^3 - 3z^2 + z - 3 = 0$ $z = i$

   b $z^3 - 6z^2 + 13z - 10 = 0$ $z = 2 + i$

   c $z^3 + 4z^2 - 10z + 12 = 0$ $z = 1 + i$

   d $2z^3 - 9z^2 + 30z - 13 = 0$ $z = 2 + 3i$

   e $2z^3 - 11z^2 + 14z + 10 = 0$ $z = 3 + i$

   f $4z^3 - 12z^2 + z + 17 = 0$ $z = 2 - \frac{1}{2}i$

★ 4 a i   How many solutions do you expect when solving $z^3 + 3z^2 - 5z + 25 = 0$?

      ii Show that $1 + 2i$ is a root of the equation $z^3 + 3z^2 - 5z + 25 = 0$ and find the other roots of the equation.

   b i   How many solutions do you expect when solving $2z^3 - 5z^2 + 12z - 5 = 0$?

      ii Show that $1 + 2i$ is a root of the equation $2z^3 - 5z^2 + 12z - 5 = 0$ and find the other roots of the equation.

   c i   How many solutions do you expect when solving $z^4 - 4z^3 + 3z^2 + 2z - 6 = 0$?

      ii Show that $1 - i$ is a root of the equation $z^4 - 4z^3 + 3z^2 + 2z - 6 = 0$ and find the other roots of the equation.

   d i   How many solutions do you expect when solving
          $2z^4 - 10z^3 + 43z^2 + 20z + 125 = 0$?

      ii Show that $3 - 4i$ is a root of the equation $2z^4 - 10z^3 + 43z^2 + 20z + 125 = 0$ and find the other roots of the equation.

Loci on the complex plane

Any complex number can be represented on an Argand diagram. A restriction can be placed on the complex number by stating a magnitude. For example, if $|z| = 3$, we would only consider complex numbers that have a magnitude of 3. (This is similar to the restriction that a pizza delivery shop states when they say that they will only deliver within a 3 mile radius of the shop.)

A set of points that follow a rule is called a **locus**. The locus of $|z| = 3$ is the set of points that have a magnitude of 3.

> The word locus is the Latin word for place. Locus relates to a set of points satisfying a particular condition often forming a line, curve or circle.

Example 4.18

If $z = x + iy$, sketch the locus which represents $|z| = 3$

Hence sketch the locus which represents $|z| \leqslant 3$

$|z| = 3$ so $|x + iy| = 3$ ———— Replace z with $x + iy$.

$|x + iy| = \sqrt{x^2 + y^2}$

$\Rightarrow \sqrt{x^2 + y^2} = 3$

$\therefore x^2 + y^2 = 9$ ———— This is an equation of a circle with a radius of 3 units and centre (0, 0).

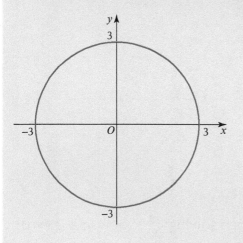

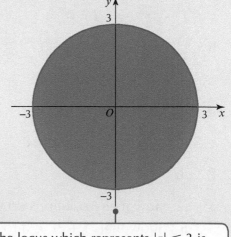

The locus which represents $|z| \leqslant 3$ is the same circle as above left with the inside shaded.

In general, if:

- $|z| = r$, the points are on the circumference of the circle
- $|z| < r$, the points are inside the circle
- $|z| > r$, the points are outside the circle.

Example 4.19

If $z = x + iy$, sketch the locus which represents $|z - 3| = 4$

$|z - 3| = 4$ so $|x + iy - 3| = 4,$ — Replace z with $x + iy$.

$|x - 3 + iy| = 4$ — Collect real and imaginary parts.

$\sqrt{(x - 3)^2 + y^2} = 4$ — Use $|x + iy| = \sqrt{x^2 + y^2}$

$(x - 3)^2 + y^2 = 16$ — Square both sides to find equation of a circle.

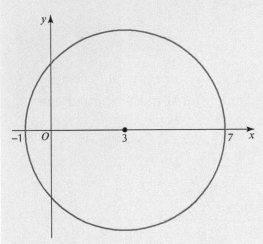

Example 4.20

If $z = x + iy$, sketch the locus which represents $|z - 2| = |z + 3i|$

$|z - 2| = |z + 3i|$ so $|x + iy - 2| = |x + iy + 3i|$

$|x - 2 + iy| = |x + (y + 3)i|$

$\sqrt{(x - 2)^2 + y^2} = \sqrt{x^2 + (y + 3)^2}$

$(x - 2)^2 + y^2 = x^2 + (y + 3)^2$

$x^2 - 4x + 4 + y^2 = x^2 + y^2 + 6y + 9$ — Multiply out the brackets.

$y = -\frac{2}{3}x - \frac{5}{6}$

This is an equation of a straight line with a gradient

of $-\frac{2}{3}$, passing through $\left(0, -\frac{5}{6}\right)$

Example 4.21

If $z = x + iy$, find the equation of the locus with $\arg z = \dfrac{\pi}{6}$

$\arg z = \dfrac{\pi}{6}$

$\tan^{-1}\dfrac{y}{x} = \dfrac{\pi}{6}$ ●————————— $\tan\theta = \dfrac{b}{a}$ where $z = a + ib$ and $\arg z = \theta$

$\dfrac{y}{x} = \tan\dfrac{\pi}{6}$

$\dfrac{y}{x} = \dfrac{1}{\sqrt{3}}$

$y = \dfrac{1}{\sqrt{3}}x$

This is the equation of a straight line with gradient $\dfrac{1}{\sqrt{3}}$ that passes through the origin.

Exercise 4H

★ 1   If $z = x + iy$, sketch the locus of z if:

    a $|z| = 5$ **b** $|z| = 2$ **c** $|z - 3| = 2$

    d $|z + i| = 4$ **e** $|z - 1 + 3i| = 4$ **f** $|z + 1 - 2i| = 5$

    g $|2z - 4 + i| = 4$ **h** $|3z + 2 - 3i| = 2$

★ 2   If $z = x + iy$, find the equation of the locus of z if:

    a $\arg z = \dfrac{\pi}{3}$ **b** $\arg z = -\dfrac{\pi}{4}$

    c $\arg z = -\dfrac{5\pi}{6}$ **d** $\arg z = \dfrac{2\pi}{3}$

3   If $z = x + iy$, sketch the locus of z if:

    a $|z| \leqslant 4$ **b** $|z + 3| \geqslant 2$

    c $|z - i| < 3$ **d** $|z - 5| \leqslant 2$

★ 4   If $z = x + iy$, state the equation of the locus of z if:

    a $|z + i| = |z - 1|$ **b** $|z - 1| = |z + 2i|$

    c $|z - 3i| = |z|$ **d** $|z + 2| = |z - i|$

A third way of writing a complex number

In Chapter 7, we will show that the series for the exponential function e^x, $x \in \mathbb{R}$, is given by:

$$e^x = 1 + x + \frac{x^2}{2!} + \frac{x^3}{3!} + \frac{x^4}{4!} + \ldots + \frac{x^r}{r!} + \ldots$$

This series also holds when x is a complex number (the proof is not included in this book). If we consider $e^{i\theta}$, then:

$$e^{i\theta} = 1 + i\theta + \frac{(i\theta)^2}{2!} + \frac{(i\theta)^3}{3!} + \frac{(i\theta)^4}{4!} + \ldots + \frac{(i\theta)^r}{r!} + \ldots$$

> Substitute $i\theta$ for x in the expansion of e^x above.

$$= 1 + i\theta - \frac{\theta^2}{2!} - \frac{i\theta^3}{3!} + \frac{\theta^4}{4!} + \ldots + \frac{i^r\theta^r}{r!} + \ldots$$

$$= 1 - \frac{\theta^2}{2!} + \frac{\theta^4}{4!} - \frac{\theta^6}{6!} + \frac{\theta^8}{8!} + \ldots + \frac{(-1)^r\theta^{2r}}{(2r)!} + \ldots$$

> Collect real and imaginary terms.

$$+ i\left(\theta - \frac{\theta^3}{3!} + \frac{\theta^5}{5!} - \frac{\theta^7}{7!} + \ldots + \frac{(-1)^r\theta^{2r+1}}{(2r+1)!} + \ldots \right)$$

$$= \cos\theta + i\sin\theta$$

> Chapter 7 will also show the series for $\sin x$ and $\cos x$.

Hence $e^{i\theta} = \cos\theta + i\sin\theta$ and, given that $z = r(\cos\theta + i\sin\theta)$ then a third way to write a complex number is $z = re^{i\theta}$ where $r = |z|$ and $\theta = \arg z$. This is the exponential form of a complex number z.

Now if we consider $\theta = \pi$:

then $e^{i\pi} = \cos\pi + i\sin\pi = -1$

that is $e^{i\pi} = -1$

This equation, attributed to Euler, combines three important numbers in mathematics, namely e, π and i in a single equation. Ian Stewart (2012) writes that 'This equation regularly come top of the list in polls for the most beautiful equations in mathematics.'

Chapter review

1   Given that $z_1 = 3 - 4i$ and $z_2 = 6 + i$, calculate the following, giving your answers in the form $a + ib$ where $a, b \in \mathbb{R}$.

    a $z_1 + z_2$ b $3z_2 - z_1$ c $z_1 z_2$

    d $\frac{z_2}{z_1}$ e $(z_2)^2$ f $\sqrt{z_1}$

2   Solve $(2a + 3i)(5 + bi) = 16 + 11i$ for a and b where $a, b \in \mathbb{R}$.

3   Draw $z_1 = 5 + 3i$ and $z_2 = -3 + 4i$ on an Argand diagram.

4   Express $z = \sqrt{3} - 3i$ in polar form.

5 Simplify the following giving your answer in the form $a + ib$ where $a, b \in \mathbb{R}$. Give your answers as exact values or, if required, to 3 significant figures.

a $5\left(\cos\dfrac{5\pi}{6} - i\sin\dfrac{5\pi}{6}\right) \times 2\left(\cos\dfrac{\pi}{4} + i\sin\dfrac{\pi}{4}\right)$

b $12\left(\cos\left(-\dfrac{\pi}{4}\right) + i\sin\left(-\dfrac{\pi}{4}\right)\right) \div 3\left(\cos\dfrac{2\pi}{7} + i\sin\dfrac{2\pi}{7}\right)$

c $\left(2\left(\cos\dfrac{3\pi}{8} + i\sin\dfrac{3\pi}{8}\right)\right)^6$

6 Express $\sin 5\theta$ in terms of $\sin\theta$.

7 Solve $z^3 = 4\sqrt{2} + 4\sqrt{2}i$ leaving your answer in polar form. Plot all the solutions on an Argand diagram.

8 Find the cube roots of unity and give your answer in the form $a + ib$, where $a, b \in \mathbb{R}$.

9 Show that $z = 2 + 3i$ is a root of the equation $z^4 - 2z^3 + 10z^2 + 6z + 65 = 0$ and hence solve the equation completely.

- I can add, subtract, multiply a complex number. ★ Exercise 4A Q3

- I can divide and find the square root of a complex number. ★ Exercise 4B Q2, Q11

- I can solve equations involving complex numbers. ★ Exercise 4B Q8

- I can use an Argand diagram. ★ Exercise 4C Q1

- I can convert a complex number from Cartesian to polar form and vice versa. ★ Exercise 4C Q5, Q6

- I can multiply and divide complex numbers in polar form. ★ Exercise 4D Q2

- I can use de Moivre's theorem with integer and fractional indices. ★ Exercise 4E Q3

- I can apply de Moivre's theorem to multiple angle trigonometric formula and to find nth roots of a complex number. ★ Exercise 4E Q8, Q11 ★ Exercise 4F Q3, Q4

- I can find the roots of a polynomial (including complex roots). ★ Exercise 4G Q4

- I can interpret geometrically equations/inequalities in the complex plane. ★ Exercise 4H Q1, Q2, Q4

5 Differential equations

This chapter will show you how to:

- solve first-order differential equations with variables separable
- solve first-order linear differential equations with non-variables separable
- solve second-order differential equations.

You should already know:

- how to differentiate products of functions
- how to differentiate quotients of functions
- how to differentiate inverse functions
- how to integrate an algebraic function which is, or can be, simplified to an expression in powers of x
- how to integrate functions of the form $f(x) = (x + q)^n$ and $f(x) = (px + q)^n$
- how to integrate functions of the form $f(x) = p \sin x$ and $f(x) = p \cos x$; $f(x) = p \sin (qx + r)$ and $f(x) = p \cos (qx + r)$
- how to solve differential equations of the form $\dfrac{dy}{dx} = Q(x)$

Differentiate first-order differential equations

Differential equations are used to calculate the growth or decay of populations over time in studies involving bacteria, and in the increase or decrease of the volume of a liquid as it enters or leaves a vessel such as a flask or a swimming pool.

Differential equations also help scientists to evaluate the weight of an isotope, after a given time, as the radioactive chemical decays according to its half-life.

In solving a differential equation we are trying to find $y = Q(x)$

First order differential equations are written with $\dfrac{dy}{dx}$ being the highest derivative involved.

In standard form: $\dfrac{dy}{dx} + P(x)y = Q(x)$

Second order differential equations are written with $\dfrac{d^2y}{dx^2}$ being the highest derivative involved.

In standard form: $a\dfrac{d^2y}{dx^2} + b\dfrac{dy}{dx} + cy = Q(x)$ or $a\dfrac{d^2y}{dx^2} + b\dfrac{dy}{dx} + cy = 0$

Differential equations can be described as either:

- variables separable
- variables non-separable

Solving first-order differential equations with variables separable

To solve first-order differential equations with variables separable:

$$\frac{dy}{dx} = f(x)g(y)$$

- first separate the terms in x and y on opposite sides of the equation
- write all terms in x and dx on one side
- write all terms in y and dy on the other side
- if fractions need to be introduced, ensure dx and dy are not written as denominators
- integrate each side:

$$\int \frac{dy}{g(y)} = \int f(x)\,dx$$

- make y the subject (if possible)
- If a point is given then substitute to find the constant of integration.

If a particular pair of values for x and y are known, then the corresponding value of c (the constant of integration) can be found and a particular solution identified. This particular pair of values for x and y are referred to as the **initial conditions**.

Example 5.1

Find the general solution of this differential equation.

$$y\frac{dy}{dx} = 7x + 2$$

$y\,dy = (7x + 2)\,dx$ — Separate the variables.

$\int y\,dy = \int (7x + 2)\,dx$ — Integrate both sides with respect to the relevant variable.

$\dfrac{y^2}{2} = \dfrac{7x^2}{2} + 2x + c$ — Only one constant of integration is needed.

$y^2 = 7x^2 + 4x + 2c$ — Double each term to eliminate the fractions.

$\quad = 7x^2 + 4x + A$ where $A = 2c$

$y = \sqrt{7x^2 + 4x + A}$ — The general solution is given when you can't evaluate the constant of integration.

Example 5.2

Find the particular solution of this differential equation.

$\frac{dy}{dx} = y + 1$ and $y = 0$ when $x = 0$

> The **initial conditions** are given, which allow the evaluation of the constant of integration.

$\frac{dy}{y+1} = dx$ —— Separate the variables.

$\int \frac{1}{y+1} dy = \int dx$

$\ln|y + 1| = x + c$

$e^{\ln|y+1|} = e^{x+c}$ —— Take exponents of each side.

$y + 1 = e^x e^c$ —— Use the laws of indices to separate the RHS.

$y = Ae^x - 1$ —— Simplify to express y in terms of x, where $A = e^c$

$0 = Ae^0 - 1$

$A = 1$ —— Substitute $x = 0$ and $y = 0$ to find A.

$y = e^x - 1$ —— Particular solution

Example 5.3

Find the particular solution of this differential equation.

$x\frac{dy}{dx} + 4\frac{dy}{dx} = y + 3$ and $y = 2$ when $x = 1$

$x\frac{dy}{dx} + 4\frac{dy}{dx} = y + 3$

$\frac{dy}{dx}(x + 4) = y + 3$ —— Factorise to get one $\frac{dy}{dx}$ term.

$\frac{dy}{y+3} = \frac{dx}{(x+4)}$

$\int \frac{1}{y+3} dy = \int \frac{1}{(x+4)} dx$

$\ln|y + 3| = \ln|x + 4| + c$

$e^{\ln|y+3|} = e^{\ln|x+4|+c}$

$e^{\ln|y+3|} = e^{\ln|x+4|}e^c$ —— The exponential function is the inverse of the logarithmic function.

$y + 3 = A(x + 4)$ —— Simplify each side, where $A = e^c$

$2 + 3 = A(1 + 4)$ —— Substitute $x = 1$ and $y = 2$ to find A.

$5A = 5 \Rightarrow A = 1$

$y + 3 = x + 4$ —— Substitute $A = 1$ and make y the subject.

$y = x + 1$

Example 5.4

A radioactive isotope of plutonium decays exponentially such that the rate of decrease of the mass of the isotope at any given time is proportional to the mass remaining at that time.

If m denotes the mass of the isotope (in grams) remaining t years after decay began, the decay is governed by a differential equation of the form:

$$\frac{dm}{dt} = -km$$

where k is a positive constant.

a Find the general solution of this differential equation, expressing m as a function of t.

b Given that this isotope has a half-life of approximately 139 years (that is, given any initial mass of the isotope, one-half will disintegrate in 139 years), show that starting with 120 grams of this isotope, the mass remaining after t years is given approximately by:

$$m = 120e^{-0.005t}$$

Hence find the mass remaining after 50 years.

a $\dfrac{dm}{dt} = -km$

$\dfrac{1}{m} dm = -k \, dt$

$\displaystyle\int \frac{1}{m} dm = -\int k \, dt$

$\ln m = -kt + c$

$e^{\ln m} = e^{c-kt}$

$m = e^c \, e^{-kt}$

$m = Ae^{-kt}$ ●——————— The general solution of the differential equation, expressing m as a function of t.

b $m = Ae^{-kt}$ ———— Use the formula from part **a**.

$\therefore 120 = Ae^{-k \times 0}$ ———— Substitute $m = 120$ and $t = 0$ to evaluate A.

$120 = Ae^0$

$A = 120$ ———— $e^0 = 1$, so $A = 120$

$m = 120e^{-kt}$

$\therefore 60 = 120 \times e^{-k \times 139}$ ———— After 139 years, half of the isotope has disintegrated. Use this information given in the question to evaluate k. Substitute $m = 60\,g$ and $t = 139$ years to evaluate k.

$e^{-k \times 139} = 0 \cdot 5$

$\ln e^{-k \times 139} = \ln 0 \cdot 5$

$-139k = \ln 0 \cdot 5$

$k = \dfrac{\ln 0 \cdot 5}{-139}$

$= 4 \cdot 98667 \times 10^{-3}$

$\therefore m = 120e^{-0 \cdot 005t}$ ———— Substitute $k \approx 0 \cdot 005$ to give the general solution of the differential equation, expressing m as a function of t.

$= 120 \times e^{-0 \cdot 005 \times 50}$ ———— Substitute $t = 50$ and evaluate m.

$= 93 \cdot 45609397$

$m = 93 \cdot 5g$ ———— Mass remaining after 50 years (3 s.f.).

Exercise 5A

1 Find the general solution of these first-order differential equations.

a $\dfrac{dy}{dx} = e^{-2x}$ 　　　　**b** $\dfrac{dy}{dx} = \sec^2 x$ 　　　　**c** $\dfrac{dy}{dx} = 5xy$

★**d** $x\dfrac{dy}{dx} = 1 - 2\dfrac{dy}{dx}$ 　　　**e** $6x + \dfrac{dy}{dx} = x^2$ 　　　★**f** $x^2 y\dfrac{dy}{dx} = 9$

g $\dfrac{dy}{dx} = 2xe^{-3y}$ 　　　　**h** $\dfrac{dy}{dx} = 3(1 + y)(1 - x)$

2 Find the particular solution these first-order differential equations.

a $\dfrac{dy}{dx} = xe^{-y}$ and $y = 0$ when $x = 0$

b $\dfrac{dx}{dt} = e^{2x} \sin 3t$ and $x = 0$ when $t = 0$

c $\dfrac{dV}{dt} = 2V(1 - t)$ and $V = 3$ when $t = 0$

★**d** $\dfrac{dP}{dt} = \dfrac{(1 - t)(1 - P^2)}{(1 - t^2)(P + 1)}$ and $P = 2$ when $t = 0$

3   The growth of bacteria in a culture at time, t, is given by the differential equation:

$$\frac{dx}{dt} = kx$$

where k is a constant and x denotes the number of bacteria present after t hours of observation.

a   Find the general solution of this differential equation, expressing x as a function of t.

b   The number of bacteria increased from 400 when $t = 0$, to 800 when $t = 2$ hours.

Find the number of bacteria present at the end of 5 hours.

Give your answer correct to the nearest whole number.

4   When a valve is opened, the rate at which the water drains from a pool is proportional to the square root of the depth of the water. This can be represented by the differential equation:

$$\frac{dh}{dt} = -k\sqrt{h}$$

where h is the depth in metres of the water and t is the time in minutes which has elapsed since the valve was opened, and k is a positive constant.

a   Find the general solution of the differential equation.

> There is no need to express h explicitly in terms of t. ⚠

b   The pool was initially 4 metres deep, and the depth of water dropped to one metre after 20 minutes of draining. How long did it take to drain the pool completely?

5   A new vlogger predicts that their viewing figures will increase according to the formula:

$$\frac{dn}{dt} = kn$$

where k is a constant and n denotes the number of viewers after t weeks.

a   Find the general solution of this differential equation, expressing n as a function of t.

b   The number of viewers when $t = 0$, is 1000. It is predicted there will be 10 925 viewers in 2 weeks' time.

i   Find the value of the constant k correct to 3 decimal places.

ii   The viewing target for this vlog is 1 000 000 viewers.

How long should it take to reach this target?

Give your answer correct to the nearest 0·1 week.

6 Newton's law of cooling states that the rate at which an object cools is proportional to the difference in temperature between the object and its surroundings.

An object cools surrounded by air at a temperature of 85 °C. The cooling of this object is modelled by the differential equation:

$$\frac{dT}{dt} = -k(T - 85)$$

where T °C is the temperature of the object after t hours of cooling and k is a constant.

a Find the general solution of this differential equation, expressing T as a function of t.

b An object cools from 125 °C to 100 °C in half an hour when surrounded by air at a temperature of 85 °C.

   i Find the temperature of this object at the end of another half hour.

   ii Find the time taken for the temperature of this object to fall to 86 °C.

7 A scientist grows a culture of bacteria in his laboratory. The number of bacteria, N, after t hours is assumed to satisfy the differential equation:

$$\frac{dN}{dt} = kN$$

where k is a constant

Initially, there are 70 bacteria present in the culture. The scientist observes that there are 197 bacteria after 3 hours.

How long will it take for the bacteria population to increase to 10 times the number initially present?

Give your answer correct to the nearest 0·1 hours.

8 Good news is spread among a large population of students. It is assumed that the rate at which the good news is spreading is proportional to the number of students who have heard the good news. N is the number of students who have heard the good news after t days. N satisfies the differential equation of the form:

$$\frac{dN}{dt} = kN$$

where k is a constant.

Two students heard the good news initially. After 5 days, 985 students had heard the good news.

a Solve the differential equation and show that $N \approx 2e^{1·24t}$

b After how many days had 250 students heard the good news?

   Give your answer correct to the nearest 0·1 days.

Solving first-order linear differential equations with variables non-separable

To solve first-order differential equations with variables non-separable:

- arrange the equation into the standard form $\dfrac{dy}{dx} + P(x)\,y = Q(x)$

- identify the integrating factor $I(x) = e^{\int P(x)\,dx}$

- multiply both sides by the integrating factor $I(x)$

- integrate both sides:

  LHS always becomes $I(x)\,y = \ldots$

  RHS always becomes $\ldots = \int I(x)\,Q(x)\,dx$

$\dfrac{dy}{dx} + P(x)\,y = Q(x)$ becomes $I(x)\,y = \int I(x)\,Q(x)\,dx$ where $I(x) = e^{\int P(x)\,dx}$

Example 5.5

Find the general solution of this first-order differential equation.

$$\frac{dy}{dx} + y = 6e^{2x}$$

$\dfrac{dy}{dx} + y = 6e^{2x}$ ⟵ The equation is in standard form $\dfrac{dy}{dx} + P(x)\,y = Q(x)$

$P(x) = 1 \Rightarrow I(x) = e^{\int P(x)\,dx}$ ⟵ Identify the integrating factor, $I(x) = e^{\int P(x)\,dx}$

$I(x) = e^{\int 1\,dx} = e^{x}$ ⟵ Evaluate the integrating factor where $P(x) = 1$

$\dfrac{dy}{dx} + P(x)\,y = Q(x)$ becomes $I(x)\,y = \int I(x)\,Q(x)\,dx$

$e^{x} \times \left(\dfrac{dy}{dx} + y \right) = e^{x} \times 6e^{2x}$ ⟵ Multiply both sides by the integrating factor, e^{x}

$e^{x} y = \int 6e^{3x}\,dx$ ⟵ Use the laws of indices to combine the e powers on the RHS.

$\quad = \dfrac{6e^{3x}}{3} + c$

$\quad = 2e^{3x} + c$ ⟵ Integrate the RHS.

$y = \dfrac{1}{e^{x}}\left(2e^{3x} + c \right)$ or $y = 2e^{2x} + \dfrac{c}{e^{x}}$ ⟵ Make y the subject.

Example 5.6

Solve this differential equation.

$\dfrac{dy}{dx} - y\cos x = 2x\,e^{\sin x}$ given that $y = 0$ when $x = \pi$

$\dfrac{dy}{dx} + y(-\cos x) = 2x e^{\sin x}$ — The equation is in standard form $\dfrac{dy}{dx} + P(x)y = Q(x)$

$P(x) = -\cos x \Rightarrow I(x) = e^{\int -\cos x\, dx}$ — Identify the integrating factor $e^{\int P(x)\,dx}$

$e^{\int -\cos x\, dx} = e^{-\sin x}$ — Evaluate the integrating factor $e^{\int -\cos x\, dx}$

$\dfrac{dy}{dx} + P(x)y = Q(x)$ becomes $I(x)\,y = \int I(x)\,Q(x)\,dx$

$\dfrac{dy}{dx} - y\cos x = 2x e^{\sin x}$ becomes $e^{-\sin x} y = \int e^{-\sin x} 2x e^{\sin x}\,dx$

$e^{-\sin x} y = \int e^{-\sin x + \sin x} 2x\,dx$

$\qquad = 2\int e^0 x\,dx$

$\qquad = 2\int x\,dx$

$\qquad = x^2 + c$

$e^{-\sin \pi}\, 0 = \pi^2 + c \Rightarrow c = -\pi^2$ — Substitute $y = 0$ and $x = \pi$ to evaluate c.

$\dfrac{y}{e^{\sin x}} = x^2 + c$

$y = e^{\sin x}(x^2 - \pi^2)$

Example 5.7

Find the particular solution of this first-order differential equation.

$3x\dfrac{dy}{dx} - y = 2x$ and $y = -1$ when $x = 1$

$3x\dfrac{dy}{dx} - y = 2x$

$\dfrac{dy}{dx} + y\left(-\dfrac{1}{3x}\right) = \dfrac{2}{3}$ — Divide each term by $3x$. The equation is now in standard form $\dfrac{dy}{dx} + P(x)y = Q(x)$

$P(x) = -\dfrac{1}{3x} \Rightarrow I(x) = e^{\int -\frac{1}{3}\cdot\frac{1}{x}dx}$ — Identify the integrating factor $e^{\int P(x)\,dx}$

$e^{\int -\frac{1}{3}\cdot\frac{1}{x}dx} = e^{-\frac{1}{3}\ln x} = e^{\ln x^{-\frac{1}{3}}} = x^{-\frac{1}{3}}$ — Evaluate the integrating factor $e^{\int -\frac{1}{3}\cdot\frac{1}{x}dx}$ and simplify using the laws of logs.

$\dfrac{dy}{dx} + P(x)\,y = Q(x)$ becomes $I(x)\,y = \int I(x)\,Q(x)\,dx$

$\dfrac{dy}{dx} - \dfrac{1}{3x}y = \dfrac{2}{3}$ becomes $x^{-\frac{1}{3}}y = \int x^{-\frac{1}{3}}\dfrac{2}{3}dx$

$x^{-\frac{1}{3}}y = \dfrac{x^{\frac{2}{3}}}{\frac{2}{3}}\cdot\dfrac{2}{3} + c \Rightarrow x^{-\frac{1}{3}}y = x^{\frac{2}{3}} + c$

$1^{-\frac{1}{3}}(-1) = 1^{\frac{2}{3}} + c \Rightarrow c = (-2))$ ●────── Substitute $y = -1$ and $x = 1$ to evaluate c.

$\dfrac{y}{x^{\frac{1}{3}}} = x^{\frac{2}{3}} - 2$

$y = x - 2x^{\frac{1}{3}}$

Exercise 5B

Find a general solution to these first-order differential equations with non-variables separable.

1 $\dfrac{dy}{dx} + \dfrac{y}{x} = x$ ★2 $\dfrac{dy}{dx} - 2y = e^{3x}$ 3 $2\dfrac{dy}{dx} + y = e^{\frac{x}{2}}$

4 $x\dfrac{dy}{dx} + y = 2\cos x$ ★5 $x\dfrac{dy}{dx} - 2y = x^3\sin x$ 6 $x\dfrac{dy}{dx} = \dfrac{3}{e^{x^2}} - 2x^2y$

Find the particular solution to these first order differential equations with non-variables separable.

★7 $\dfrac{dy}{dx} = 5x - \dfrac{3y}{x}$ and $y = 3$ when $x = 1$

8 $\dfrac{dy}{dx} = x^2\sin x + \dfrac{2y}{x}$ and $y = 0$ when $x = 2\pi$

9 $2x\dfrac{dy}{dx} - 6y = 2x^4\cos x$ and $y = 0$ when $x = \pi$

10 a Use the identity $\tan x = \dfrac{\sin x}{\cos x}$ to rewrite the integral $\int\tan x\,dx$

    Use a suitable substitution to show that $\int\tan x\,dx = \ln(\sec x) + c$

  b Find the general solution of the differential equation $\cos x\dfrac{dy}{dx} + y\sin x = \cos^2 x$ expressing y explicitly in terms of x.

11 A garden centre sells young plants which are suitable for hedging.

The growth in metres after planting (the increase in height), after t years is modelled by the differential equation:

$$\frac{dG}{dt} = \frac{25k - G}{25}$$

where k is a constant and $G = 0$ when $t = 0$

a   Express G in terms of t and k.

b   After 5 years, the plant has grown 0·6 metres.

Find the value of k correct to 3 decimal places.

c   Stated on the plant's label are the words:

'After 10 years this plant will be about 1 metre tall.'

Is this claim justified?

d   The initial height of this plant was 0·3 metres.

What is the likely long-term height of this plant?

Solving second-order differential equations

There are two types of second-order differential equations:

- **homogeneous** $\Rightarrow a\dfrac{d^2y}{dx^2} + b\dfrac{dy}{dx} + cy = 0$ RHS = 0

- **heterogeneous** $\Rightarrow a\dfrac{d^2y}{dx^2} + b\dfrac{dy}{dx} + cy = Q(x)$ RHS = a function

(Heterogeneous differential equations are also known as non-homogeneous differential equations.)

For both types, we need to form and solve the **auxiliary equation** $ak^2 + bk + c = 0$

The roots of this quadratic determine the solution of the differential equation.

Homogeneous second-order linear differential equations

Homogeneous $\Rightarrow a\dfrac{d^2y}{dx^2} + b\dfrac{dy}{dx} + cy = 0$ (i.e. RHS = 0)

The auxiliary equation $ak^2 + bk + c = 0$ can have:

- 2 real and distinct roots, or

- 2 real and equal roots, or

- complex conjugate roots.

2 real and distinct roots

When the auxiliary equation has 2 real and distinct roots m_1 and m_2, then the solution is of the form:

$y = Ae^{m_1 x} + Be^{m_2 x}$ where A and B are constants.

2 real and equal roots

When the auxiliary equation has 2 real and equal roots p, then the solution is of the form:

$y = (Ax + B)e^{px}$ where A and B are constants.

Complex conjugate roots

When the auxiliary equation has 2 complex conjugate roots $k = p \pm qi$, then the solution is of the form:

$y = e^{px}(A\sin qx + B\cos qx)$ where A and B are constants.

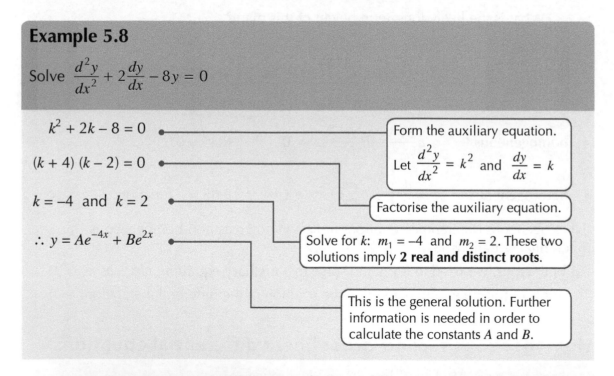

Example 5.8

Solve $\dfrac{d^2 y}{dx^2} + 2\dfrac{dy}{dx} - 8y = 0$

$k^2 + 2k - 8 = 0$ — Form the auxiliary equation.

Let $\dfrac{d^2 y}{dx^2} = k^2$ and $\dfrac{dy}{dx} = k$

$(k + 4)(k - 2) = 0$ — Factorise the auxiliary equation.

$k = -4$ and $k = 2$ — Solve for k: $m_1 = -4$ and $m_2 = 2$. These two solutions imply **2 real and distinct roots**.

$\therefore y = Ae^{-4x} + Be^{2x}$ — This is the general solution. Further information is needed in order to calculate the constants A and B.

Example 5.9

Solve $\dfrac{d^2 y}{dx^2} - 6\dfrac{dy}{dx} + 9y = 0$

$k^2 - 6k + 9 = 0$
$(k - 3)(k - 3) = 0$
$k = 3$ (repeated root)
$\therefore y = (Ax + B)e^{3x}$ — This is the general solution.

Example 5.10

Solve $\dfrac{d^2y}{dx^2} + 4\dfrac{dy}{dx} + 13y = 0$

$k^2 + 4k + 13 = 0$

$k = \dfrac{-b \pm \sqrt{b^2 - 4ac}}{2a}$

$ = \dfrac{-4 \pm \sqrt{-36}}{2} = \dfrac{-4 \pm \sqrt{i^2 36}}{2} = \dfrac{-4 \pm 6i}{2}$

$k = -2 \pm 3i$ ●————

> Compare with $p \pm qi$: $p = (-2)$ and $q = 3$.
> This solution implies **complex roots**.

$\therefore\ y = e^{px}(A\sin qx + B\cos qx)$

$\ y = e^{-2x}(A\sin 3x + B\cos 3x)$

Example 5.11

Solve $\dfrac{d^2y}{dx^2} - 4\dfrac{dy}{dx} + 3y = 0$ when $x = 0$, $y = 1$ and $\dfrac{dy}{dx} = 5$

$k^2 - 4k + 3 = 0$

$(k - 3)(k - 1) = 0$

$k = 3$ and $k = 1$

$\therefore\ y = Ae^{3x} + Be^{x}$ ●————

> This is the general solution.
> Use it to calculate A and B.

$\dfrac{dy}{dx} = 3Ae^{3x} + Be^{x}$ ●————

> Differentiate both sides.

$5 = 3Ae^{0} + Be^{0}$ ●————

> Substitute $x = 0$ and $\dfrac{dy}{dx} = 5$

$5 = 3A + B$

$y = Ae^{3x} + Be^{x}$

$1 = Ae^{0} + Be^{0}$ ●————

> Substitute $x = 0$ and $y = 1$

$1 = A + B$ ①

$5 = 3A + B$ ②

$4 = 2A$ ②−①

$A = 2$

$B = -1$ ●————

> Substitute $A = 2$ into ①

$y = Ae^{3x} + Be^{x}$ ●————

> General solution

$y = 2e^{3x} - e^{x}$ ●————

> Particular solution

Example 5.12

Solve $\dfrac{d^2y}{dx^2} + 4\dfrac{dy}{dx} + 4y = 0$ when $x = 0$, $y = 1$ and $\dfrac{dy}{dx} = 2$

$k^2 + 4k + 4 = 0$

$(k + 2)(k + 2) = 0$

$k = (-2)$ or $k = (-2)$

$\therefore\ y = (Ax + B)e^{-2x}$ ← Use the general solution to calculate A and B. Substitute $x = 0$ and $y = 1$.

$1 = (B)e^0 \Rightarrow B = 1$

$\dfrac{dy}{dx} = \dfrac{d}{dx}(Ax + 1)e^{-2x}$

$u = Ax + 1 \qquad v = e^{-2x}$

$\dfrac{du}{dx} = A \qquad\qquad \dfrac{dv}{dx} = -2e^{-2x}$

$\dfrac{dy}{dx} = Ae^{-2x} - 2e^{-2x}(Ax + 1)$ ← Differentiate the general solution. Substitute $x = 0$ and $\dfrac{dy}{dx} = 2$ and evaluate A.

$2 = Ae^0 - 2e^0(A \times 0 + 1)$

$2 = A - 2$

$A = 4$

$y = (Ax + B)e^{-2x}$

$y = (4x + 1)e^{-2x}$

Example 5.13

Solve $\dfrac{d^2y}{dx^2} - 4\dfrac{dy}{dx} + 5y = 0$ when $x = 0$, $y = 2$ and $\dfrac{dy}{dx} = 1$

$k^2 - 4k + 5 = 0$

$k = \dfrac{-b \pm \sqrt{b^2 - 4ac}}{2a}$

$= \dfrac{-(-4) \pm \sqrt{-4}}{2 \times 1} = \dfrac{4 \pm \sqrt{i^2 4}}{2}$

$k = 2 \pm i$

$y = e^{px}(A\sin qx + B\cos qx)$ where $k = p \pm qi$

$y = e^{2x}(A\sin x + B\cos x)$ ← Use the general solution to calculate A and B. Substitute $x = 0$ and $y = 2$

$2 = e^0(A\sin 0 + B\cos 0)$

$B = 2$

$$\frac{dy}{dx} = \frac{d}{dx}\left(e^{2x}\left(A\sin x + 2\cos x\right)\right)$$

$$u = e^{2x} \qquad\qquad v = A\sin x + 2\cos x$$

$$\frac{du}{dx} = 2e^{2x} \qquad\qquad \frac{dv}{dx} = A\cos x - 2\sin x$$

$$\frac{dy}{dx} = 2e^{2x}\left(A\sin x + 2\cos x\right) + e^{2x}\left(A\cos x - 2\sin x\right) \quad\bullet\!\!-\!\!-\!\!-$$

> Differentiate the general solution.

$$1 = 2e^{0}\left(A\sin 0 + 2\cos 0\right) + e^{0}\left(A\cos 0 - 2\sin 0\right) \quad\bullet\!\!-\!\!-$$

> Substitute $x = 0$ and $\frac{dy}{dx} = 1$

$$1 = 2(2) + A$$

$$A = -3$$

$$y = e^{2x}\left(A\sin x + B\cos x\right)$$

$$y = e^{2x}\left(2\cos x - 3\sin x\right)$$

Exercise 5C

1   Find the general solution for each of these second-order differential equations.

a $\dfrac{d^2y}{dx^2} - 11\dfrac{dy}{dx} + 18y = 0$

b $\dfrac{d^2y}{dx^2} + 2\dfrac{dy}{dx} + y = 0$

c $\dfrac{d^2y}{dx^2} - 2\dfrac{dy}{dx} - 35y = 0$

d $\dfrac{d^2y}{dx^2} + 6\dfrac{dy}{dx} + 10y = 0$

★e $\dfrac{d^2y}{dx^2} - 8\dfrac{dy}{dx} + 16y = 0$

★f $\dfrac{d^2y}{dx^2} + 4\dfrac{dy}{dx} + 8y = 0$

★2   Find the particular solution for each of these second-order differential equations.

a $\dfrac{d^2y}{dx^2} + 5\dfrac{dy}{dx} + 6y = 0$ where $\dfrac{dy}{dx} = 2, y = 1$ and $x = 0$

b $\dfrac{d^2y}{dx^2} + 4y = 0$ where $\dfrac{dy}{dx} = 2, y = 1$ and $x = 0$

c $\dfrac{d^2y}{dx^2} - 10\dfrac{dy}{dx} + 25y = 0$ where $\dfrac{dy}{dx} = 5, y = 2$ and $x = 0$

d $\dfrac{d^2y}{dx^2} + 2\dfrac{dy}{dx} - 3y = 0$ where $\dfrac{dy}{dx} = 7, y = 3$ and $x = 0$

Heterogeneous second-order linear differential equations

Heterogeneous $\Rightarrow a\dfrac{d^2y}{dx^2} + b\dfrac{dy}{dx} + cy = Q(x)$ (i.e. RHS = a non-zero function)

To solve second-order linear differential equations where $Q(x) \neq 0$:

- solve the auxiliary equation $ak^2 + bk + c = 0$ to find the **complementary function** (CF)

- find the form of the **particular integral** (PI). This is usually the same form as $Q(x)$, however see Examples 5.18 and 5.19 for exceptions

- find $\dfrac{dy}{dx}$, $\dfrac{d^2y}{dx^2}$ and compare with the original second-order differential equation

- state the solution $y = \textbf{CF} + \textbf{PI}$

Finding the complementary function
The table shows the form of the complementary function according to the roots.

Type of roots	Form of roots	Complementary function
2 real and distinct roots	$k = m_1$ and m_2	$y = Ae^{m_1 x} + Be^{m_2 x}$
real and equal roots	$k = p$	$y = (Ax + B)e^{px}$
complex conjugate roots	$k = p \pm qi$	$y = e^{px}(A\sin qx + B\cos qx)$

Finding the particular integral
The table shows the different forms of the particular integral relating to the form of the RHS function, $Q(x)$.

$$a\dfrac{d^2y}{dx^2} + b\dfrac{dy}{dx} + cy = Q(x)$$

$Q(x)$	Particular integral
constant	p
linear	$px + q$
quadratic	$px^2 + qx + r$
trigonometric function	$p\sin x + q\cos x$
exponential	pe^x

Example 5.14

Solve $\dfrac{d^2y}{dx^2} - \dfrac{dy}{dx} - 2y = 4x + 1$

$k^2 - k - 2 = 0$ ●————————————— Form the auxiliary function and solve for k.

$(k - 2)(k + 1) = 0$ ●————————————— Substitute $m_1 = 2$ and $m_2 = -1$.

$k = 2$ and $k = (-1)$ ●————————————— These two solutions imply **2 real and distinct roots**.

$CF = Ae^{m_1 x} + Be^{m_2 x} \Rightarrow Ae^{2x} + Be^{-x}$

$PI = px + q$ ●————————————— $Q(x) = 4x + 1$, linear function.

$\dfrac{dy}{dx} = p$ and $\dfrac{d^2y}{dx^2} = 0$

$\dfrac{d^2y}{dx^2} - \dfrac{dy}{dx} - 2y = 4x + 1$

$0 - p - 2(px + q) = 4x + 1$ ●————————————— Substitute $\dfrac{d^2y}{dx^2} = 0$, $\dfrac{dy}{dx} = p$ and $y = px + q$

$-p - 2px - 2q = 4x + 1$ ●————————————— Simplify and compare LHS with RHS.

$-2px = 4x$ ●————————————— Equate equivalent terms and solve for p and q.

$p = -2$

$-(-2) - 2q = 1$

$q = \dfrac{1}{2}$

$y = \dfrac{1}{2} - 2x$ ●————————————— Substitute $p = -2$ and $q = \frac{1}{2}$ to find the **particular integral**.

$y = PI + CF$ ●————————————— General solution

$y = \dfrac{1}{2} - 2x + Ae^{2x} + Be^{-x}$

Example 5.15

Solve $\dfrac{d^2y}{dx^2} - 4\dfrac{dy}{dx} + 4y = 2x^2 + 1$

$k^2 - 4k + 4 = 0$

$(k - 2)(k - 2) = 0$

$k = 2$ and $k = 2$

CF $= (Ax + B)e^{px} \Rightarrow (Ax + B)e^{2x}$

PI try $y = px^2 + qx + r$ •————————

> $Q(x) = 2x^2 + 1$, quadratic function

$\dfrac{dy}{dx} = 2px + q$ and $\dfrac{d^2y}{dx^2} = 2p$

$\dfrac{d^2y}{dx^2} - 4\dfrac{dy}{dx} + 4y = 2x^2 + 1$

$2p - 4(2px + q) + 4(px^2 + qx + r) = 2x^2 + 1$ •——

> Substitute $\dfrac{d^2y}{dx^2} = 2p$, $\dfrac{dy}{dx} = 2px + q$ and $y = px^2 + qx + r$

$4px^2 + x(4q - 8p) + 2p - 4q + 4r = 2x^2 + 1$ •——

> Simplify and compare LHS with RHS.

$4px^2 = 2x^2$

$p = \dfrac{1}{2}$ •————————

> Equate x^2 terms and solve for p.

$4q - 8p = 0$ •————————

> Equate x terms and solve for q.

$4q = 4$

$q = 1$

$2p - 4q + 4r = 1$ •————————

> Equate numerical terms and solve for r.

$4r = 4$

$r = 1$

$y = px^2 + qx + r$

$y = \dfrac{x^2}{2} + x + 1$

$y = \textbf{PI} + \textbf{CF}$ •————————

> General solution

$y = \dfrac{x^2}{2} + x + 1 + (Ax + B)e^{2x}$

Example 5.16

Solve $\dfrac{d^2y}{dx^2} + 16y = e^{-2x}$

$k^2 + 16 = 0$

$k = \sqrt{-16}$

$k = \pm 4i$

CF $= y = e^{px}(A\sin qx + B\cos qx) \Rightarrow A\sin 4x + \cos 4x$

PI try $y = pe^{-2x}$ •————————————⟶ $Q(x) = e^{-2x}$, exponential function

$\dfrac{dy}{dx} = -2pe^{-2x}$ and $\dfrac{d^2y}{dx^2} = 4pe^{-2x}$

$\dfrac{d^2y}{dx^2} + 16y = e^{-2x}$ •————————⟶ The original second-order differential equation.

$4pe^{-2x} + 16pe^{-2x} = e^{-2x}$ •————⟶ Substitute $\dfrac{d^2y}{dx^2} = 4pe^{-2x}$ and $y = pe^{-2x}$

$20p = 1$

$\quad p = \frac{1}{20}$

$y = pe^{px}$

$\quad = pe^{-2x}$

$y = $ PI $+$ CF

$y = A\sin 4x + B\cos 4x + \frac{1}{20}e^{-2x}$

Example 5.17

Solve $\dfrac{d^2y}{dx^2} - 4\dfrac{dy}{dx} + 3y = 10\sin x$

$k^2 - 4k + 3 = 0$

$(k-3)(k-1) = 0$

$k = 3$ and $k = 1$

CF $= Ae^{px} + Be^{px} \Rightarrow Ae^{3x} + Be^{x}$

PI try $y = p\sin x + q\cos x$ •————⟶ $Q(x) = 10\sin x$, trigonometric function

$\dfrac{dy}{dx} = p\cos x - q\sin x$ and $\dfrac{d^2y}{dx} = -p\sin x - q\cos x$

$\dfrac{d^2y}{dx^2} - 4\dfrac{dy}{dx} + 3y = 10\sin x$

$-p\sin x - q\cos x - 4(p\cos x - q\sin x) + 3(p\sin x + q\cos x) = 10\sin x$

Substitute $y = p\sin x + q\cos x$,

$$\frac{d^2y}{dx^2} = -p\sin x - q\cos x \text{ and } \frac{dy}{dx} = p\cos x - q\sin x$$

$-p\sin x - q\cos x - 4p\cos x + 4q\sin x + 3p\sin x + 3q\cos x = 10\sin x$

$(-p + 4q + 3p)\sin x + (-q - 4p + 3q)\cos x = 10\sin x$

$2p + 4q = 10 \text{ and } -4p + 2q = 0$

$2p + 4q = 10 \text{ and } -2p + q = 0$

$5q = 10$

$q = 2$

$2p = 10 - 8$

$p = 1$

$y = p\sin x + q\cos x$

$\quad = \sin x + 2\cos x$

$y = \text{PI} + \text{CF}$

$y = Ae^{3x} + Be^{x} + \sin x + 2\cos x$

When the right hand side of the equation has the same form as the complementary function, we must modify our choice of particular integral.

Example 5.18

Solve $\dfrac{d^2y}{dx^2} - \dfrac{dy}{dx} - 6y = 2e^{3x}$.

$k^2 - k - 6 = 0$

$(k - 3)(k + 2) = 0$

$k = 3 \text{ and } k = -2$

$\text{CF} = Ae^{3x} + Be^{-2x}$

$\text{PI try } y = pxe^{3x}$

As pe^{3x} is the same form as the CF, we multiply by x and use pxe^{3x} instead.

$\dfrac{dy}{dx} = pe^{3x} + 3pxe^{3x}$

Differentiate using the product rule.

$\dfrac{d^2y}{dx^2} = 3pe^{3x} + 3pe^{3x} + 9pxe^{3x}$

$\quad = 6pe^{3x} + 9pxe^{3x}$

$$6pe^{3x} + 9pxe^{3x} - (pe^{3x} + 3pxe^{3x}) - 6pxe^{3x} = 2e^{3x}$$

$$5pe^{3x} = 2e^{3x}$$

The terms involving xe^{3x} cancel.

$$p = \frac{2}{5}$$

$$y = \text{PI} + \text{CF}$$

$$y = Ae^{3x} + Be^{-2x} + \frac{2}{5}xe^{3x}$$

Example 5.19

Solve $\dfrac{d^2y}{dx^2} - 2\dfrac{dy}{dx} + y = e^x$.

$$k^2 - 2k + 1 = 0$$

$$(k - 1)(k - 1) = 0$$

$$k = 1 \text{ and } k = 1$$

$$\text{CF} = (Ax + B)e^x$$

$$\text{PI try } y = px^2e^x$$

Both pe^x and pxe^x are the same form as the CF, so we multiply by x^2 and use px^2e^x instead.

$$\frac{dy}{dx} = 2pxe^x + px^2e^x$$

Differentiate using the product rule.

$$\frac{d^2y}{dx^2} = 2pe^x + 2pxe^x + 2pxe^x + px^2e^x$$

$$= px^2e^x + 4pxe^{3x} + 2pe^3$$

$$px^2e^x + 4pxe^x + 2pe^x - 2(2pxe^x + px^2e^x) + px^2e^x = e^x$$

$$2pe^x = e^x$$

The terms involving xe^x and x^2e^x cancel.

$$p = \frac{1}{2}$$

$$y = \text{PI} + \text{CF}$$

$$y = (Ax + B)e^x + \frac{1}{2}x^2e^x$$

When additional information is given it may be possible to determine any constants and write down the specific solution.

Example 5.20

Solve $\dfrac{d^2y}{dx^2} - 4\dfrac{dy}{dx} + 4y = e^x$ where $x = 0$, $y = 2$ and $\dfrac{dy}{dx} = 1$

$k^2 - 4k + 4 = 0$

$(k - 2)(k - 2) = 0$

$k = 2$

CF $= (Ax + B)e^{px} \Rightarrow (Ax + B)e^{2x}$

PI try $y = pe^x$ ●━━━━━━━━━━━━━━━ $Q(x) = e^x$, exponential function

$\dfrac{dy}{dx} = pe^x$ and $\dfrac{d^2y}{dx^2} = pe^x$

$\dfrac{d^2y}{dx^2} - 4\dfrac{dy}{dx} + 4y = e^x$

$pe^x - 4pe^x + 4pe^x = e^x$ ●━━━━ Substitute $\dfrac{d^2y}{dx^2} = pe^x$, $\dfrac{dy}{dx} = pe^x$ and $y = pe^x$

$pe^x = e^x$

$p = 1$

$y = pe^x$

$= e^x$

$y = $ PI $+$ CF

$y = e^x + (Ax + B)e^{2x}$

$2 = e^0 + (0 + B)e^0$ ●━━━━━━━━━━━━━━━ Solve for B.

$B = 1$

$\dfrac{dy}{dx} = \dfrac{d}{dx}\left(e^x + (Ax + 1)e^{2x}\right)$

$u = e^{2x} \qquad v = Ax + 1$

$\dfrac{du}{dx} = 2e^{2x} \qquad \dfrac{dv}{dx} = A$

$\dfrac{dy}{dx} = e^x + 2e^{2x}(Ax + 1) + Ae^{2x}$ ●━━━ Differentiate the general solution.

$1 = e^0 + 2e^0(0 + 1) + Ae^0$

$1 = 1 + 2 + A$

$A = -2$

$y = e^x + (Ax + B)e^{2x}$

$y = e^x + (1 - 2x)e^{2x}$

Choosing the form of the complementary function

Use this summary to help decide the form of the complementary function in the questions in Exercise 5D.

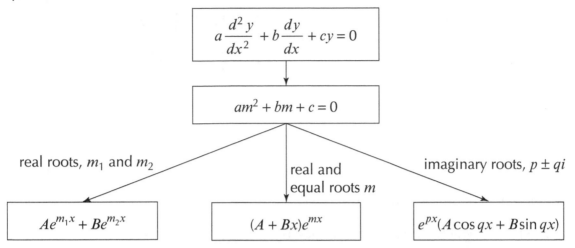

$$a\frac{d^2 y}{dx^2} + b\frac{dy}{dx} + cy = 0$$

$$am^2 + bm + c = 0$$

real roots, m_1 and m_2

real and equal roots m

imaginary roots, $p \pm qi$

$$Ae^{m_1 x} + Be^{m_2 x}$$

$$(A + Bx)e^{mx}$$

$$e^{px}(A\cos qx + B\sin qx)$$

Exercise 5D

1 Find the general solution for each of these second-order differential equations.

★ a $\dfrac{d^2 y}{dx^2} - \dfrac{dy}{dx} - 2y = 4x + 2$

★ b $\dfrac{d^2 y}{dx^2} + 6\dfrac{dy}{dx} + 9y = -e^{2x}$

★ c $2\dfrac{d^2 y}{dx^2} + 2\dfrac{dy}{dx} + 5y = 25x$

★ d $\dfrac{d^2 y}{dx^2} - 10\dfrac{dy}{dx} + 25y = 32e^x$

e $\dfrac{d^2 y}{dx^2} - 8\dfrac{dy}{dx} + 12y = 8\sin x + 11\cos x$

f $5\dfrac{d^2 y}{dx^2} - 6\dfrac{dy}{dx} + 9y = 9\sin x - 6\cos x$

2 Find the particular solution for each of these second-order differential equations.

★ a $\dfrac{d^2 y}{dx^2} - 8\dfrac{dy}{dx} + 16y = e^{2x}$ given that, when $x = 0$, $y = \dfrac{5}{4}$ and $\dfrac{dy}{dx} = -\dfrac{1}{2}$

★ b $\dfrac{d^2 y}{dx^2} + 2\dfrac{dy}{dx} + 5y = 10x - 1$ given that, when $x = 0$, $y = 2$ and $\dfrac{dy}{dx} = -1$

★ c $\dfrac{d^2 y}{dx^2} - y = 6\sin x - 4\cos x$ given that, when $x = 0$, $y = 3$ and $\dfrac{dy}{dx} = 0$

d $5\dfrac{d^2 y}{dx^2} + 10\dfrac{dy}{dx} + 10y = 15\cos x + 20\sin x$ given that, when $x = 0$, $y = 2$

and $\dfrac{dy}{dx} = 1$

Chapter review

1   Solve these first-order differential equations with variables separable.

a $y\dfrac{dy}{dx} = 4x + 3$

b $(2x - 1)\dfrac{dy}{dx} = 3y$

c $xe^{3y}\dfrac{dy}{dx} = 4x^2$

d $\dfrac{dy}{dx} = 2x\cos^2 y$

e $\dfrac{dV}{dt} = \dfrac{2V}{t}$ and $V = 1$ when $t = 0$

f $\sqrt{y}\dfrac{dy}{dx} = x(3x + 4)$ and $y = 0$ when $x = 0$

2   Express $\dfrac{2x + 1}{x(x + 1)}$ in partial fractions and hence find the general solution of the
differential equation $x(x + 1)\dfrac{dy}{dx} = y(2x + 1)$ expressing y explicitly in terms of x.

3   The rate at which a population of mice is increasing at any given time is assumed
to be proportional to the number of mice.

a m denotes the number of mice after t months. Write down a differential
equation which satisfies m and find the general solution, expressing m as a
function of t.

b   When time $t = 0$ there are 6 mice. It takes 6 months for a number of mice to
double. After how many months will there be 90 mice?

Give your answer correct to the nearest 0·1 month.

4   When taken from the oven, the temperature, $T°C$, of a bowl of pasta after
t minutes of cooling is such that:

$$\dfrac{dT}{dt} = -kT$$

where k is a constant

The bowl of pasta was at a temperature of 70°C after one minute of cooling. After
a further minute of cooling, the pasta reached a temperature of 49°C.

a   Find the temperature of the pasta on leaving the oven.

Give your answer correct to the nearest °C.

b   How long, to the nearest 0·1 minute, will the pasta take to cool down to room
temperature? (Room temperature = 20°C.)

5   a   Express $\dfrac{1}{x(x + 1)}$ in partial fractions.

b   The spread of a disease in a large population can be modelled with the
differential equation:

$$\dfrac{dx}{dt} = kx(x + 1)$$

where k is a constant.

Find the general solution of the differential equation, expressing x explicitly as a function of t, where t is measured in days.

c   Given that $x = \dfrac{1}{1000}$ (0·1%) when $t = 0$, and $x = \dfrac{1}{100}$ (1%) when $t = 5$,

   i   verify that about 10·9% of the population was infected after 10 days

   ii   calculate the number of days in which 25% of the population will become infected.

6   Solve these first-order differential equations with variables non-separable.

a $\dfrac{dy}{dx} + \dfrac{y}{x} = \sin x$

b $3x\dfrac{dy}{dx} - 3y = 2x$

c $\dfrac{dy}{dx} = x^2 \sin x + \dfrac{2y}{x}$

d $\dfrac{dy}{dx} - y\sin x = 3e^{-\cos x}$ and $y = 0$ when $x = 0$

e $\dfrac{dy}{dx} = e^x - \dfrac{y}{x}$ and $y = 1$ when $x = 1$

f $\dfrac{dy}{dx} = y - 2e^x$ and $y = 0$ when $x = 0$

7   Solve these second-order differential equations.

a $\dfrac{d^2y}{dx^2} - 2\dfrac{dy}{dx} + 2y = 0$

b $\dfrac{d^2y}{dx^2} - 12\dfrac{dy}{dx} + 35y = 0$

c $\dfrac{d^2y}{dx^2} + 12\dfrac{dy}{dx} + 36y = 108$

d $\dfrac{d^2y}{dx^2} - 2\dfrac{dy}{dx} + 37y = 80e^{3x}$ given that, when $x = 0$, $y = 3$ and $\dfrac{dy}{dx} = 0$

e $\dfrac{d^2y}{dx^2} + 3\dfrac{dy}{dx} - 10y = 97\sin x + 9\cos x$ given that, when $x = 0$, $y = 0$

   and $\dfrac{dy}{dx} = (-2)$

- I can solve first-order differential equations with variables separable.
  ★ Exercise 5A Q1d, Q1f, Q2d

- I can solve first-order linear differential equations with non-variables separable.
  ★ Exercise 5B Q2, Q5 and Q7

- I can solve second-order differential equations. ★ Exercise 5C Q1e, Q1f, Q2a–d ★ Exercise 5D Q1a–d, Q2a–c

6 Applying algebraic skills and calculus skills to properties of functions

This chapter will show you how to:

- find the equations of vertical, horizontal and oblique asymptotes of the graphs of rational functions
- use second derivatives to investigate stationary points and points of inflection
- find extreme values of functions on given intervals
- determine when a function is even or odd
- identify points of discontinuity for a given function
- sketch graphs of rational functions
- sketch graphs of functions related to a known graph.

You should already know:

- the meaning of a limit of a function
- how to add, subtract and simplify algebraic fractions
- the meaning of the degree of a polynomial
- how to find the quotient and remainder when dividing one polynomial by another
- how to differentiate basic functions and apply the product and chain rules
- the graphs of basic functions such as x^2, x^3, $\frac{1}{x}$, $\sin x$, $\cos x$, $\tan x$, e^x and $\ln x$.

Asymptotes

An **asymptote** of the graph of a function $f(x)$ is a straight line such that the distance between the graph and the straight line tends to 0 as either $x \to \pm\infty$ or $f(x) \to \pm\infty$. Knowledge of the asymptotes of a function can be very useful, as they help to describe 'large-scale' behaviour of the function.

Many types of functions may have asymptotes; however, here we will be primarily concerned with the calculation of asymptotes of **rational functions**, that is, functions which may be expressed as an algebraic fraction $\frac{g(x)}{h(x)}$ where both $g(x)$ and $h(x)$ are polynomials.

Vertical asymptotes

Consider the graph of the function $f(x) = \dfrac{1}{x-1}$ as shown. When approaching the value $x = 1$ from above, the value of $f(x)$ is positive and its absolute value gets larger, tending towards $+\infty$. When approaching the value $x = 1$ from below, the value of $f(x)$ is negative and its absolute value gets larger, tending towards $-\infty$.

x	1·5	1·25	1·1	1·01	1·001	0·999	0·99	0·9	0·75
$\dfrac{1}{x-1}$	2	4	10	100	1000	−1000	−100	−10	4

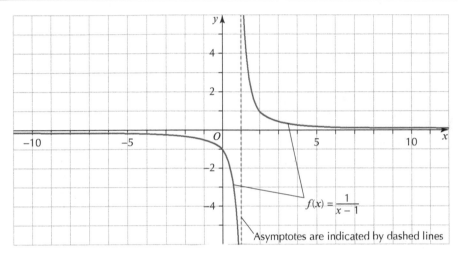

$f(x) = \dfrac{1}{x-1}$

Asymptotes are indicated by dashed lines

As the value of x gets closer and closer to 1, the denominator of $\dfrac{1}{x-1}$ gets closer to 0, so the absolute value of the fraction gets larger. At $x = 1$, $f(x)$ is not defined, as the denominator is equal to 0. The line with equation $x = 1$ is a **vertical asymptote**.

It is possible for the graph of a function to have more than one asymptote. For example, the graph $y = \tan x$ has vertical asymptotes $x = \ldots, \dfrac{-5\pi}{2}, \dfrac{-3\pi}{2}, \dfrac{-\pi}{2}, \dfrac{\pi}{2}, \dfrac{3\pi}{2}, \dfrac{5\pi}{2}, \ldots$

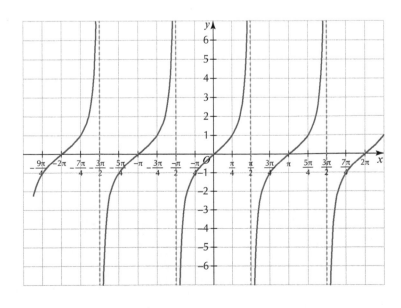

The line $x = a$ is a vertical asymptote of the graph of a function $f(x)$ if one of the following statements hold:

- $\lim\limits_{x \to a^+} f(x) = \pm\infty$

- $\lim\limits_{x \to a^-} f(x) = \pm\infty$

Finding vertical asymptotes of a rational function involves calculating roots of the denominator of the function. It is important to cancel any common factors of the numerator and denominator before calculating the roots of the denominator. Not only does this often make calculation of the roots easier, it avoids the possibility that a root of the denominator does **not** lead to a vertical asymptote (see Example 6.3).

Example 6.1

Find the equations of any vertical asymptotes of the graph $y = \dfrac{1}{x - 2}$

$x - 2 = 0$ if and only if $x = 2$

$\Rightarrow x = 2$ is the only root of the denominator. •————

> Set the denominator equal to 0, then solve for x.

$x = 2$ is not a root of the numerator •————

$\Rightarrow x = 2$ is a vertical asymptote.

> The numerator is the constant function 1, so has no roots.

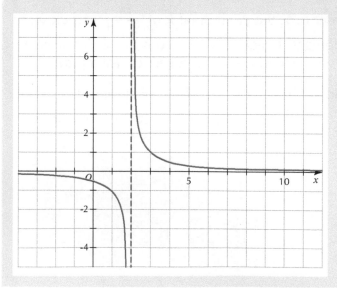

Example 6.2

Find the equations of any vertical asymptotes of the graph $y = \dfrac{4x}{x^2 - x - 2}$

$x^2 - x - 2 = (x - 2)(x + 1)$ — Factorise the quadratic to find the roots.

$(x - 2)(x + 1) = 0$ if and only if $x = 2$ or $x = -1$

$\Rightarrow$ neither $x = 2$ nor $x = -1$ are roots of the numerator $4x$ — Remember to check that the roots of the denominator are not also roots of the numerator.

$\Rightarrow$ vertical asymptotes are $x = 2$ and $x = -1$

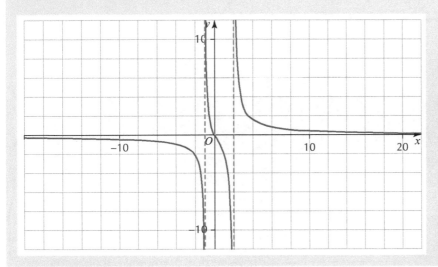

Example 6.3

Find the equations of any vertical asymptotes of the graph $y = \dfrac{-2x + 2}{x^2 + 2x - 3}$

$y = \dfrac{-2x + 2}{x^2 + 2x - 3}$

$= \dfrac{-2(x - 1)}{(x - 1)(x + 3)}$

$= \dfrac{-2}{x + 3}$ — Cancel any common factors of the numerator and the denominator.

$x + 3 = 0$ if and only if $x = -3$

$x = -3$ is not a root of the numerator

$\Rightarrow$ vertical asymptote is $x = -3$

(continued)

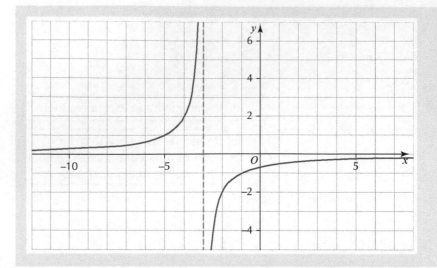

Example 6.4

Find the equations of any vertical asymptotes of the graph of the function

$$f(x) = \frac{x}{x+1} + \frac{1}{x+1}$$

$$f(x) = \frac{x}{x+1} + \frac{1}{x+1}$$

$$= \frac{x+1}{x+1} = 1$$

⇒ There are no vertical asymptotes.

> Be careful with functions given as a sum of rational functions. Although both $\frac{x}{x+1}$ and $\frac{1}{x+1}$ have vertical asymptotes, their sum does not. Write the sum as a single fraction.

Exercise 6A

1 For each of the following rational functions, find the equations of any vertical asymptotes.

a $f(x) = \dfrac{1}{x-1}$

b $f(x) = \dfrac{1}{x+3}$

c $f(x) = -\dfrac{2}{x-4}$

d $f(x) = \dfrac{1}{x^2-1}$

e $f(x) = \dfrac{x}{x^2-4x+4}$

f $f(x) = \dfrac{2x^2}{3x^2+8x-3}$

g $f(x) = \dfrac{4}{x^3+x^2-2x}$

h $f(x) = \dfrac{x^2+1}{x^3-3x-2}$

2   For the following graphs, decide if any vertical asymptotes exist. If so, give the equations of the asymptotes to 2 decimal places.

a $y = \dfrac{1}{x^2 - 5x + 1}$

b $y = -\dfrac{3}{2x^2 + 4x + 2}$

c $y = \dfrac{4}{-x^2 - 7x - 3}$

d $y = \dfrac{5}{x^2 + x + 6}$

★ 3   Find the equations of the vertical asymptotes, should any exist, for the following graphs.

a $y = \dfrac{8x}{x^2 + x}$

b $y = \dfrac{3x - 3}{x^2 - 3x + 2}$

c $y = \dfrac{2x^2 - 3x}{5x}$

d $y = \dfrac{x^2 + 3x - 4}{x^2 + 2x - 3}$

e $y = \dfrac{6x - 1}{x^2 - 9}$

f $y = \dfrac{x - 1}{x^3 - x^2 + 2x - 2}$

g $y = \dfrac{x^2 - 4}{x^4 - 3x^2 - 4}$

h $y = \dfrac{x^3 - 2x^2 - x + 2}{x^4 - x^3 - 3x^2 + x + 2}$

4   The graph of $y = \dfrac{3x + 6}{x^3 + 3x^2 - 4}$ is shown below. $x = -2$ is a root of the

denominator, and is a root of the numerator. Explain why the line $x = -2$ still appears as a vertical asymptote.

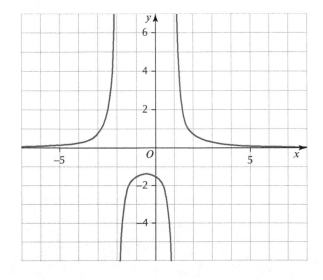

5 Find the equations of any vertical asymptotes of these functions.

a $\quad f(x) = \dfrac{4}{x - 2} + \dfrac{x}{x - 2}$

b $\quad f(x) = \dfrac{2x}{x - 2} - \dfrac{4}{x - 2}$

Horizontal asymptotes

A line $y = a$ is a **horizontal asymptote** for a function $f(x)$ if $\lim\limits_{x \to \infty} f(x) = a$ or $\lim\limits_{x \to -\infty} f(x) = a$.

A rational function will have a horizontal asymptote if the degree of its numerator is less than or equal to the degree of its denominator.

Example 6.5

Find the equations of any horizontal asymptotes for the graph of $f(x) = \dfrac{1}{3x} + 2$

As $x \to \infty$, $\dfrac{1}{3x} \to 0$, so $\lim\limits_{x \to \infty} f(x) = 0 + 2 = 2$

$\Rightarrow y = 2$ is a horizontal asymptote.

As $x \to -\infty$, $\dfrac{1}{3x} \to 0$,

> Calculation of horizontal asymptotes involves calculation of the limit of the function as $x \to \pm\infty$

so $\lim\limits_{x \to -\infty} f(x) = 0 + 2 = 2$

Therefore the line $y = 2$ is a horizontal asymptote for the graph as both $x \to \infty$ and $x \to -\infty$

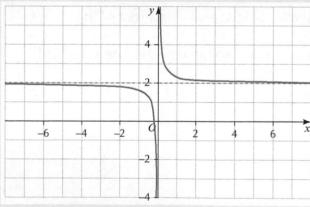

For rational functions, the limit (if it exists) as $x \to \infty$ must be equal to the limit as $x \to -\infty$. However, this need not be the case for non–rational functions.

Example 6.6

Find the equations of any horizontal asymptotes for the graph of $f(x) = \dfrac{1}{e^x + 1}$

As $x \to \infty$, $e^x \to \infty$

$\lim\limits_{x \to \infty} \dfrac{1}{e^x + 1} = 0$

$\Rightarrow y = 0$ is a horizontal asymptote for the graph as $x \to \infty$

As $x \to -\infty$, $e^x \to 0$

$$\lim_{x \to -\infty} \frac{1}{e^x + 1} = \frac{1}{0 + 1} = 1$$

$\Rightarrow y = 1$ is a horizontal asymptote for the graph as $x \to -\infty$

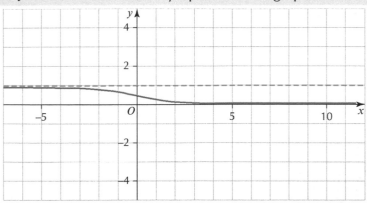

Example 6.7

Find the equations of any horizontal asymptotes for the graph of the function

$$g(x) = \frac{2x^2 - 3x - 1}{3x^2 + 3}$$

Method 1

$$\frac{2x^2 - 3x - 1}{3x^2 + 3} = \frac{\dfrac{2x^2 - 3x - 1}{x^2}}{\dfrac{3x^2 + 3}{x^2}}$$

> Divide both the numerator and denominator by the highest power of x that appears. This makes the limits easier to calculate.

$$= \frac{\dfrac{2x^2}{x^2} - \dfrac{3x}{x^2} - \dfrac{1}{x^2}}{\dfrac{3x^2}{x^2} + \dfrac{3}{x^2}}$$

$$= \frac{2 - \dfrac{3}{x} - \dfrac{1}{x^2}}{3 + \dfrac{3}{x^2}}$$

$$\lim_{x \to \infty} \frac{3}{x} = 0$$

$$\lim_{x \to \infty} \frac{1}{x^2} = 0$$

$$\lim_{x \to \infty} \frac{3}{x^2} = 0$$

$$\Rightarrow \lim_{x \to \infty} 2 - \frac{3}{x} - \frac{1}{x^2} = 2$$

and

$$\Rightarrow \lim_{x \to \infty} 3 + \frac{3}{x^2} = 3$$

$$\Rightarrow \lim_{x \to \infty} g(x) = \frac{2}{3}$$

$$\Rightarrow y = \frac{2}{3} \text{ is a horizontal asymptote as } x \to \infty$$

$$\lim_{x \to -\infty} \frac{3}{x} = 0$$

$$\lim_{x \to -\infty} \frac{1}{x^2} = 0$$

$$\lim_{x \to -\infty} \frac{3}{x^2} = 0$$

$$\Rightarrow \lim_{x \to -\infty} 2 - \frac{3}{x} - \frac{1}{x^2} = 2$$

and

$$\Rightarrow \lim_{x \to -\infty} 3 + \frac{3}{x^2} = 3$$

$$\Rightarrow \lim_{x \to -\infty} g(x) = \frac{2}{3}$$

$$\Rightarrow y = \frac{2}{3} \text{ is a horizontal asymptote as } x \to -\infty$$

Method 2

$$
\begin{array}{r}
\frac{2}{3} \\
3x^2 + 3 \overline{\big)\, 2x^2 - 3x - 1} \\
-\,(2x^2 \qquad + 2) \\
\hline
-3x - 3
\end{array}
$$

> Use long division to find the quotient and remainder when the numerator is divided by the denominator.

Therefore:

$$g(x) = \frac{2}{3} + \frac{-3x - 3}{3x^2 + 3}$$

$$= \frac{2}{3} - \frac{x + 1}{x^2 + 1}$$

> Cancel the common factor of 3 in the algebraic fraction.

$$\text{as } x \to \infty, \frac{x + 1}{x^2 + 1} \to 0$$

$$\text{as } x \to -\infty, \frac{x + 1}{x^2 + 1} \to 0$$

$$\Rightarrow g(x) \to \frac{2}{3} \text{ as } x \text{ tends toward both } \infty \text{ and } -\infty$$

$$\Rightarrow y = \frac{2}{3} \text{ is a horizontal asymptote.}$$

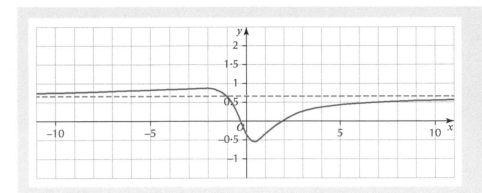

Example 6.8

Find the equations of all asymptotes for the graph of the function

$$f(x) = \frac{x}{x^2 - 1} + 3$$

Vertical asymptotes

$\dfrac{x}{x^2 - 1} = \dfrac{x}{(x - 1)(x + 1)}$ and $(x - 1)(x + 1) = 0$ if and only if $x = 1$ or $x = -1$

$\Rightarrow \dfrac{x}{x^2 - 1}$ has vertical

asymptotes $x = 1$ and $x = -1$

$\Rightarrow f(x)$ also has vertical
asymptotes $x = 1$ and $x = -1$

> The graph of $f(x)$ is the graph of $y = \dfrac{x}{x^2 - 1}$ shifted **vertically upwards by 3 units** so the **vertical** asymptotes of $f(x)$ will be equal to those of $\dfrac{x}{x^2 - 1}$

Horizontal asymptotes

As $x \to \infty$, $\dfrac{x}{x^2 - 1} \to 0$,

so $\lim\limits_{x \to \infty} f(x) = 0 + 3 = 3$

$\Rightarrow y = 3$ is a horizontal
asymptote.

As $x \to -\infty$, $\dfrac{x}{x^2 - 1} \to 0$,

so $\lim\limits_{x \to -\infty} f(x) = 0 + 3 = 3$

$\Rightarrow y = 3$ is a horizontal
asymptote for the graph as
$x \to \infty$ and $x \to -\infty$

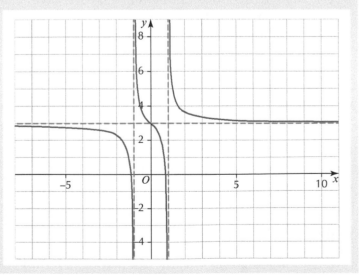

Exercise 6B

1 For each of these rational functions, find the equations of any horizontal asymptotes.

a $f(x) = \dfrac{2}{x}$

b $f(x) = \dfrac{1}{x} + 4$

c $f(x) = \dfrac{5}{x + 1}$

d $f(x) = -\dfrac{1}{x} - 2$

e $f(x) = \dfrac{3}{x^2 + 4} - 9$

f $f(x) = \dfrac{2}{x^3 - x} + 1$

★ 2 For each of these rational functions, find the equations of any horizontal asymptotes.

a $f(x) = \dfrac{x}{x + 2}$

b $g(x) = \dfrac{x}{x^2 + 2}$

c $r(t) = \dfrac{t^2 + 3}{2t^2}$

d $s(t) = \dfrac{t^3}{t^2 + t} - 3$

3 For each of these rational functions, find the equations of any horizontal asymptotes.

a $y(x) = \dfrac{1}{x} + \dfrac{1}{x - 2}$

b $y(x) = \dfrac{x^2}{3x^2} - \dfrac{2}{x^2 + 1}$

4 A physics student is asked to find an equation to model the velocity of a skydiver in freefall.

The student produces the equation $v(t) = \dfrac{108t}{2t + 11}$ where v is the velocity and t is the time in seconds after the skydiver begins their descent.

a Calculate the initial acceleration of the skydiver, given by the derivative $v'(t)$.

b Calculate the terminal velocity of the skydiver, i.e. the velocity the skydiver will tend towards as time increases.

c After 15 seconds of freefall, a skydiver will usually have reached 99% of their terminal velocity. Given this fact, do you think the given equation is an accurate model of the velocity of the skydiver? Justify your answer.

Oblique asymptotes

A rational function $f(x)$ will have an **oblique** asymptote (or **slant** asymptote) when the degree of its numerator is exactly one larger than the degree of its denominator.

In this situation, as x tends to infinity or negative infinity, $f(x)$ will be approximated by a non-constant linear function.

As a straight line, the equation of an oblique asymptote will take the form $y = mx + c$. There are two common methods for finding the equations of oblique asymptotes.

Method 1

1 First find the gradient m by calculating $m = \lim\limits_{x \to \infty} \dfrac{f(x)}{x}$ or $m = \lim\limits_{x \to -\infty} \dfrac{f(x)}{x}$

2   Use the value of m to find c, by calculating $c = \lim\limits_{x \to \infty} (f(x) - mx)$ or
$c = \lim\limits_{x \to -\infty} (f(x) - mx)$.

This method can also be used to find asymptotes of non-rational functions.

Method 2

1   Write the rational function $f(x)$ in the form $f(x) = \dfrac{g(x)}{h(x)}$ where $g(x)$ and $h(x)$ are polynomials.

2   Use long division or synthetic division to find the quotient and remainder when $g(x)$ is divided by $h(x)$. If the quotient is a linear function, then this will be an oblique asymptote of the graph of $f(x)$.

This method can be quicker than Method 1, but is limited to rational functions.

Note that if a rational function $f(x)$ has an oblique asymptote as $x \to \infty$, it will have the same oblique asymptote as $x \to -\infty$. However, this need not be the case for non-rational functions, so it is good practice to deal with the two limits separately.

Example 6.9

Find the equation of the oblique asymptote to the graph of $f(x) = \dfrac{2x^3 - x + 3}{x^2 - 4}$

Method 1

$$\frac{f(x)}{x} = \frac{2x^3 - x + 3}{x(x^2 - 4)}$$

$$= \frac{2 - \dfrac{1}{x^2} + \dfrac{3}{x^3}}{1 - \dfrac{4}{x^2}}$$

> Divide numerator and denominator by the highest power of x, so x^3 in this case.

$$\Rightarrow \lim_{x \to \infty} \frac{f(x)}{x} = \frac{2}{1} = 2$$

and

$$\lim_{x \to -\infty} \frac{f(x)}{x} = 2$$

$$f(x) - 2x = \frac{2x^3 - x + 3}{x^2 - 4} - 2x$$

$$= \frac{2x^3 - x + 3}{x^2 - 4} - 2x\frac{\left(x^2 - 4\right)}{\left(x^2 - 4\right)}$$

> Express this as a single fraction to make evaluating the limits easier.

$$= \frac{2x^3 - x + 3 - 2x\left(x^2 - 4\right)}{x^2 - 4}$$

(continued)

$$= \frac{2x^3 - x + 3 - 2x^3 + 8x}{x^2 - 4}$$

$$= \frac{7x + 3}{x^2 - 4}$$

$$= \frac{\dfrac{7}{x} + \dfrac{3}{x^2}}{1 - \dfrac{4}{x^2}}$$

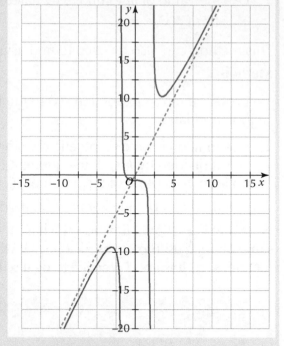

$$\Rightarrow \lim_{x \to \infty} (f(x) - 2x) = 0$$

and

$$\lim_{x \to -\infty} (f(x) - 2x) = 0$$

$\therefore f(x)$ has an oblique asymptote of $y = 2x$ as x tends to both ∞ and $-\infty$

Method 2

$$
\begin{array}{r}
2x \\
x^2 - 4 \overline{\smash{\big)}\, 2x^3 - x + 3} \\
-\,(2x^3 - 8x) \\
\hline
7x + 3
\end{array}
$$

> If the quotient was not a linear function, the graph of $f(x)$ would have no oblique asymptotes.

$$f(x) = \frac{2x^3 - x + 3}{x^2 - 4} = 2x + \frac{7x + 3}{x^2 - 4}$$

As $x \to \pm\infty, \dfrac{7x + 3}{x^2 - 4} \to 0$

$\therefore$ As $x \to \pm\infty$ the graph of $f(x)$ will approach the straight line $y = 2x$.

$\Rightarrow f(x)$ has an oblique asymptote of $y = 2x$

Example 6.10

Find the equation of any oblique asymptotes for the function $g(x) = \dfrac{-x^2 + x + 1}{2x + 4}$

Method 1

$$\frac{g(x)}{x} = \frac{-x^2 + x + 1}{x(2x + 4)}$$

$$= \frac{-x^2 + x + 1}{2x^2 + 4x}$$

$$= \frac{-1 + \dfrac{1}{x} + \dfrac{1}{x^2}}{2 + \dfrac{4}{x}}$$

$$\Rightarrow \lim_{x \to \infty} \frac{g(x)}{x} = \lim_{x \to \infty} \frac{-1 + \dfrac{1}{x} + \dfrac{1}{x^2}}{2 + \dfrac{4}{x}}$$

$$= \frac{-1 + 0 + 0}{2 + 0}$$

$$= -\frac{1}{2}$$

Similarly, $\displaystyle\lim_{x \to -\infty} \frac{g(x)}{x} = -\frac{1}{2}$

Now:

$$g(x) - \left(-\frac{1}{2}x\right) = \frac{-x^2 + x + 1}{2x + 4} - \left(-\frac{1}{2}x\right)$$

$$= \frac{-x^2 + x + 1}{2x + 4} + \frac{1}{2}x$$

$$= \frac{2\left(-x^2 + x + 1\right)}{2(2x + 4)} + \frac{x(2x + 4)}{2(2x + 4)}$$

$$= \frac{2\left(-x^2 + x + 1\right) + x(2x + 4)}{2(2x + 4)}$$

$$= \frac{-2x^2 + 2x + 2 + 2x^2 + 4x}{2(2x + 4)}$$

$$= \frac{6x + 2}{2(2x + 4)}$$

$$= \frac{3x + 1}{2x + 4}$$

$$= \frac{3 + \dfrac{1}{x}}{2 + \dfrac{4}{x}}$$

(continued)

$$\therefore \lim_{x \to \infty} g(x) - \frac{1}{2}x = \frac{3}{2}$$

and

$$\lim_{x \to -\infty} g(x) - \frac{1}{2}x = \frac{3}{2}$$

$\therefore g(x)$ has an oblique asymptote of $y = -\frac{1}{2}x + \frac{3}{2}$ as x tends to both ∞ and $-\infty$

Method 2

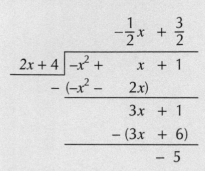

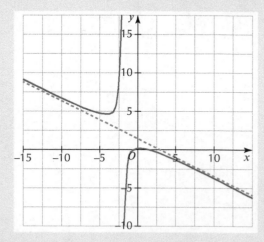

$$\frac{-x^2 + x + 1}{2x + 4} = -\frac{1}{2}x + \frac{3}{2} - \frac{5}{2x + 4}$$

As $x \to \pm\infty$, $\dfrac{5}{2x + 4} \to 0$

$\therefore f(x)$ will approach the straight line $y = -\frac{1}{2}x + \frac{3}{2}$

$\Rightarrow y = -\frac{1}{2}x + \frac{3}{2}$ is an oblique asymptote of the graph of $f(x)$.

Identifying asymptotes of rational functions

We may not know in advance whether a rational function has any asymptotes, or what type of asymptotes these might be. In this situation, we can compare the degrees of the numerator and denominator of our function to identify whether any asymptotes exist, and their type.

Before using these properties to make conclusions about asymptotes, it is important that you write your rational function as a single fraction and then simplify as far as possible.

Properties of $f(x) = \dfrac{g(x)}{h(x)}$	Notes on asymptotes
$h(x)$ has real roots	$f(x)$ has vertical asymptotes of the form $x = $ constant
$\deg(g(x)) > \deg(h(x)) + 1$	$f(x)$ has no horizontal or oblique asymptotes
$\deg(g(x)) = \deg(h(x)) + 1$	$f(x)$ has oblique asymptotes of the form $y = mx + c$
$\deg(g(x)) \leqslant \deg(h(x))$	$f(x)$ has horizontal asymptotes of the form $y = $ constant

Example 6.11

Decide whether the function $f(x) = \dfrac{2x^2 + 3x + 1}{2x^2 - x - 1} + \dfrac{1}{x}$ has any asymptotes. If any asymptotes exist, identify their type.

$$f(x) = \frac{2x^2 + 3x + 1}{2x^2 - x - 1} + \frac{1}{x}$$

$$= \frac{(2x + 1)(x + 1)}{(2x + 1)(x - 1)} + \frac{1}{x}$$

$$= \frac{(x + 1)}{(x - 1)} + \frac{1}{x}$$

> Simplify the first fraction.

$$= \frac{x(x + 1)}{x(x - 1)} + \frac{x - 1}{x(x - 1)}$$

> Write the function as a single algebraic fraction.

$$= \frac{x(x + 1) + (x - 1)}{x(x - 1)}$$

$$= \frac{x^2 + x + x - 1}{x(x - 1)}$$

$$= \frac{x^2 + 2x - 1}{x^2 - x}$$

> As numerator and denominator share no common factors you cannot simplify further.

The denominator has real roots, so $f(x)$ will have vertical asymptotes.

The degree of the numerator is equal to the degree of the denominator, so $f(x)$ will have a horizontal asymptote.

> After simplifying as much as possible you can make conclusions regarding the type of asymptotes.

Exercise 6C

★ 1   Determine the equations of any oblique asymptotes of the graphs of the following functions.

a $f(x) = \dfrac{2x^2 - 2}{x}$

b $f(x) = \dfrac{3 - x^2}{x}$

c $f(x) = \dfrac{4x^2}{x + 1}$

d $f(x) = \dfrac{x^3}{2x^2 - 1}$

e $f(x) = \dfrac{x - 2x^3}{x^2}$

f $f(x) = \dfrac{1 - 2x^2}{3x + 1}$

2   Determine the equations of any oblique asymptotes of the graphs of the following functions.

a $f(x) = \dfrac{1}{x + 3} + \dfrac{x^2}{x - 1}$

b $f(x) = \dfrac{5x^3 + 2}{x - x^2} - 2$

c $f(x) = \dfrac{2x^3 + x^2}{2x - 3} - x^2$

d $f(x) = \dfrac{3x - x^4}{(3x - 1)(x^2 + 1)} + \dfrac{4x - 3}{x^2 - 1}$

e $f(x) = \dfrac{1}{x + 1} + \dfrac{1}{x + 2} + \dfrac{x^2}{x + 3} - \dfrac{6x}{x + 4}$

f $f(x) = \dfrac{x(x - 1)(x - 2)(x + 2)}{3 + x - x^2 + 4x^3}$

3 Decide whether the following functions have any asymptotes. If so, identify their type.

a $f(x) = \dfrac{6}{x + 2}$

b $f(x) = \dfrac{2}{x^2 + 1}$

c $g(x) = \dfrac{-3x}{x + 1}$

d $d(t) = \dfrac{t - t^2}{2 - t}$

e $m(n) = \dfrac{n^3}{n^2 - 1} + 3$

f $q(x) = \dfrac{x^4}{x^2 + 6}$

4 Determine the equations of all asymptotes of the following functions.

a $f(x) = \dfrac{x + 1}{(x - 1)^2}$

b $r(\theta) = \dfrac{2\theta - \theta^2}{\theta^2 + \theta}$

c $p(x) = \dfrac{4x^4 - 3}{x^3}$

d $g(x) = \dfrac{x^2 + 4x + 4}{x^2 + x - 2}$

e $g(x) = \dfrac{1}{3x^2 + 1} - 2$

f $h(k) = \dfrac{2 - k^3}{2k^2 - 1}$

Stationary points, points of inflection, maxima and minima

Given a function $f(x)$ and a point c in the domain of f, we say that f is **stationary at a point** c if $f'(c) = 0$, that is, if the gradient of f is equal to 0 when $x = c$. A point on the graph of $f(x)$ at which $f(x)$ is stationary is called a **stationary point** (or **critical point**).

There are three types of stationary point:

- local minimum
- local maximum
- point of horizontal inflection.

When investigating a particular function, determining its stationary points and their nature is key.

A point $(x, f(c))$ is a **local minimum** for $f(x)$ if there exists a positive real number ε such that $f(c) \leqslant f(x)$ for all x with the distance ε of c.

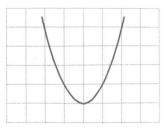

A point $(x, f(c))$ is a **local maximum** for $f(x)$ if there exists a positive real number ε such that $f(c) \geqslant f(x)$ for all x with the distance ε of c.

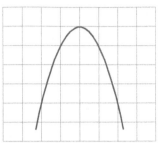

A point $(x, f(c))$ is a **point of horizontal inflection** for $f(x)$ if $f'(c) = 0$ but the gradient of $f(x)$ does not change sign when moving through the stationary point from left to right (or right to left).

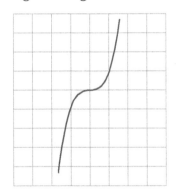 or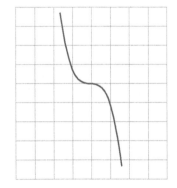

The second derivative test

At Higher level you investigated stationary points and their nature using the first derivative and by constructing a nature table. The **second derivative test** is an alternative method for investigating stationary points, which makes use of the second derivative of a function.

Given a function $f(x)$, first identify all stationary points using the first derivative $f'(x)$. Then, calculate the second derivative $f''(x)$, and evaluate at each stationary point a. The sign of $f''(a)$ gives the following information about the nature of the stationary point at a.

- If $f''(a) > 0$, then a is a local minimum.
- If $f''(a) < 0$, then a is a local maximum.
- If $f''(a) = 0$, then further investigation is needed.

Example 6.12

Find any stationary points of the graph of the function $f(x) = x^3 - 3x$, and use the second derivative test to determine their nature.

$f'(x) = 3x^2 - 3$ and $3x^2 - 3 = 0$ if and only if $x^2 = 1$, so if and only if $x = 1$ or $x = -1$

> Find all values of x for which $f'(x) = 0$.

$f(1) = 1^3 - 3 \times 1 = -2 \Rightarrow$ stationary point at $(1, -2)$

$f(-1) = (-1)^3 - 3 \times (-1) = 2 \Rightarrow$ stationary point at $(-1, 2)$

$f''(x) = 6x$

$f''(-1) = -6 < 0$

$\Rightarrow (-1, 2)$ is a local maximum

$f''(1) = 6 > 0$

$\Rightarrow (1, -2)$ is a local minimum

> Substitute the x-coordinates of the stationary points into $f''(x)$. If the value is positive or negative you can make an immediate conclusion about the nature of the stationary point.

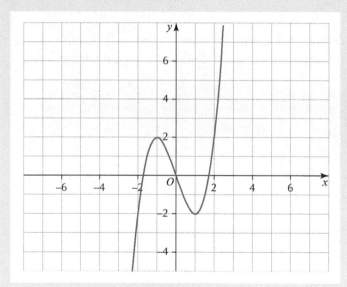

Example 6.13

Find all stationary points of the graph of the function $f(x) = x^4 - \dfrac{4x^3}{3}$ and determine their nature.

$f'(x) = 4x^3 - 4x^2$ and $4x^3 - 4x^2 = 4x^2(x - 1) = 0$ if and only if $x = 0$ or $x = 1$

$f(0) = 0$

$\Rightarrow (0, 0)$ is a stationary point

$f(1) = 1^4 - \dfrac{4 \times 1^3}{3} = -\dfrac{1}{3}$

$\Rightarrow \left(1, -\dfrac{1}{3}\right)$ is a stationary point

$f''(x) = 12x^2 - 8x$

$f''(1) = 12 \times 1^2 - 8 \times 1 = 4 > 0$

$\Rightarrow (1, -\frac{1}{3})$ is a local minimum

$f''(0) = 12 \times 0^2 - 8 \times 0 = 0$ •

> You need to investigate either $f(x)$ or $f'(x)$ for values just above and below $x = 0$ to determine the nature of this stationary point.

just below $x = 0$, $f'(x) < 0$

just above $x = 0$, $f'(x) < 0$

$\Rightarrow f(x)$ is decreasing on either side of this stationary point

$\therefore (0, 0)$ is a point of horizontal inflection. •

> Alternatively you could have substituted values into $f(x)$ to show that $f(x) > 0$ just below $x = 0$ and $f(x) < 0$ just above $x = 0$. Further methods for investigating points of inflection will be studied later in this chapter.

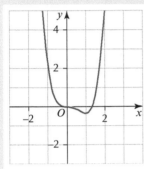

Exercise 6D

1 For each of the following functions, identify all stationary points and use the second derivative test to determine their nature.

   a $f(x) = x^2 - 3$

   b $f(x) = 3x - 2x^2 + 5$

   c $f(x) = \dfrac{x^3}{3} - 4x + 2$

   d $f(x) = x^3 - 9x^2 + 15x$

   e $f(x) = \dfrac{x^4}{4} + \dfrac{2x^3}{3} - \dfrac{x^2}{2} - 2x$

   f $f(x) = x^2 - \dfrac{1}{x}$

★ 2 For each of the following functions, identify all stationary points and use the second derivative test, along with further investigation if necessary, to determine their nature.

   a $f(x) = \dfrac{x^3}{3} - 2$

   b $f(x) = \dfrac{x^3}{3} - x^2 + x$

   c $f(x) = x^2 e^x$

   d $f(x) = x^3 e^x + 1$

   e $f(x) = x^2 e^{-x}$

3 Consider the function $h(\theta) = \sin\theta - \cos\theta$. Identify all stationary points of $h(\theta)$ and use the second derivative test to determine which are maxima and which are minima.

> You might want to use the identity $\dfrac{\sin\theta}{\cos\theta} = \tan\theta$

Concavity and points of inflection

The second derivative tells us the **concavity** of a function.

A real-valued function $f(x)$ is said to be **concave upward** on an interval D if, for all a, b, c in D with $a < c < b$, the point $(c, f(c))$ lies below the straight line joining the points $(a, f(a))$ and $(b, f(b))$.

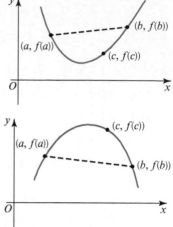

A function $f(x)$ is said to be **concave downward** on an interval D if, for all a, b, c in D with $a < c < b$, the point $(c, f(c))$ lies above the straight line joining the points $(a, f(a))$ and $(b, f(b))$.

Given a function $f(x)$ and a point a, the second derivative gives us the following information about the concavity of $f(x)$ at the point a.

- If $f''(a) > 0$ then $f(x)$ is concave upward at the point a.

- If $f''(a) < 0$ then $f(x)$ is concave downward at the point a.

In general, a **point of inflection** occurs at any point where $f(x)$ changes from concave upward to concave downward, or vice versa.

Another way of visualising this is that if a point of inflection occurs at $x = a$, then the tangent to $f(x)$ at a will cross the graph of $f(x)$ at a.

Points of inflection do not need to be stationary points. To find points of inflection for the graph of $f(x)$ we must calculate the points at which the second derivative $f''(x)$ changes sign from positive to negative, or vice versa.

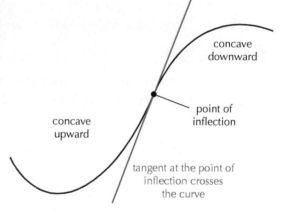

For a point of inflection to occur when $x = a$, it is necessary that either $f''(a) = 0$ or $f''(a)$ is undefined. However, it is possible that $f''(a) = 0$ or $f''(a)$ is undefined but a is not a point of inflection.

Example 6.14

Identify any points of inflection for the graph of the function $f(x) = x^3 - x$.

$f'(x) = 3x^2 - 1$ and $f''(x) = 6x$ (Calculate the second derivative $f''(x)$.)

$6x = 0$ if and only if $x = 0$ (Identify any points at which $f''(x) = 0$.)

If $x < 0$ then $6x < 0$

If $x > 0$ then $6x > 0$

∴ $f''(x)$ changes sign when $x = 0$

∴ There is a point of inflection at $x = 0$.

> You must check that $f''(x)$ changes sign at this point to show a point of inflection occurs.

$f(0) = 0^3 - 0 = 0$

⟹ (0, 0) is a point of inflection for the graph of $f(x)$.

> After calculating the x-coordinate, don't forget to find the y-coordinate.

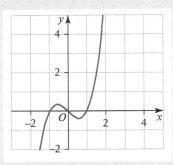

> Notice that the point of inflection is not a stationary point of the graph. Such points are **non-horizontal** or **tilted** points of inflection.

Example 6.15

Let $f(x) = \dfrac{x^4}{4} - \dfrac{3x^2}{2} + 2x$

a Find all stationary points and determine their nature.

b Identify any further points of inflection.

a $f'(x) = x^3 - 3x + 2$

By inspection, $x = 1$ is a root of $x^3 - 3x + 2$

⟹ $f'(x) = (x - 1)(x^2 + x - 2) = (x - 1)(x - 1)(x + 2)$

> Alternatively long division or synthetic division could be used here.

∴ $f'(x) = 0$ if and only if $x = 1$ or $x = -2$

> Identify all stationary points using the first derivative.

$f(1) = \dfrac{1^4}{4} - \dfrac{3 \times 1^2}{2} + 2 \times 1 = \dfrac{3}{4}$

∴ $\left(1, \dfrac{3}{4}\right)$ is a stationary point

$f(-2) = \dfrac{(-2)^4}{4} - \dfrac{3 \times (-2)^2}{2} + 2 \times (-2) = -6$

∴ (-2, -6) is a stationary point

$f''(x) = 3x^2 - 3$

> Calculate the second derivative and apply the second derivative test.

$f''(-2) = 9 > 0$

∴ (-2, -6) is a local minimum

$f''(1) = 0$

When x is just below 1, $f''(x) < 0$

When x is just above 1, $f''(x) > 0$

> This shows the concavity of f changes at $x = 1$. You could have used the first derivative test to determine the nature of this stationary point.

∴ $\left(1, \dfrac{3}{4}\right)$ is a point of horizontal inflection.

b $3x^2 - 3 = 3(x^2 - 1) = 0$ if and only if $x = 1$ or $x = -1$

$f''(x) > 0$ when x is just below -1

$f''(x) < 0$ when x is just above -1

$$f(-1) = \frac{(-1)^4}{4} - \frac{3 \times (-1)^2}{2} + 2 \times (-1) = -\frac{13}{4}$$

$\therefore \left(-1, -\frac{13}{4}\right)$ is a point of inflection.

> To identify any further points of inflection, find all values of x for which $f''(x) = 0$.

> There will be a further point of inflection when $x = -1$, which is not a stationary point.

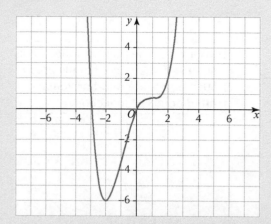

Example 6.16

Show that the function $f(x) = \dfrac{1}{(x - 2)}$ has no points of inflection.

$$f'(x) = \frac{-1}{(x - 2)^2} \qquad f''(x) = \frac{2}{(x - 2)^3}$$

The equation $f''(x) = 0$ has no solutions.

$f''(2)$ is undefined, but $f(2)$ is also undefined.

Therefore f(x) has no points of inflection.

> Remember to consider the possibility that the second derivative is undefined. Here, the original function is also undefined when $x = 2$, so there cannot be a point of inflection.

Example 6.17

Find any points of inflection for the function $f(x) = x^{\frac{1}{3}}$.

$$f'(x) = \frac{1}{3}x^{-\frac{2}{3}} \qquad f''(x) = -\frac{2}{9}x^{-\frac{5}{3}}$$

$$= -\frac{2}{9^3\sqrt{x^5}}$$

$f'(0)$ is undefined, but $f(0)$ is defined.

$f''(-1) = \dfrac{2}{9} > 0$

$f'(1) = -\dfrac{2}{9} < 0$

$f(0) = 0$

> We need to investigate the point where the second derivative is undefined.

> Check the concavity of the function when x is just below and just above 0. Since the concavity changes, we have a point of inflection.

Therefore there is a point of inflection at $(0, 0)$.

Exercise 6E

★ 1 For each of the following real-valued functions, find all points of inflection should any exist.

a $f(x) = x^3 - 2$
b $f(x) = (x + 3)^3$
c $f(x) = x^4 + 8x^3 + 18x^2$

d $f(x) = x^4 + 2x^3 + 18x^2$
e $f(x) = \dfrac{x}{x^2 - 1}$
f $f(x) = 3 - 2x^{\frac{1}{5}}$

★ 2 For each of the following real-valued functions, find all points of inflection in the interval $[-2\pi, 2\pi]$.

a $g(x) = \sin x$
b $h(x) = 3\cos 2x - 1$

c $r(\theta) = \tan\left(\theta - \dfrac{\pi}{6}\right)$
d $s(t) = 2\sin t + 2\cos t$

3 Show that the function $f(x) = \dfrac{x - 3}{x + 2}$ has no points of inflection.

4 Find any points of inflection of the graph of the function $f(x) = \dfrac{x}{\ln x}$ and

determine whether or not they are points of horizontal inflection.

This question is quite tricky. After differentiating twice and simplifying, you should get

$$f''(x) = \dfrac{-4x^3 + 24x}{\left(x^2 + 2\right)^3}$$

5 Find all points of inflection of the graph of

$$f(x) = \dfrac{x^3}{x^2 + 2} + 4 \ .$$

Finding maxima and minima of functions over given intervals

We may be required to calculate maximum or minimum values of functions (sometimes referred to as **extreme values** or **extrema**), either over their whole domain or over a particular interval.

A function $f(x)$ has a **global** (or **absolute**) **maximum** at $x = a$ if $f(a) \geqslant f(x)$ for all x in the domain of f.

Similarly, $f(x)$ has a **global** (or **absolute**) **minimum** at $x = a$ if $f(a) \leqslant f(x)$ for all x in the domain of f.

In general, given a real-valued function $f(x)$, the maximum and minimum values of $f(x)$ over a closed interval $[a, b]$ can occur at the following points:

• stationary points

• the end points a or b

• points at which the derivative $f'(x)$ is not defined.

All three possibilities must be considered to ensure that the correct maximum and minimum values are identified.

Example 6.18

Find the maximum and minimum values of the function $f(x) = x^3 - 3x$ over the closed interval $\left[-\dfrac{3}{2}, 3\right]$.

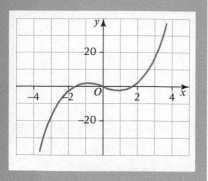

$f'(x) = 3x^2 - 3$ is defined at every point in the interval $\left[-\dfrac{3}{2}, 3\right]$.

$3x^2 - 3 = 0$ if and only if $x = 1$ or $x = -1$

$\therefore$ stationary points at $x = 1$ and $x = -1$

$f(1) = -2 \Rightarrow (1, -2)$ is a stationary point

$f''(x) = 6x$ so $f''(1) = 6 > 0$

$\Rightarrow (1, -2)$ is a local minimum

$f(-1) = 2 \Rightarrow (-1, 2)$ is a stationary point

$f''(-1) = -6 < 0$

$\Rightarrow (-1, 2)$ is a local maximum •————

> Calculate the value of $f(x)$ at any stationary points. The local maxima and minima may not be extrema of $f(x)$ over the given interval.

$f\left(-\dfrac{3}{2}\right) = \left(-\dfrac{3}{2}\right)^3 - 3 \times \left(-\dfrac{3}{2}\right) = \dfrac{9}{8}$

$f(3) = 3^3 - 3 \times 3 = 18$ •————

> Calculate the value of $f(x)$ at the end points of the interval before making conclusions about the maximum and minimum values.

$\Rightarrow$ Maximum value of f over $\left[-\dfrac{3}{2}, 3\right]$ is 18 (occurring at the end point $x = 3$) and the minimum value is -2 (occurring at the stationary point when $x = -1$).

Example 6.19

Calculate the maximum and minimum values of the function

$f(x) = \begin{cases} -x + 1, & x < 0 \\ x + 1, & x \geqslant 0 \end{cases}$ for $-2 \leqslant x \leqslant 1$

When $-2 < x < 0$, $f(x) = -x + 1$ so $f'(x) = -1$

$\therefore$ cannot have $f'(x) = 0$ for $-2 < x < 0$

$\therefore$ no stationary points in this interval

(continued)

When $0 < x < 1$, $f(x) = x + 1$ so $f'(x) = 1$

$\therefore$ cannot have $f'(x) = 0$ for $0 < x < 1$

$\therefore$ no stationary points in this interval

$f(-2) = 3$

$f(1) = 2$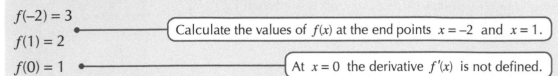

Calculate the values of $f(x)$ at the end points $x = -2$ and $x = 1$.

$f(0) = 1$ At $x = 0$ the derivative $f'(x)$ is not defined.

$\therefore$ Maximum value of f for $-2 \leqslant x \leqslant 1$ is 3, at the end point $x = -2$

and minimum value is 1, at $x = 0$.

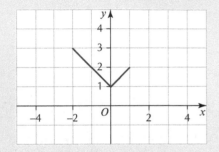

Exercise 6F

★ 1   Find the maximum and minimum values of the following real-valued functions on the given interval. Give your answers to 2 decimal places.

   a $f(x) = 3x - 4$ on the interval $[-5, 3]$.

   b $f(x) = 2x - x^2 + 6$ on the interval $[-10, 10]$.

   c $f(x) = 5x^2 + x - 4$ on the interval $[-2, 4]$.

   d $f(x) = x^3 + x^2 - 2x - 2$ on the interval $[-2, 2]$.

   e $f(x) = 2x^3 - x^2 + 3$ on the interval $[-1, 1]$.

   f $f(x) = x^4 - x^2$ on the interval $\left[-\frac{1}{2}, \frac{3}{2}\right]$.

★ 2   Find the maximum and minimum values of the following real-valued functions on the given interval. Give your answers to 2 decimal places.

   a $f(x) = \dfrac{1}{x^2 + 1}$ on the interval $[-1, 2]$

   b $f(x) = \dfrac{e^x}{x}$ on the interval $[0{\cdot}4, 3{\cdot}2]$

   c $f(x) = x \ln x$ on the interval $[0{\cdot}1, 2]$

   d $f(x) = e^x \cos x$ on the interval $[-5, 5]$

   e $f(x) = \sin x \cos x$ on the interval $\left[-3\pi, \ -\dfrac{\pi}{2}\right]$

   f $f(x) = e^x \sin^2 x$ on the interval $[0, 4]$

★ 3   Find the extrema of the following functions on the given interval.

   a $f(x) = \begin{cases} -2x, & x < 0 \\ 2x, & x \geqslant 0 \end{cases}$ on the interval $[-1, 3]$

b $\quad f(x) = \begin{cases} 3x + 1, & x < 0 \\ -3x + 2, & x \geqslant 0 \end{cases}$ on the interval $[-1, 2]$

c $\quad f(x) = \begin{cases} x + 1, & x < 0 \\ -x + 1, & 0 \leqslant x < 1 \\ x - 1, & x \geqslant 1 \end{cases}$ on the interval $\left[-2, \frac{3}{2}\right]$

d $\quad f(x) = \begin{cases} 2x, & x \leqslant 1 \\ x^2, & x > 1 \end{cases}$ on the interval $\left[0, \frac{3}{2}\right]$

4 What can be said about any maximum or minimum values for the function $f(x) = \dfrac{1}{x}$ on the interval $[-1, 1]$?

Even and odd functions

Certain functions display types of symmetry. This can be useful when analysing the behaviour of a function or sketching its graph.

- A function f is an **even** function if $f(-x) = f(x)$ for all x in the domain of f. Graphs of even functions display reflectional symmetry across the vertical axis. For example, $\cos x$ is an even function.

- A function f is an **odd** function if $f(-x) = -f(x)$ for all x in the domain of f. Graphs of odd function display rotational symmetry of 180° around the origin. For example, $\sin x$ is an odd function.

Note that many functions are neither even nor odd.

Example 6.20

Prove that $f(x) = x^2$ with domain $\mathbb{R}$ is an even function.

Let x be any real number.

$f(-x) = (-x)^2 = x^2 = f(x)$ ⟶ Substitute $-x$ into f and show that it equals $f(x)$.

$\therefore f(x) = x^2$ is an even function.

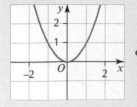

The reflectional symmetry across the y-axis can be seen from the graph.

Example 6.21

Show that $r(t) = 2t^3 - t$ with domain $\mathbb{R}$ is an odd function.

Let t be any real number.

$r(-t) = 2(-t)^3 - (-t) = -2t^3 + t = -(2t^3 - t) = -r(t)$

$\therefore r(t)$ is an odd function.

(*continued*)

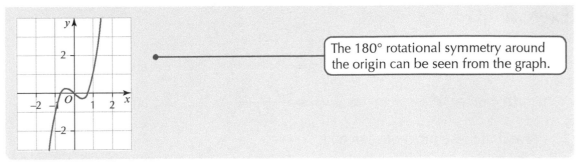

The 180° rotational symmetry around the origin can be seen from the graph.

Functions that are neither even nor odd

To show that a function $f(x)$ is **not** even, we must produce an element a in the domain of f such that $f(-a) \neq f(a)$.

Similarly, to show that a function $f(x)$ is **not** odd, we must produce an element a in the domain of f such that $f(-a) \neq -f(a)$.

If 0 is an element of the domain of function f, then for f to be odd it must be the case that $f(0) = 0$ (since otherwise the graph of f would not display rotational symmetry about the origin). Therefore a quick way to show a function f is not odd is to show that $f(0) \neq 0$.

Example 6.22

Show that the function $s(t) = t^2 - 3t + 2$ is neither even nor odd.

$s(0) = 0^2 - 3 \times 0 + 2 = 2 \neq 0$
$\therefore s$ cannot be an odd function.
$s(-1) = (-1)^2 - 3 \times (-1) + 2 = 6$
$s(1) = 1^2 - 3 \times 1 + 2 = 0$ •————— Calculate the values for $s(-1)$ and $s(1)$.
$\therefore s$ cannot be an even function.

Sums and products of even and odd functions

These properties of sums and products of even and odd functions are useful:

- the sum of two even functions is an even function
- the sum of two odd functions is an odd function
- the product of two even functions is an even function
- the product of two odd functions is an even function
- the product of an even function with an odd function is an odd function.

Example 6.23

Show that $h(x) = 3x^3\cos x - x$ is an odd function.

$3(-x)^3 = -3x^3 \therefore 3x^3$ is an odd function
$\cos(-x) = \cos x \therefore \cos x$ is an even function
$\Rightarrow 3x^3\cos x$ is an odd function •————— odd × even = odd
As the function x is an odd function, •——— It is also possible to prove directly that $h(x)$
$h(x)$ is also a sum of odd functions.   is an odd function. See Exercise 6H Q4.
Therefore $h(x)$ is also an odd function.

Exercise 6G

1 Prove that these functions are even.

a $f(x) = 5x^2$

b $g(x) = -3$

c $h(x) = 3\cos 3x$

d $r(t) = 6t^6 - 3t^4 + t^2$

e $s(t) = \dfrac{4t^3 - 2t}{6t}$

f $d(\theta) = 8\theta \sin 2\theta$

2 Prove that these functions are odd.

a $f(x) = -\dfrac{x^3}{2}$

b $q(x) = x^5 + 4x^7$

c $h(\theta) = 3\tan\theta + \sin\theta$

d $s(t) = t^2(t - t^3)$

e $f(x) = -\dfrac{5x^4 - x^2}{x^3 - 2x}$

f $v(t) = 2t^3\cos 3t$

★ 3 For each of these functions decide if $f(x)$ is even, odd or neither. In each case provide justification for your answer.

a $f(x) = \dfrac{x^3}{4} + \sin x$

b $f(x) = x^2 + x^5 - 2$

c $f(x) = \dfrac{x}{x^3 - 4x} + 5$

d $f(x) = \sin^2 x + \cos^2 x$

e $f(x) = \sin^2 x - \cos x - x$

f $f(x) = x^{35} + x^{33}$

4 Prove directly that $h(x) = 3x^3\cos x - x$ is an odd function that is, prove that $h(-x) = -h(x)$ for all real numbers x.

5 Can you think of a real-valued function which is both even and odd?

6 Prove that the following properties hold.

a The sum of two even functions is an even function.

b The sum of two odd functions is an odd function.

> ⚠ See Chapter 11 for more about proofs.

c The product of two even functions is an even function.

d The product of two odd functions is an even function.

e The product of an even function with an odd function is an odd function.

7 A function $f(x)$ is defined over the real numbers. Part of the graph of f is shown below.

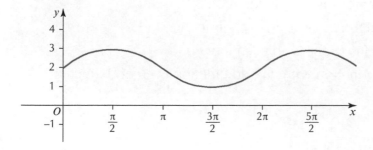

a Sketch the complete graph assuming that f is an even function.

b Sketch the complete graph assuming that f is an odd function.

Continuity

If $f(x)$ is a real-valued function with domain D, then f **is continuous at the point** $c \in D$ if $\lim_{x \to c} f(x) = f(c)$.

If f is continuous at every point in its domain, we say that f is **continuous**.

If this condition does not hold at some point $c \in D$, we call c a **point of discontinuity**, and say f is **discontinuous** (or not continuous).

Discontinuous function	Graph	Reason for discontinuity
$f(x) = \dfrac{1}{x}$		$f(x)$ is not defined when $x = 0$
$f(x) = \tan x$		$f(x)$ is not defined when $x = \ldots, -\dfrac{3\pi}{2}, -\dfrac{\pi}{2}, \dfrac{\pi}{2}, \dfrac{3\pi}{2}, \ldots$
$f(x) = \begin{cases} \dfrac{1}{x}, & x < 0 \\ 2, & x \geqslant 0 \end{cases}$		$\lim_{x \to 0^-} f(x)$ does not exist

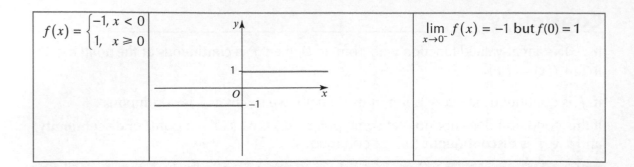

$f(x) = \begin{cases} -1, & x < 0 \\ 1, & x \geq 0 \end{cases}$

$\lim\limits_{x \to 0^-} f(x) = -1$ but $f(0) = 1$

Example 6.24

Identify all points of discontinuity for this function:

$$f(x) = \begin{cases} \dfrac{1}{x+1}, & x < -1 \\ 3, & -1 \leq x < 0 \\ x + 3, & 0 \leq x < 1 \\ x, & x \geq 1 \end{cases}$$

At each point, give a reason for the discontinuity.

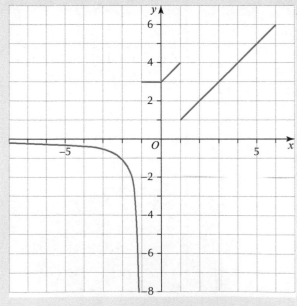

Sketching the graph is often useful to get an idea where any 'jumps' occur.

There is a point of discontinuity at $x = -1$, as $\lim\limits_{x \to -1^-} f(x)$ does not exist.

There is a point of discontinuity at $x = 1$, as $\lim\limits_{x \to 1^-} f(x) = 4$ but $f(1) = 1$.

Exercise 6H

★ 1 For each of the following real-valued functions, identify all points of discontinuity, should any exist. Give the reason for the discontinuity at each such point.

a $f(x) = \dfrac{1}{x+5}$

b $f(x) = \dfrac{2x}{1-x^2}$

c $f(x) = \begin{cases} -2x, & x < 1 \\ x-3, & x \geqslant 1 \end{cases}$

d $f(x) = \begin{cases} x^3, & x < 1 \\ \dfrac{1}{x}, & x \geqslant 1 \end{cases}$

e $f(x) = \begin{cases} 2, & x < 0 \\ 4x, & x \geqslant 0 \end{cases}$

f $f(x) = \begin{cases} x, & x \leqslant 0 \\ \ln x, & x > 0 \end{cases}$

g $f(x) = \begin{cases} x, & x < -1 \\ -1, & -1 \leqslant x < 0 \\ x-1, & x \geqslant 0 \end{cases}$

h $f(x) = \begin{cases} -x, & x < 0 \\ x^2, & 0 \leqslant x < 1 \\ 2x, & x \geqslant 1 \end{cases}$

Sketching graphs of functions

The ability to sketch graphs is an essential skill, and follows a few basic techniques:

• find any y-intercept and x-intercepts

• determine any stationary points and their nature

• find any non-horizontal points of inflection

• find any asymptotes

• investigate the behaviour of the function near any asymptotes and as $x \to \pm\infty$

• determine if $f(x)$ is odd, even or neither.

The sketched graph must show any intercepts, stationary points, points of inflection and asymptotes.

Example 6.25

a Sketch the graph of $f(x) = \dfrac{2}{x^2 - 1}$

b Use your answer to part **a** to determine the values of k for which the equation $f(x) = k$ has no solutions.

a **Intercepts**

$f(0) = \dfrac{2}{0^2 - 1} = -2$

∴ y-intercept is –2 ←———— Set $x = 0$ to determine the y-intercept.

$\dfrac{2}{x^2 - 1}$ is always non-zero

∴ there are no x-intercepts.

Stationary points

$f(x) = 2(x^2 - 1)^{-1}$

$f'(x) = -\dfrac{2}{(x^2 - 1)^2} \times 2x = -\dfrac{4x}{(x^2 - 1)^2}$ ←———— Rewrite $f(x)$ and then use the chain rule to calculate $f'(x)$.

(continued)

$-\dfrac{4x}{\left(x^2-1\right)^2}$ if and only if $-4x=0$, which occurs if and only if $x=0$

$f(0)=-2$ ●————————————— You have already calculated $f(0)$ above.

∴ $(0,-2)$ is a stationary point

$f''(x)=-\dfrac{4}{\left(x^2-1\right)^2}+\dfrac{8x}{\left(x^2-1\right)^3}\times 2x=\dfrac{16x^2}{\left(x^2-1\right)^3}-\dfrac{4}{\left(x^2-1\right)^2}$ ● Use the chain and product rules to calculate $f''(x)$.

$f''(0)=\dfrac{\left(16\times 0^2\right)}{\left(0^2-1\right)^3}-\dfrac{4}{\left(0^2-1\right)^2}=0-4=-4,<0$

∴ $(0,-2)$ is a local maximum

Points of inflection

$\dfrac{16x^2}{\left(x^2-1\right)^3}-\dfrac{4}{\left(x^2-1\right)^2}=0$ ●————————— Find any values of x for which $f''(x)=0$.

$\Leftrightarrow\quad 16x^2-4(x^2-1)=0$

$\Leftrightarrow\quad 16x^2-4x^2+4=0$

$\Leftrightarrow\quad\quad 12x^2+4=0$

$\Leftrightarrow\quad\quad\quad 3x^2+1=0$

$\Leftrightarrow\quad\quad\quad\quad x^2=-\dfrac{1}{3}$

There are no real values of x for which $x^2=-\dfrac{1}{3}$

∴ Graph of $f(x)$ has no points of inflection.

Asymptotes

Roots of the denominator x^2-1 are $x=1$ and $x=-1$

⇒ vertical asymptotes at $x=1$ and $x=-1$

As $x\to -1^-,f(x)\to\infty$ ●——————— Determine the behaviour of $f(x)$ around the vertical asymptotes.

As $x\to -1^+,f(x)\to -\infty$

As $x\to 1^-,f(x)\to -\infty$

As $x\to 1^+,f(x)\to\infty$

$f(x)$ will have a horizontal asymptote. ●——— The degree of the numerator is less than the degree of the denominator.

As $x\to\infty,f(x)\to 0$, and as $x\to -\infty,f(x)\to 0$

$y=0$ is a horizontal asymptote

As $x\to\infty,f(x)$ will be positive, and as $x\to -\infty,f(x)$ will be positive. ●——— Determine the behaviour of $f(x)$ around the horizontal asymptote.

(continued)

Symmetry

$$f(-x) = \frac{2}{(-x)^2 - 1} = \frac{2}{x^2 - 1} = f(x)$$

$\therefore f$ is an even function.

> f is even so the graph has reflectional symmetry across the y-axis. This will help when sketching the graph.

Sketching the graph

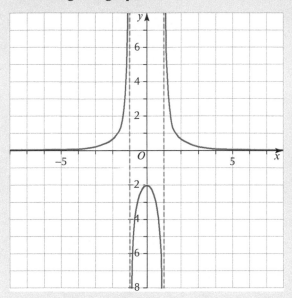

> First mark on any asymptotes and important points. Mark the asymptotes using dashed lines.

> Complete the sketch by drawing the curve as smoothly as possible. Use the asymptotes and symmetry of the function as a guide.

b By inspecting the graph, if $-2 < k \leq 0$, then there are no values of x for which $f(x) = k$.

$\therefore f(x) = k$ has no solutions for $-2 < k \leq 0$

Example 6.26

Sketch the graph of the function $f(x) = \dfrac{x^2}{2x - 3} + 1$

Intercepts

$$f(0) = \frac{0^2}{2 \times 0 - 3} + 1 = 1$$

$\therefore y$-intercept is 1

$$\frac{x^2}{2x - 3} + 1 = 0$$

> Find any values of x for which $f(x) = 0$.

$\Leftrightarrow \quad x^2 + 2x - 3 = 0$

$\Leftrightarrow \quad (x + 3)(x - 1) = 0$

which occurs if and only if $x = -3$ or $x = 1$

$\therefore x$-intercepts at $x = -3$ and $x = 1$

(continued)

Stationary points

$$f'(x) = \frac{2x}{(2x-3)} - \frac{x^2}{(2x-3)^2} \times 2 = \frac{2x}{(2x-3)} - \frac{2x^2}{(2x-3)^2}$$

$$\frac{2x}{(2x-3)} - \frac{2x^2}{(2x-3)^2} = 0$$ ← Find any values of x for which $f'(x) = 0$.

$\Leftrightarrow \quad 2x(2x-3) - 2x^2 = 0$
$\Leftrightarrow \quad 4x^2 - 6x - 2x^2 = 0$
$\Leftrightarrow \quad 2x^2 - 6x = 0$
$\Leftrightarrow \quad x^2 - 3x = 0$
$\Leftrightarrow \quad x(x-3) = 0$

$\Rightarrow$ stationary points at $x = 0$ and $x = 3$

$f(0) = 1$

$\therefore$ (0, 1) is a stationary point

$$f(3) = \frac{3^2}{(2 \times 3 - 3)^2} + 1 = 4$$

$\therefore$ (3, 4) is a stationary point

$$f''(x) = \frac{2}{(2x-3)} - \frac{4x}{(2x-3)^2} - \frac{4x}{(2x-3)^2} + \frac{8x^2}{(2x-3)^3}$$

$$= \frac{2}{(2x-3)} - \frac{8x}{(2x-3)^2} + \frac{8x^2}{(2x-3)^3}$$

$$f''(0) = \frac{2}{(2 \times 0 - 3)} - \frac{8 \times 0}{(2 \times 0 - 3)^2} + \frac{\left(8 \times 0^2\right)}{(2 \times 0 - 3)^3}$$

$$= \frac{2}{(-3)} - 0 + 0$$

$$= -\frac{2}{3} < 0$$

$\therefore$ (0, 1) is a local maximum

$$f''(3) = \frac{2}{(2 \times 3 - 3)} - \frac{8 \times 3}{(2 \times 3 - 3)^2} + \frac{\left(8 \times 3^2\right)}{(2 \times 3 - 3)^3}$$

$$= \frac{2}{3} - \frac{24}{9} + \frac{72}{27}$$

$$= \frac{2}{3} > 0$$

$\therefore$ (3, 4) is a local minimum

Points of inflection

$$f''(x) = \frac{2}{(2x-3)} - \frac{8x}{(2x-3)^2} + \frac{8x^2}{(2x-3)^3} = 0$$

$\Leftrightarrow \qquad 2(2x-3)^2 - 8x(2x-3) + 8x^2 = 0$

$\Leftrightarrow 2(4x^2 - 12x + 9) - 16x^2 + 24x + 8x^2 = 0$

$\Leftrightarrow \quad 8x^2 - 24x + 18 - 16x^2 + 24x + 8x^2 = 0$

$\Leftrightarrow \qquad\qquad\qquad\qquad\qquad 18 = 0$

∴ There are no points of inflection.

Asymptotes

$$f(x) = \frac{x^2}{2x-3} + 1 = \frac{x^2}{2x-3} + \frac{2x-3}{2x-3} = \frac{x^2 + 2x - 3}{2x-3}$$

> Write the function as a single fraction before calculating asymptotes.

The denominator has a single root of $x = \frac{3}{2}$, so there is a vertical asymptote $x = \frac{3}{2}$.

As $x \to \frac{3}{2}^-, f(x) \to -\infty$

As $x \to \frac{3}{2}^+, f(x) \to \infty$

The graph of $f(x)$ will have an oblique asymptote.

> The degree of the numerator is exactly one larger than the degree of the denominator.

$$
\begin{array}{r}
\frac{1}{2}x + \frac{7}{4} \\
2x - 3 \enclose{longdiv}{x^2 + 2x - 3} \\
-\left(x^2 - \frac{3}{2}x\right) \\
\hline
\frac{7}{2}x - 3 \\
-\left(\frac{7}{2}x - \frac{21}{4}\right) \\
\hline
\frac{9}{4}
\end{array}
$$

So the graph has an oblique asymptote of $y = \frac{1}{2}x + \frac{7}{4}$

Symmetry

The function $2x - 3$ is neither even nor odd. Consequently $\frac{x^2}{(2x-3)}$ and then $f(x)$ are neither even nor odd.

(continued)

Sketching the graph

- Mark any asymptotes and important points, such as maxima and minima, and x- and y-intercepts

- Sketch freehand to fit the known asymptotes and points.

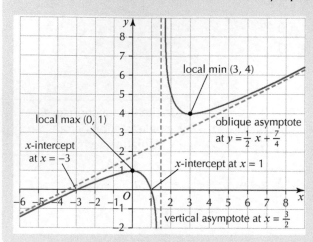

Exercise 6I

★ 1 Sketch the graphs of these rational functions.

a $\quad f(x) = x^3 - x + 1$

b $\quad f(x) = \dfrac{1}{x - 2}$

c $\quad f(x) = x^3 + x^2 + x$

d $\quad f(x) = \dfrac{1}{4 - x^2}$

e $\quad f(x) = \dfrac{x}{4 - x}$

f $\quad f(x) = \dfrac{x^2}{x + 1}$

g $\quad f(x) = \dfrac{x^4 - 3}{x}$

h $\quad f(x) = \dfrac{x^4 - 3x}{x}$

Reflections, translations and scaling

If the graph of a function $f(x)$ is known, we can determine the graphs of the following related functions:

- $-f(x)$
- $f(-x)$
- $f(x) \pm a$
- $f(x \pm b)$
- $kf(x)$
- $f(cx)$

If a, b, c and k are positive constants, the graphs of these functions are reflections, translations or enlargements/reductions of the graph of $f(x)$. You may have studied this previously in Higher maths.

You should already know the graphs of some basic functions, such as x^2, x^3, $\dfrac{1}{x}$, $\sin x$, $\cos x$, $\tan x$, e^x and $\ln x$.

Graph of –$f(x)$

The graph of –$f(x)$ is the **reflection** of the graph of $f(x)$ **in the x-axis**.

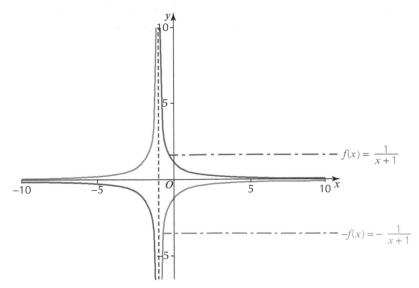

Any vertical asymptotes of $f(x)$ are also vertical asymptotes of –$f(x)$.

Graph of $f(-x)$

The graph of $f(-x)$ is the **reflection** of the graph of $f(x)$ **in the y-axis**.

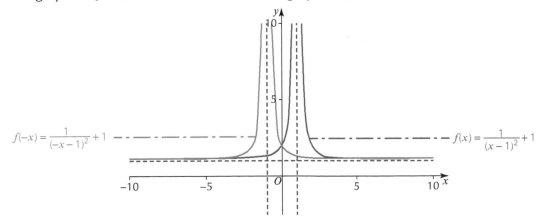

Any horizontal asymptotes of $f(x)$ are also asymptotes of $f(-x)$. Any vertical asymptotes will be reflected in the y-axis.

Graph of $f(x) \pm a$

Adding or subtracting a positive constant a to a function $f(x)$ gives a **vertical translation** of the graph of $f(x)$. To produce the graph of $f(x) + a$ we shift the graph of $f(x)$ **up** a units. To produce the graph of $f(x) - a$ we shift the graph of $f(x)$ **down** a units.

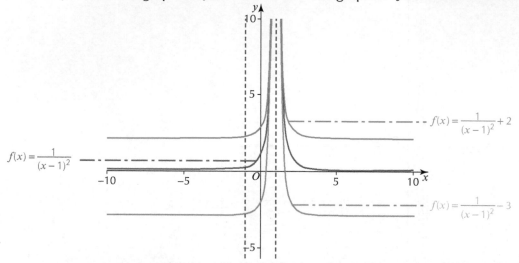

Any horizontal asymptotes are affected by the vertical translation. Any vertical asymptotes are unaffected.

Graph of $kf(x)$

Multiplying $f(x)$ by a positive constant k will **vertically scale** the graph of $f(x)$. If $k > 1$ then the graph of $f(x)$ will be stretched vertically. If $k < 1$ then the graph of $f(x)$ will be compressed vertically.

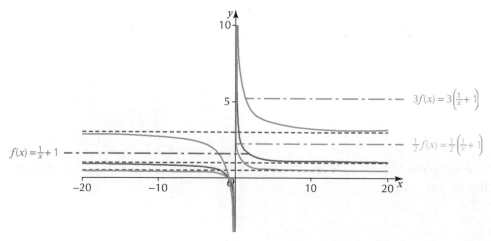

Vertical asymptotes are unaffected by vertical scaling. Horizontal asymptotes (other than $y = 0$) will be affected.

Graph of $f(x \pm b)$

Adding or subtracting a positive constant b to the argument of the function $f(x)$ will **translate** the graph of $f(x)$ **horizontally**. To produce the graph of $f(x + b)$ we shift the graph of $f(x)$ to the **left** b units. To produce the graph of $f(x - b)$ we shift the graph of $f(x)$ to the **right** b units.

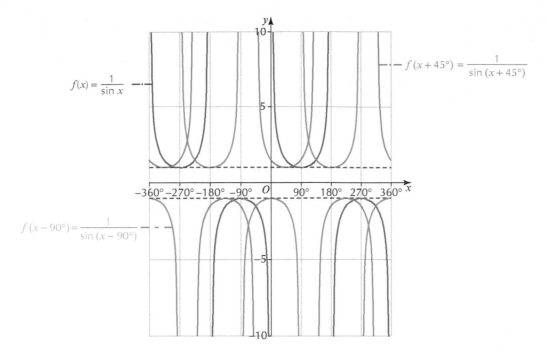

Vertical asymptotes are affected by horizontal translations. Any horizontal asymptotes are unaffected.

Graph of $f(cx)$

Multiplying the argument of $f(x)$ by a positive constant c will **scale** the graph of $f(x)$ **horizontally**.

If $c > 1$ then the graph of $f(x)$ will be stretched horizontally.

If $c < 1$ then the graph of $f(x)$ will be compressed horizontally.

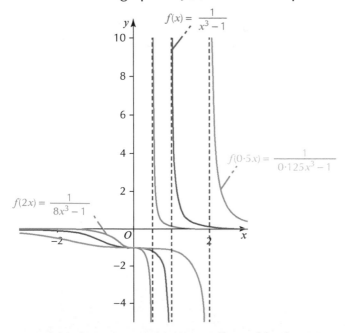

Horizontal asymptotes are unaffected by horizontal scaling, but vertical asymptotes (other than $x = 0$) will be affected.

Example 6.27

The graph of the function $f(x) = \dfrac{x^2 + 2x - 4}{x^2 - 4}$ is shown. By modifying this graph, sketch the graphs of these functions.

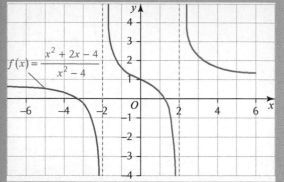

a $\quad g(x) = \dfrac{x^2 + 2x - 4}{x^2 - 4} + 2$

b $\quad h(x) = \dfrac{(x + 1)^2 + 2(x + 1) - 4}{(x + 1)^2 - 4}$

c $\quad r(x) = \dfrac{-x^2 - 2x + 4}{x^2 - 4}$

a $\quad g(x) = f(x) + 2$

b $\quad h(x) = f(x + 1)$

c $\quad r(x) = -f(x)$

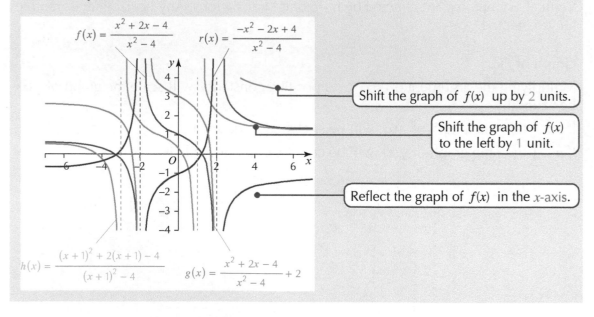

Shift the graph of $f(x)$ up by 2 units.

Shift the graph of $f(x)$ to the left by 1 unit.

Reflect the graph of $f(x)$ in the x-axis.

Example 6.28

The graph of $f(x) = \dfrac{x}{x^2 - 9}$ is shown.

Sketch the graph of the function

$3f(x - 1) + 2.$

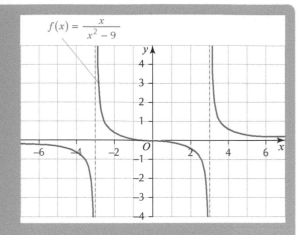

The graph of $f(x)$ has been shifted to the right by 1 unit, then vertically scaled by a factor of 3, then shifted up 2 units.

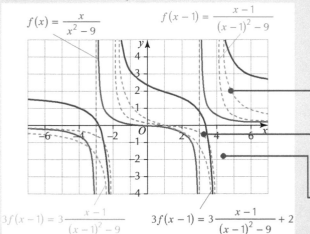

$f(x) = \dfrac{x}{x^2 - 9}$ $\quad f(x - 1) = \dfrac{x - 1}{(x - 1)^2 - 9}$

$3f(x - 1) = 3\dfrac{x - 1}{(x - 1)^2 - 9}$ $\quad 3f(x - 1) = 3\dfrac{x - 1}{(x - 1)^2 - 9} + 2$

When multiple modifications of a graph are required, it is important they are applied in the correct order. This order follows the usual order in which you perform arithmetic operations.

First apply the horizontal translation. $f(x - 1)$ shifts the graph to the **right by 1** unit.

Then apply the vertical scaling. $3f(x - 1)$ scales the graph by a factor of **3**.

Finally, apply the vertical translation. $3f(x - 1) + 2$ shifts the graph **up 2 units**.

Exercise 6J

1   Let $f(x) = x^2$. For each of these functions, first state how the graph is related to the graph of $f(x)$, then sketch the graph.

   a $f(x) + 3$ **b** $f(x - 2)$ **c** $f(x + 3) - 5$

   d $f(2x)$ **e** $-3f(-x + 1)$ **f** $2f(x) - 1$

2   Let $f(x) = \cos x$. For each of these functions, first state how the graph is related to the graph of $f(x)$, then sketch the graph.

   a $3f(x) - 2$ **b** $f(-x) + 1$ **c** $-f\left(x + \dfrac{\pi}{2}\right)$

   d $f\left(2x - \dfrac{\pi}{4}\right)$ **e** $4f\left(\dfrac{x}{3}\right) + 1$

★ 3   In each of these cases, first identify a function whose graph you are familiar with, then modify it to sketch the graph of the given function.

   a $f(x) = 2(x - 1)^2$ **b** $f(x) = 3x^3 - 3$

   c $f(x) = -3\sin 2x + 4$ **d** $y(x) = 4\cos\left(3x - \dfrac{\pi}{2}\right) + 1$

e $g(x) = (x + 3)^3 - 2$ f $h(x) = e^{x-1}$

g $d(a) = -e^{2a}$ h $r(\theta) = 2\tan\left(3x + \dfrac{\pi}{4}\right)$

i $s(t) = -\ln t + 1$ j $f(x) = \dfrac{1}{x + 2} - 6$

★ 4   Suppose that $f(x) = \dfrac{1 - x^2}{x}$

    a   Sketch the graph of $f(x)$, clearly showing any intercepts, stationary points and asymptotes.

    b   Use your answer to part **a** to produce graphs of the following functions.

      i $\dfrac{1 - x^2}{x} + 2$

      ii $\dfrac{1 - 4x^2}{2x}$

      iii $\dfrac{-x^2 - 2x}{x + 1}$

Graphs of inverse functions and modulus functions
Graph of $f^{-1}(x)$

Given the graph of $f(x)$, the graph of the inverse function $f^{-1}(x)$ is the **reflection** of the graph of $f(x)$ **in the line** $y = x$.

Recall that one method for calculating the inverse of a function $f(x)$ involves setting $y = f(x)$ and then rearranging to make x the subject of the equation, so the roles of x and y are 'swapped'. To achieve this effect on the graph of $f(x)$, the x- and y-axes must be swapped – a reflection of the graph in the line $y = x$ produces this effect.

Example 6.29

The graph of the function $f(x) = 3x - 1$ is shown. Sketch the graph of $f^{-1}(x)$, clearly marking any points of intersection with the graph of $f(x)$.

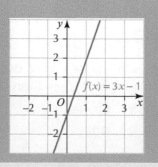

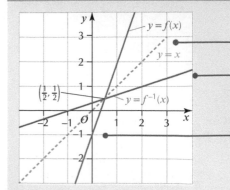

First sketch the line $y = x$ as a guide.

Reflect the graph of $f(x)$ in the line $y = x$ to produce the graph of $f^{-1}(x)$.

The y-intercept $(0, -1)$ for $y = f(x)$ is reflected to become the x-intercept $(-1, 0)$ for $y = f^{-1}(x)$.

(continued)

$$x = 3x - 1$$
$$2x - 1 = 0$$
$$2x = 1$$
$$x = \tfrac{1}{2}$$

Points of intersection occur when $y = x$, so we must solve the equation $x = f(x)$. Note that we do not have to explicitly calculate $f^{-1}(x)$.

The point $x = \tfrac{1}{2}$ lies on the line $y = x$, so the y-coordinate must also equal $\tfrac{1}{2}$.

$\therefore$ Single point of intersection $\left(\tfrac{1}{2}, \tfrac{1}{2}\right)$.

Example 6.30

The graph $y = x^3$ is shown. Sketch the graphs of $y = x^3$ and $y = \sqrt[3]{x}$ on the same diagram, clearly marking any points of intersection.

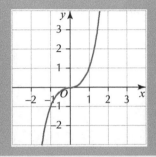

If $f(x) = x^3$, then $f^{-1}(x) = \sqrt[3]{x}$

$\Rightarrow$ the graph $y = \sqrt[3]{x}$ is the reflection of the graph $y = x^3$ in the line $y = x$.

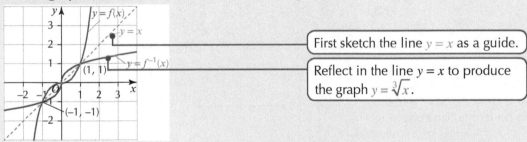

First sketch the line $y = x$ as a guide.

Reflect in the line $y = x$ to produce the graph $y = \sqrt[3]{x}$.

$$x = x^3$$
$$x^3 - x = 0$$
$$x(x^2 - 1) = 0$$
$$x(x - 1)(x + 1) = 0$$

To find any points of intersection we must solve $x = x^3$.

Factorise the cubic to find all solutions.

$\Rightarrow x = -1, 0$ or 1

Points of intersection lie on the line $y = x$, they must be $(-1, -1)$, $(0, 0)$ and $(1, 1)$.

Graph of $|f(x)|$

Given a function $f(x)$, the modulus function $|f(x)|$ is defined as follows:

$$|f(x)| = \begin{cases} f(x) \text{ if } f(x) \;\geqslant 0 \\ -f(x) \text{ if } f(x) < 0 \end{cases}$$

When $f(x) \geqslant 0$ the graph of $|f(x)|$ will equal the graph of $f(x)$

When $f(x) < 0$ the graph of $|f(x)|$ will be the reflection of the graph of $f(x)$ in the x-axis.

In particular, since $|f(x)|$ is always non-negative, the graph of $|f(x)|$ will always lie on or above the x-axis.

Example 6.31

Given that $f(x) = \dfrac{1}{x + 3}$ sketch the graph of $|f(x)|$.

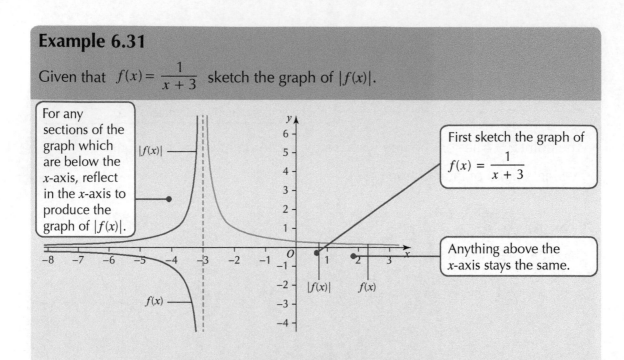

For any sections of the graph which are below the x-axis, reflect in the x-axis to produce the graph of $|f(x)|$.

First sketch the graph of $f(x) = \dfrac{1}{x + 3}$

Anything above the x-axis stays the same.

Example 6.32

Given that $f(x) = \dfrac{1}{x^2}$, sketch the graph of $|f(x) - 1|$

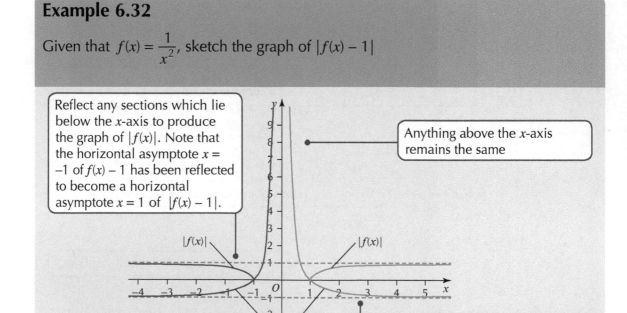

Reflect any sections which lie below the x-axis to produce the graph of $|f(x)|$. Note that the horizontal asymptote $x = -1$ of $f(x) - 1$ has been reflected to become a horizontal asymptote $x = 1$ of $|f(x) - 1|$.

Anything above the x-axis remains the same

First sketch the graph of $f(x) - 1$, which is the graph $y = x^2$ shifted down by 1 unit.

Use this checklist when you are asked to sketch the graph of a function $f(x)$.

- **Intercepts** – determine all points where the curve crosses an axis
- **Stationary points** – find all points where $f'(x) = 0$
- **Points of inflection** – find all points where $f''(x) = 0$ or $f''(x)$ is undefined
- **Asymptotes** – potentially vertical, horizontal or oblique
- **Symmetry** – is the function even, odd or neither
- **Sketch** – make sure to annotate clearly

Exercise 6K

1 Given the following graphs of functions $f(x)$, sketch the graph of $f^{-1}(x)$ in each case. Clearly mark any known points.

a
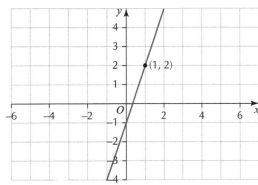

b
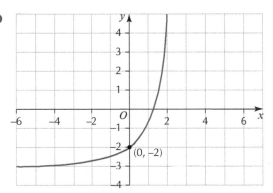

c
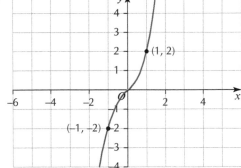

★ 2 For each of the following functions $f(x)$, sketch the graphs of $f(x)$ and the inverse function $f^{-1}(x)$ on the same diagram, clearly marking any intercepts with the axes.

  a $f(x) = 2x - 2$
  b $f(x) = x^3 + 2$
  c $f(x) = (1 - x)^3$
  d $f(x) = e^x - 4$
  e $f(x) = 2 - e^x$
  f $f(x) = \dfrac{1}{x}$

★ 3 For each of the following functions $f(x)$, sketch the graphs of $f(x)$ and the inverse function $f^{-1}(x)$ on the same diagram, clearly marking any points where the graphs meet.

  a $f(x) = \dfrac{x}{2} + 1$
  b $f(x) = x^3 + x$
  c $f(x) = -2x^3$
  d $f(x) = e^x - 1$

4 In each case the graph of a function $f(x)$ is given. Sketch the graph of $|f(x)|$, clearing marking any known points.

a

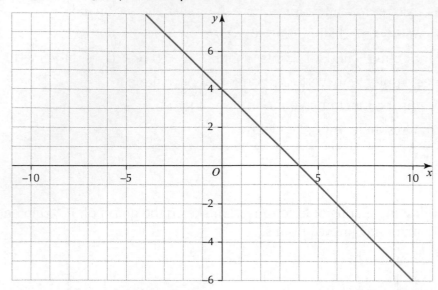

b

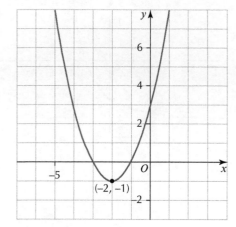

c

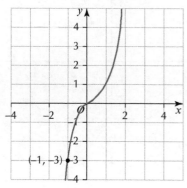

d

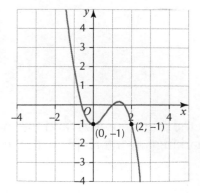

e

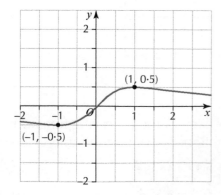

f

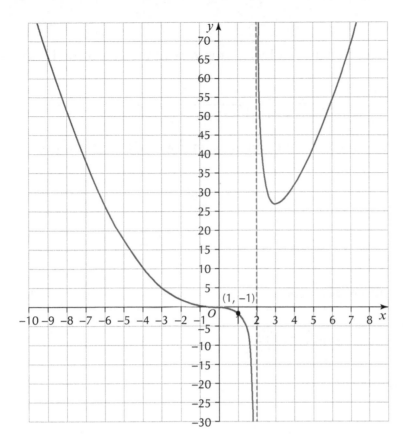

★ 5 In each of the following cases, sketch the graph of $|f(x)|$

a $f(x) = x - 5$

b $f(x) = 4 - 3x$

c $f(x) = 9 - x^2$

d $f(x) = (x + 1)^2 - 4$

e $f(x) = \dfrac{1}{x}$

f $f(x) = -x^3 - x^2$

g $f(x) = e^x - 1$

h $f(x) = \cos x$

6 Let $f(x) = \sin x$. Sketch the graphs of the following functions.

a $|f(x)| + 2$

b $|f(x) + 2|$

c $3|f(x)|$

d $|3f(x)|$

7 Sketch the graphs of

a $|\ln x|$

b $\ln |x|$

8 Sketch the graph of $|\sec x|$.

9 Suppose that $f(x) = \dfrac{1}{x} + 1$

a Sketch the graph of $f(x)$, clearly marking any intercepts and asymptotes.

Sketch the graphs of:

b $f^{-1}(x)$

c $|f(x)|$

d $|f^{-1}(x)|$

e If you were asked to sketch the graph of $|f(x)|^{-1}$, what would the problem be?

Chapter review

1   Determine the equations of any vertical asymptotes of the graphs of the following functions.

   a $f(x) = \dfrac{3}{x^2 - 4}$
   b $h(x) = \dfrac{x^2 - 2x - 3}{x^3 - x^2 - 5x - 3}$
   c $g(x) = \dfrac{2x}{x + 1} + \dfrac{1}{x}$

2   Determine the equations of any horizontal asymptotes of the graphs of the following functions.

   a $f(x) = -\dfrac{3}{x + 3}$
   b $f(x) = -\dfrac{3}{x + 3} - 2$

   c $r(t) = \dfrac{2t^2}{1 - t^2}$
   d $s(q) = \dfrac{(3q^2 - 1)(q + 1)}{(2q - 3)(2q^2 + 1)}$

3   Determine the equations of any oblique asymptotes of the graphs of the following functions.

   a $f(x) = -\dfrac{3x^2}{x + 1}$
   b $d(v) = \dfrac{5 - 2v - v^2}{2 - v}$
   c $p(t) = \dfrac{2t(t + 3)(t + 2)}{3t^2 + 1}$

4   Find the equations of all asymptotes of the graph $y = \dfrac{x^3 + 6x^2 + 11x + 6}{x^2 - x - 2}$

5   Find all stationary points of the graph $y = \dfrac{x^4}{4} + x^3$ and use the second derivative test to determine their nature.

6   Consider the function $y = \dfrac{x^3}{3} - x^2 - 3x$

   a   Find all stationary points of the graph of $f(x)$, and determine their nature.
   b   Find any points of inflection of the graph of $f(x)$.

7   Find the maximum and minimum values of the following functions over the given interval.

   a $f(x) = 3x^3 - \dfrac{15x^2}{2} + 18x$ over the interval $[1, 4]$.

   b $f(x) = -2e^x \cos x$ over the interval $[-3, 2]$.

   c $f(x) = \begin{cases} x, & x < -1 \\ -x, & x \geqslant 1 \end{cases}$ over the interval $[-5, 5]$.

8   Determine whether the following functions are even, odd or neither. In each case provide justification for your answer.
   a $f(x) = 4 - 3x^4$
   b $g(x) = \cos x - \dfrac{1}{x}$
   c $h(x) = 2x^3 \tan x$

9   Identify any points of discontinuity for the following real-valued functions.

   a $f(x) = -\dfrac{2x^2}{x^2 - 4x - 5}$
   b $f(x) = \begin{cases} 2x - 1, & x < 1 \\ 0, & 1 \leqslant x < 2 \\ x - 2, & x \geqslant 2 \end{cases}$
   c $f(x) = \begin{cases} x^3, & x < 0 \\ x^2, & 0 \leqslant x < 2 \\ x, & x \geqslant 2 \end{cases}$

10 Sketch the graphs of the following rational functions, clearly marking any intercepts, stationary points, points of inflection and asymptotes.

a $f(x) = \dfrac{2}{(x-1)^2}$ b $f(x) = (x+2)^3 - 1$ c $f(x) = \dfrac{x^2}{x-3}$

11 Suppose that $f(x) = x^3$.

a Sketch the graph of $f(x)$.

b For each of the following functions, state the effect on the graph of $f(x)$ and then sketch the graph of the function.

   i $-3f(x)$ ii $f(x+4)$ iii $2f(-x) - 6$

12 Suppose that $f(\theta) = \sin\theta$.

a Sketch the graph of $f(x)$.

b For each of the following functions, state the effect on the graph of $f(x)$ and then sketch the graph of the function.

   i $f\left(\theta + \dfrac{\pi}{6}\right)$ ii $-f(2\theta) + 3$ iii $4f\left(-\dfrac{\theta}{2}\right)$

13 For each of the following functions $f(x)$, sketch the graph of the inverse function $f^{-1}(x)$.

a $f(x) = 5x - 3$ b $f(x) = (x-2)^3$ c $f(x) = e^{x-1}$

14 Sketch the following functions.

a $\left|x^3\right| + 1$ b $\left|x^3 + 1\right|$ c $\left|x^2 - 5x + 6\right|$

d $|\tan x|$ e $|\ln(x+1)|$ f $\left|\dfrac{4}{x^2 - 4}\right|$

g $\left|-\dfrac{1}{x^2 - 9} - 2\right|$ ... an interval D if, for all a, b, c in D with $a < c < b$, the point...

- I can find equations of vertical asymptotes. ★ Exercise 6A Q3
- I can find equations of horizontal asymptotes. ★ Exercise 6B Q2
- I can find equations of oblique asymptotes. ★ Exercise 6C Q1
- I can use the second derivative test to determine the nature of stationary points. ★ Exercise 6D Q2
- I can use the second derivative to find points of inflection. ★ Exercise 6E Q1, Q2
- I can find maxima and minima of functions over given intervals. ★ Exercise 6F Q1–Q3
- I can determine if a function is even or odd. ★ Exercise 6G Q3
- I find points of discontinuity for functions. ★ Exercise 6H Q1
- I can sketch graphs of rational functions. ★ Exercise 6I Q1
- I can modify the graph of a function to produce graphs of related functions. ★ Exercise 6J Q3, Q4
- I can sketch graphs of inverse and modulus functions. ★ Exercise 6K Q2, Q3, Q5

7 Sequences and series

This chapter will show you how to:

- find and use the general term of an arithmetic sequence
- find the sum to n terms of an arithmetic series
- find and use the general term of a geometric sequence
- find the sum to n terms of a geometric series and the sum to infinity of a geometric series
- use the Maclaurin expansion to find a power series
- use sigma notation and apply summation formulae.

You should already know:

- that a sequence is an ordered list of numbers that follow a set rule
- that each number in a sequence is called a term and the terms in a sequence can be represented as $u_1, u_2, u_3, \ldots, u_n$ for the 1st, 2nd, 3rd and nth term respectively
- how to use a formula to calculate the nth term of a sequence.

Arithmetic sequences

An **arithmetic sequence** is one in which each term differs from the previous term by a constant amount. Examples of arithmetic sequences are:

- 1, 3, 5, 7, 9, … (consecutive terms differ by 2 each time)
- 29, 24, 19, 14, 9, 4, −1, … (consecutive terms differ by −5 each time).

The constant amount is the **common difference** denoted by d. The 1st term of a sequence is denoted by a and so:

$u_1 = a$

$u_2 = a + d$

$u_3 = a + 2d$

$u_4 = a + 3d$

$\vdots \quad \vdots \quad \vdots$

$u_n = a + (n - 1)d$

> The nth term (general term) of an arithmetic sequence is
> $u_n = a + (n - 1)d$
> where a is the 1st term
> d is the common difference, that is $d = u_2 - u_1 = u_3 - u_2 = u_n - u_{n-1}$

Example 7.1

Find the nth term of the sequence 3, 7, 11, 15, . . .

Hence find the 100th term.

Method 1

$a = 3$ — The 1st term of the sequence is 3.

$d = u_2 - u_1$

$\quad = 7 - 3 = 4$

$d = 4$

$u_n = a + (n - 1)d$ — General term of an arithmetic sequence.

$\quad = 3 + (n - 1) \times 4$ — Substitute values for a and d.

$\quad = 3 + 4n - 4$

$\quad = 4n - 1$

Hence $u_{100} = 4 \times 100 - 1 = 399$

Method 2

Term number (n)	1	2	3	4
Sequence	3	7	11	15

$+4 \quad +4 \quad +4$

$u_n = 4n + \ldots$ — The sequence adds 4 each time, so the formula starts with $u_n = 4n + \ldots$ (since the 4 times table involves a sequence of numbers where you add 4 to the previous number to get next number).

the 4 times table is 4, 8, 12, 16, ...

our sequence is 3, 7, 11, 15, ... — We need to subtract 1 from the 4 times table to get our sequence.

nth term of $u_n = 4n - 1$

Although this method can be used for finding the nth term, remembering that $u_n = a + (n - 1)d$ is more useful as it can be used in more types of examples.

Example 7.2

For a given arithmetic sequence, $u_{10} = 38$ and $d = 3$

Find the 1st term in the sequence and the formula for the nth term.

From the question, $u_{10} = 38$, $d = 3$ and $n = 10$

$u_n = a + (n - 1)d$ — Formula for nth term of an arithmetic sequence.

$38 = a + 9 \times 3$ — Substitute $u_{10} = 38$, $d = 3$ and $n = 10$

(continued)

$a = 11$ •───────────────────── Solve for a.

$\Rightarrow$ 1st term in the sequence is 11

$\Rightarrow$ the formula for the nth term is

$u_n = a + (n - 1)d$

$\quad = 11 + (n - 1) \times 3$

$\quad = 11 + 3n - 3$

$u_n = 8 + 3n$

Example 7.3

The 3rd term in an arithmetic series is 14 and the 5th term is 38.

Find a formula for u_n and write down the first four terms of the sequence.

$u_3 = 14$ and $u_5 = 38$.

$u_n = a + (n - 1)d,$

$\Rightarrow u_3 = a + 2d$ and $u_5 = a + 4d$

$\quad a + 2d = 14 \qquad ①$

and $a + 4d = 38 \qquad ②$

$\qquad 2d = 24$ •───────────────── Subtract ① from ②.

$\qquad d = 12$

$\qquad a = 14 - 2 \times 12$ •────── Substitute $d = 12$ into ① to find a.

$\qquad a = -10$

Hence $u_n = a + (n - 1)d$

$\qquad = -10 + 12(n - 1)$

$\qquad = 12n - 22$

First four terms in the sequence are:

$u_1 = -10;\ u_2 = 2;\ u_3 = 14;\ u_4 = 26.$

Example 7.4

The first three terms of an arithmetic sequence are $3x$, 20 and $7x$.

Find the value of x and hence state the first three numerical values of the sequence.

$d = u_2 - u_1 = u_3 - u_2$ •──────── The sequence is an arithmetic sequence.

$\Rightarrow 20 - 3x = 7x - 20$

$\qquad 10x = 40$

$\qquad x = 4$

First three terms of the arithmetic sequence

are 12, 20 and 28. •───────────── Substitute $x = 4$ into $3x$ and $7x$

Exercise 7A

This exercise should be completed without the use of a calculator.

1 Check that these sequences are arithmetic sequences. If they are, state the values of the first term (a) and the common difference (d).

  a 8, 11, 14, 17, …   b 2, 4, 8, 16, …   c 21, 19, 17, 15, …

  d 1, 1, 2, 3, 5, 8, …   e 5, 4·7, 4·4, 4·1, …   f $\dfrac{7}{6}, 1, \dfrac{5}{6}, \dfrac{2}{3}, …$

  g −3, −7, −11, −15, …   h 1, −1, 1, −1, …

★ 2 Find the nth term for each arithmetic sequence.

  a 5, 8, 11, 14, …   b −9, −5, −1, 3, …   c 10, 8, 6, 4, 2, …

  d 240, 150, 60, −30, …   e $x, 4x, 7x, 10x, …$ f $6x, 2x, −2x, −6x, −10x, …$

★ 3 Find the 1st term in each arithmetic sequence where the common difference d and a term in the sequence are given.

  a $d = 3$ and $u_4 = 31$ b $d = −4$ and $u_5 = −19$

  c $d = −\dfrac{1}{2}$ and $u_8 = 5·5$ d $d = 1·4$ and $u_4 = −1$

4 How many terms are in these arithmetic sequences?

  a 5, 9, 13, 17, …, 121   b 1, 2·25, 3·5, 4·75, …, 44·75

  c 100, 95, 90, …, −900   d $x + 9y, x + 6y, x + 3y, …, x − 57y$

5 For each arithmetic sequence, find the 1st term a and the common difference d.

  a the 5th term is 19 and the 10th term is 39

  b the 3rd term is 6 and the 9th term is −36

  c $u_3 = 20$ and $u_7 = 12$

  d $u_5 = 21$ and $u_{10} = 41$

  e $u_4 = 20$ and $u_9 = 8$

6 The 4th term of an arithmetic sequence is 19·5 and the 16th term is 49·5.
  Find the 51st term.

7 The 4th term of an arithmetic sequence is 27 and the 9th term is 12.
  Find the 27th term.

8 The 5th term of an arithmetic sequence is 25 and the 9th term is 9.

  a Find a and d.

  b Hence find, algebraically, the first term to become negative.

9 The 1st term in an arithmetic sequence is 98 and the difference between two consecutive terms is −3·5.

  a Find a value of the 100th term in the sequence.

  b Find, algebraically, the term in the sequence that has a value of 0.

10 The first three terms of an arithmetic sequence are -3, x and x^2 respectively.

Find the values of x and hence state the possible first three terms of the sequence.

11 A boy saves some money over a period of a year. He saves 1p in the first week, 6p in the second week, 11p in the third week and so on over a whole year.

How much will he save in the last week of the year?

Sum to n terms of an arithmetic series

A series is a sum, where the terms of the sum form a sequence. An arithmetic series is formed by adding together the terms of an arithmetic sequence.

Gauss's solution

When he was 10 years old, the German mathematician Carl Friedrich Gauss was asked by his teacher to find the sum of the first 100 whole numbers, that is, the answer to $1 + 2 + 3 + \ldots + 100$.

Rather than completing the task as a straightforward addition sum, Gauss solved it in this way:

Consider the sum $\qquad S = 1 + 2 + 3 + \ldots + 98 + 99 + 100$

then reverse the sum so that $\quad S = 100 + 99 + 98 + \ldots + 3 + 2 + 1$

Adding the two sums gives

$2S = 101 + 101 + 101 + \ldots + 101 + 101 + 101$ ●————

> Since $1 + 100 = 101$ and $2 + 99 = 101$, ...

$2S = 100 \times 101$

$S = \dfrac{100 \times 101}{2}$

$S = 5050$

The first 100 whole numbers form an arithmetic sequence (with $a = 1$ and $d = 1$). If we use the same ideas that Gauss used for any arithmetic series, we can find a formula for the sum of n terms of an arithmetic series.

Consider the sum:

$$S_n = \qquad a \qquad + \qquad (a + d) \qquad + \ldots + \quad (a + (n - 2)d) \; + \quad (a + (n - 1)d) \; ●$$
$$S_n = (a + (n - 1)d) \quad + \quad (a + (n - 2)d) \quad + \ldots + \qquad (a + d) \qquad + \qquad a$$
$$2S_n = (2a + (n - 1)d) \quad + \quad (2a + (n - 1)d) \quad + \ldots + \quad (2a + (n - 1)d) \; + \; (2a + (n - 1)d)$$
$$2S_n = n(2a + (n - 1)d)$$
$$S_n = \tfrac{1}{2}n\left[2a + (n - 1)d\right]$$

> Remember the nth term of an arithmetic sequence is $a + (n - 1)d$.

The formula for the sum of an arithmetic series is

$$S_n = \frac{1}{2}n\big[2a + (n-1)d\big]$$ ● —————— This is provided in the exam formulae list.

where a is the 1st term of the series

d is the common difference of the series

n is the number of terms in the series

The formula can also be expressed as

$$S_n = \frac{1}{2}n\big[a + L\big]$$

where L is the last term in the series.

Example 7.5

Find the sum of the first twenty terms of the arithmetic series that starts
$3 + 7 + 11 + 15 + \dots$

Method 1

$$S_n = \frac{1}{2}n\big[2a + (n-1)d\big]$$ ● —————— Formula for calculating the sum of n terms of an arithmetic series.

$$\Rightarrow S_{20} = \frac{1}{2}(20)(2 \times 3 + 19 \times 4)$$ ● —————— Since $a = 3$, $d = 4$ and $n = 20$.

$$= 10(6 + 76)$$

$$= 820$$

Method 2 using $S_n = \frac{1}{2}n\big[a + L\big]$

We need to know the value of the last term ($L = u_{20}$) to be able to use this formula. Use the formula $u_n = a + (n-1)d$

$$L = u_{20} = 3 + 19 \times 4 = 79$$ ●

$$\Rightarrow S_n = \frac{1}{2}n\big[a + L\big]$$ ● —————— Alternative formula for calculating the sum of n terms of an arithmetic series.

$$S_{20} = \frac{1}{2}(20)(3 + 79)$$ ● —————— Since $a = 3$, $L = 79$ and $n = 20$.

$$= 10 \times 82$$

$$= 820$$

Example 7.6

Find the sum of this series:

$36 + 31 + 26 + 21 + \ldots + (-109)$

Method 1

$u_n = a + (n-1)d$

To use $S_n = \frac{1}{2}n\left[2a + (n-1)d\right]$, we need to know how many terms are in the series.

$-109 = 36 + (n-1)(-5)$

$-109 = 41 - 5n$

Since $u_n = -109$, $a = 36$ and $d = -5$.

$-150 = -5n$

$n = 30$

$\Rightarrow$ 30 terms are in the series $36 + 31 + 26 + 21 + \ldots + -109$

$S_n = \frac{1}{2}n\left[2a + (n-1)d\right]$

Formula for calculating the sum of n terms of an arithmetic series.

so $S_{30} = \frac{1}{2}(30)(2 \times 36 + 29 \times -5)$

Since $a = 36$, $d = -5$ and $n = 30$.

$= 15(72 - 145)$

$= -1095$

Method 2 using $S_n = \frac{1}{2}n\left[a + L\right]$

$S_n = \frac{1}{2}n\left[a + L\right]$

$S_{30} = \frac{1}{2}(30)(36 - 109)$

Find $n = 30$, using the same method as in Method 1 above.

$= 15 \times -73$

$= -1095$

Example 7.7

How many terms of the arithmetic series $5 + 8 + 11 + \ldots$ are required to make a total of $16\,274$?

$S_n = \frac{1}{2}n\left[2a + (n-1)d\right]$

$16\,274 = \frac{1}{2}(n)(2 \times 5 + 3(n-1))$

$S_n = 16\,274$, $a = 5$ and $d = 3$.

$32\,548 = n(3n + 7)$

$3n^2 + 7n - 32\,548 = 0$

Multiply out brackets and equate to zero.

$(3n + 316)(n - 103) = 0$

The quadratic formula could also be used here if the quadratic is not easily factorised.

$\Rightarrow n = -\dfrac{316}{3}$ or 103

Since $n > 0$, then $n = 103$.

$\therefore$ 103 terms are needed to make the total of $16\,274$

Example 7.8

The 5th term of an arithmetic series is 14.

The sum of the first three terms is –3.

a   Calculate the 1st term a and the common difference d.

b   Given that the nth term of the series is greater than 282, find the least possible value of n.

a $u_5 = 14$ and $S_3 = -3$

$u_5 = a + 4d$ *Use the formula for the nth term of an arithmetic series: $u_n = a + (n-1)d$.*

$S_n = \frac{1}{2}n[2a + (n-1)d]$, then $S_3 = \frac{3}{2}[2a + 2d] = 3a + 3d$

$\quad a + 4d = 14$ ①

$\quad 3a + 3d = -3$ ②

$\quad d = 5$ and $a = -6$ *Solve ① and ② simultaneously to find a and d.*

b $u_n = a + (n-1)d$

$\quad -6 + 5(n-1) > 282$ *Since $d = 5$, $a = -6$ and $u_n > 282$*

$\quad\quad\quad\quad 5n > 293$

$\quad\quad\quad\quad\quad n > 58{\cdot}6$

so the least possible value of n is 59

Exercise 7B

 **1**   Find the required sum for each arithmetic series.

     a $1 + 3 + 5 + 7 + ..., S_{20}$ b $7 + 11 + 15 + 19 + ..., S_{15}$

     c $25 + 22 + 19 + 16 + ..., S_{40}$ d $8 - 1 - 10 - 19 - ..., S_{17}$

     e $7 + 10 + 13 + ... + 91$ f $13{\cdot}5 + 15 + 16{\cdot}5 + ... + 81$

     g $(x + 3) + (2x + 4) + (3x + 5) + ..., S_{20}$

     h $(2x + 1) + (5x - 1) + (8x - 3) + ... + (68x - 43)$

2   Calculate how many terms are required in each series to make the given sum.

     a $5 + 7 + 9 + 11 + ... = 725$

     b $27 + 19 + 11 + 3 + ... = -629$ *The prime factorisation of 629 is $629 = 17 \times 37$.*

     c $48 + 45 + 42 + 39 + ... = 0$

     d $(x - 8) + (3x - 5) + (5x - 2) + ... = 1156x + 1411$

3   Find the 1st term in the arithmetic series with the given common difference and S_n.

     a $d = 2$ and $S_{15} = 360$ b $d = -\frac{1}{2}$ and $S_{23} = -11{\cdot}5$

     c $d = 2{\cdot}5$ and $S_{38} = 2023{\cdot}5$ d $d = -\frac{1}{3}$ and $S_{21} = 35$

4   For each arithmetic series, calculate the 1st term a and the common difference d.

   a   The 2nd term of an arithmetic series is 12.

       The sum of the first ten terms is 295.

   b   The 7th term of an arithmetic series is 10.

       The sum of the first twenty terms is 410.

   c   The 8th term of an arithmetic series is -1.

       The sum of the first 29 terms is -87.

★ 5   The sum of the first six terms of an arithmetic series is 126 and the sum of the first sixty terms is 7740. Calculate S_{54}.

6   An arithmetic series has a 1st term of 5 and a common difference of 3. Calculate n such that S_n is greater than 450.

7   An arithmetic series has a 1st term of 20 and a common difference of -3. Calculate n such that $S_n \leqslant 0$.

8   Find the sum of the first 100 odd numbers.

9   Find the sum of the multiples of 3 between 100 and 200.

10 Find the sum of the first n even numbers.

11 The first three terms of an arithmetic sequence, expressed in base 8 are:

   12_8, 26_8 and 42_8

   Find the sum of the first nine terms of the sequence and express your answer in base 8.

   > ⚠ Writing numbers in different bases is covered in Chapter 10.

12 Given that $S_n = 8n - n^2$, find u_1, u_2 and u_3. Hence find a formula for u_n.

   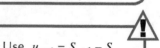

   > ⚠ Use $u_{n+1} = S_{n+1} - S_n$

13 The terms of an arithmetic sequence are given by $u_k = 11 - 2k$, $k \geqslant 1$

   a   Obtain a formula for S_n.

   b   Find the value(s) of n for which $S_n = 21$.

14 A girl saves money over a year (52 weeks). She saves 5p in week 1, 10p in week 2, 15p in week 3 and so on. How much does she save over the whole year?

15 Richard makes annual payments into a savings scheme. In the first year, he pays £400 into the scheme and the amount he pays in increases by £50 each year.

   a   Calculate the amount Richard pays into the scheme in the sixth year.

   b   Find a formula showing the total amount, in pounds, that Richard pays into the scheme during the first n years.

   Julia also makes annual payments into a similar saving scheme. She initially saves £300 with the amount she saves increasing by £70 each year.

   c   After how many years of paying into the saving scheme will Richard and Julia have the same amount of money saved?

   > ⚠ Find a formula for the total amount saved after n years for Julia and then use this and your answer for part **b** to help you answer this.

16 Jeremy started work 25 years ago. In the first year, his annual salary was £15 000. His annual salary increased by £1800 each year. This continued until he reached his maximum annual salary of £36 600 in year n. His annual salary has not increased since.

    a How many years was Jeremy working for before he reached his maximum annual salary?

    b Calculate the total amount that Jeremy earned in his 25 years working.

17 Daniel played a computer game. Every time he destroyed a monster, he scored points. As the game progressed, it was more difficult to destroy the monsters so more points were scored for each further monster destroyed. He scored 50 points for the first monster, 80 points for his second, 110 points for the third and so on. Daniel destroyed 23 monsters. How many points did he score in the game?

18 A primary class are decorating their classroom with triangle patterns made from straws.

   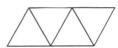

Each pattern contains 2 straws more than the pattern before.

    a The series continues until they make the pattern with 20 triangles. How many straws are required to make all 20 patterns?

    b The teacher has 1000 straws altogether. How many triangles are in the pattern she could make? (Remember she has to make the pattern with one triangle, then the pattern with two triangles, then the pattern with three triangles etc.)

Geometric sequences

These sequences are examples of **geometric sequences**:

- 1, 2, 4, 8, 16, … (previous term is doubled to get the next term)

- 400, 100, 25, 6·25, 1·5625, … (previous term is multiplied by $\frac{1}{4}$ to get the next term)

- 2, −10, 50, −250, 1250, … (previous term is multiplied by −5 to get the next term)

To obtain a term in a geometric sequence, multiply the previous term by the same amount each time.

Consider the sequence

$a, ar, ar^2, ar^3, …, ar^{n-2}, ar^{n-1}, …$

If we look at each term of the sequence, we have

$u_1 = a$

$u_2 = u_1 r \quad = ar$

$u_3 = u_2 r \quad = ar^2$

$u_4 = u_3 r \quad = ar^3$

$\vdots \qquad \vdots \qquad \vdots$

$u_n = u_{n-1} r = ar^{n-1}$

The sequence is a geometric sequence, because each term is multiplied by r to obtain the next term. The nth term of the geometric sequence is ar^{n-1}.

In general, we say that the nth term of a geometric sequence is

$u_n = ar^{n-1}$

where a is the 1st term in the sequence and r is the common ratio.

The nth term of a geometric sequence is

$u_n = ar^{n-1}$ for all $n \in \mathbb{N}$

where: a is the 1st term

r is the common ratio, that is $r = \dfrac{u_2}{u_1} = \dfrac{u_3}{u_2} = \ldots = \dfrac{u_n}{u_{n-1}}$

Example 7.9

Find an expression for the nth term and evaluate the 15th term of the geometric sequence 2, 6, 18, 54, …

$a = 2$ •———————————————————— The 1st term of the sequence is 2.

$r = \dfrac{u_2}{u_1} = \dfrac{6}{2} = 3$ •———— Always check that both $\dfrac{u_2}{u_1}$ and $\dfrac{u_3}{u_2}$ give the same value.

$r = \dfrac{u_3}{u_2} = \dfrac{18}{6} = 3$

$r = 3$

$u_n = ar^{n-1}$ •——————————— General term of a geometric sequence is ar^{n-1}.

$u_n = 2 \times 3^{n-1}$

$\therefore u_{15} = 2 \times 3^{14} = 9\,565\,938$

Example 7.10

The 3rd term of a geometric sequence is $4\sqrt{2}$ and the 6th term is 16. Find an expression for the nth term.

$u_3 = 4\sqrt{2}$ and $u_6 = 16$

$u_n = ar^{n-1}$

so $u_3 = ar^2$ and $u_6 = ar^5$

$ar^2 = 4\sqrt{2}$ ①

$ar^5 = 16$ ②

$\dfrac{ar^5}{ar^2} = \dfrac{16}{4\sqrt{2}}$ •———————————————— Divide ② by ①.

so $r^3 = 2\sqrt{2}$

$r = \sqrt[3]{2\sqrt{2}}$

$\quad = \left(2^{\frac{3}{2}}\right)^{\frac{1}{3}}$

$\quad = 2^{\frac{1}{2}}$

$\quad = \sqrt{2}$

$a = 2\sqrt{2}$ •————————— | Substitute $r = \sqrt{2}$ into equation 1 to find a. |

$u_n = ar^{n-1}$

$\quad = 2\sqrt{2} \times \left(\sqrt{2}\right)^{n-1}$

$\quad = \left(\sqrt{2}\right)^{n+2}$ •————————— | In this case the formula can be simplified further. |

Example 7.11

What is the 1st term in the geometric sequence 2, 4, 8, 16, 32, … to exceed half a million?

$a = 2$ •————————————————————— | The 1st term of the sequence is 2. |

$r = \dfrac{u_2}{u_1} = \dfrac{4}{2} = 2$

$r = \dfrac{u_3}{u_2} = \dfrac{8}{4} = 2$ •————— | Always check that both $\dfrac{u_2}{u_1}$ and $\dfrac{u_3}{u_2}$ give the same value. |

$r = 2$

$u_n = ar^{n-1}$ •————————— | General term of a geometric sequence is ar^{n-1} |

$u_n = 2 \times 2^{n-1}$

We want $u_n > 500\,000$

$\therefore 2 \times 2^{n-1} > 500\,000$

$2^{n-1} > 250\,000$ •————————— | Divide both sides by 2. |

$(n-1)\log_e 2 > \log_e 250\,000$ •————— | Taking logs of both sides and dealing with the power of $(n-1)$. |

$n - 1 > \dfrac{\log_e 250\,000}{\log_e 2}$

$\quad n - 1 > 17{\cdot}9315\ldots$

$\quad\quad n > 18{\cdot}9315\ldots$

Hence the first term to exceed half a million is the 19th term.

> ⚠ It is good practice to check your working here. Calculate the 18th and 19th term – is the 19th term the first one to exceed half a million?
>
> Using the calculator, $u_{18} = 262\,144$ and $u_{19} = 524\,288$ so the 19th term is the first one to exceed half a million.

Exercise 7C

1   Check that each sequence is a geometric sequence. For each geometric sequence, state the value of the 1st term (a) and the common ratio (r).

a   1, 2, 4, 8, …

b   4, 7, 10, 13, …

c   256, 128, 64, 32, …

d   1, −3, 9, −27, …

e $\dfrac{1}{3}, \dfrac{1}{6}, \dfrac{1}{12}, \dfrac{1}{24}, \ldots$

f   125, −50, 20, −8, …

g $\dfrac{1}{2}, \dfrac{3}{4}, \dfrac{5}{8}, \dfrac{7}{16}, \ldots$

2   For each geometric sequence, find the value of the term specified in the square brackets.

a   1, 3, 9, 27, … [8th term]

b $1, \dfrac{1}{2}, \dfrac{1}{4}, \dfrac{1}{8}, \ldots$ [10th term]

c   8, −4, 2, −1, … [15th term]

d   45, 15, 5, $\dfrac{5}{3}$, … [10th term]

e   3, 15, 75, … [7th term]

f $\dfrac{3}{2}$, 6, 24, … [8th term]

3   Find the 1st term in each geometric sequence where the common ratio and a term in the sequence is given.

a $r = 6$ and the 4th term is 432

b $r = \dfrac{3}{2}$ and the 5th term is 4050

c $r = -\dfrac{1}{2}$ and $u_{10} = -1$

d $r = \dfrac{3}{4}$ and $u_5 = \dfrac{81}{512}$

4   How many terms are in each geometric sequence?

a   3, 6, 12, 24, …,1536

b   100, 50, 25, 12.5, …, $\dfrac{25}{2048}$

c   4, −12, 36, −108, … −78 732

d   600, −300, 150, −75, …, $\dfrac{75}{32}$

★ 5   For each geometric sequence, find the 1st term a and the common ratio r where:

a   the 3rd term is 18 and the 6th term is 486

b   the 2nd term is −125 and the 5th term is 1

c $u_3 = 9$ and $u_6 = 30{\cdot}375$

d $u_3 = -32$ and $u_5 = -128$

6   The numbers x^2, $4x$ and $x - 4$ form the first three terms of a geometric sequence with $r > 0$. Find x.

7   The numbers $x - 1$, 6 and $2x^2$ form the first three terms of a geometric sequence. Find x.

★ 8   Find the 1st term in the geometric sequence 3, 6, 12, 24, … to exceed 3 million.

9 What is the 1st term in the geometric sequence 2000, 1000, 500, 125, ... that is smaller than 1?

Remember to switch the inequality sign when dividing by a negative number.

10 I invest £100 at a rate of 4% per annum. How long will it take to double my investment?

The amount in the bank account at the end of each year forms a geometric sequence.

11 If I invest £A at a rate of 6% per annum, how long will it take me to triple my money?

12 Bradley is training for the Tour de France. On day 1 of training, he cycles 20 miles. He cycles every other day on his training programme. He increases the distance he cycles each time by 15%. On the Tour de France they cycle about 100 miles a day. After how many days training will Bradley cycle more than 100 miles in a day?

Sum to n terms of a geometric series

A **geometric series** is formed by adding together the terms of a geometric sequence.

The terms of a geometric sequence are $a, ar, ar^2, ar^3, \ldots, ar^{n-1}$

Let S_n be the geometric series such that

$$S_n = a + ar + ar^2 + \ldots + ar^{n-2} + ar^{n-1}$$
$$\Rightarrow rS_n = ar + ar^2 + ar^3 + \ldots + ar^{n-1} + ar^n$$

Multiply each term by r.

so

$$S_n - rS_n = (a + ar + ar^2 + \ldots + ar^{n-2} + ar^{n-1}) - (ar + ar^2 + ar^3 + \ldots + ar^{n-1} + ar^n)$$
$$S_n - rS_n = a - ar^n$$
$$S_n(1 - r) = a(1 - r^n)$$
$$S_n = \frac{a(1 - r^n)}{1 - r}$$

The formula for the sum of a geometric series is

This is provided in the exam formulae list.

$$S_n = \frac{a(1 - r^n)}{1 - r} = \frac{a(r^n - 1)}{r - 1}, r \neq 1$$

where a is the 1st term of the series

r is the common ratio of the series

n is the number of terms in the series.

If r was 1, then the series would be easy to sum without the need for a formula as we would have $u_1 = a, u_2 = a, u_3 = a, \ldots, u_n = a \therefore S_n = na$

When $r < 1$, it is often easier to use the $S_n = \dfrac{a(1 - r^n)}{1 - r}$ form of the formula for the sum of a geometric series

When $r > 1$, it is often easier to use the $S_n = \dfrac{a(r^n - 1)}{r - 1}$ form of the formula for the sum of a geometric series

Example 7.12

Find the sum of the first 10 terms of the geometric series $3 + 6 + 12 + ...$

$$S_n = \frac{a(r^n - 1)}{r - 1}$$ — Formula for calculating the sum of n terms of a geometric series.

$a = 3$

$n = 10$

$r = \frac{u_2}{u_1} = \frac{6}{3} = 2$ — Check $r = \frac{u_3}{u_2} = \frac{12}{6} = 2$

so $S_{10} = \frac{3(2^{10} - 1)}{2 - 1}$

$= 3069$ — Calculate

Example 7.13

Find the sum of the geometric series $1024 - 512 + 256 - 128 + ... + 1$

Method 1

$a = 1024$ — We need to calculate how many terms are in the geometric series, i.e. we need to find n.

$r = \frac{-512}{1024} = \frac{256}{-512} = -\frac{1}{2}$

$u_n = 1$ — State what is known.

$u_n = ar^{n-1}$

$1024 \times \left(-\frac{1}{2}\right)^{n-1} = 1$ — Substitute values of a, r and u_n into general term of geometric series.

$\left(-\frac{1}{2}\right)^{n-1} = \frac{1}{1024}$

$(-2)^{n-1} = 1024$ — Use index laws.

$(-1)^{n-1} \times (2)^{n-1} = 1024$ — 2^{n-1} is always positive and 1024 (on the RHS) is positive, so $(-1)^{n-1}$ must be positive. A negative number to a given power can only be positive if the power is even, hence n must be odd.

$2^{n-1} = 1024$

$(n - 1)\log_e 2 = \log_e 1024$

$n - 1 = \frac{\log_e 1024}{\log_e 2}$ — Take logs of both sides and use logarithmic laws to bring the power of 2 in front of $\log_e 2$.

$n - 1 = 10$

$n = 11$ — Calculate n

Hence

$$S_n = \frac{a\left(1 - r^n\right)}{1 - r}$$

$$= \frac{1024\left(1 - \left(-\frac{1}{2}\right)^{11}\right)}{1 - \left(-\frac{1}{2}\right)}$$

$$= \frac{1024\left(1 + \frac{1}{2048}\right)}{1 + \frac{1}{2}}$$

$$= 683$$

Method 2

$a = 1024$, $r = \frac{-512}{1024} = \frac{256}{-512} = -\frac{1}{2}$ and $u_n = 1$

$$S_n = \frac{a\left(1 - r^n\right)}{1 - r}$$ — Formula for sum of n terms of a geometric series.

$$= \frac{a - ar^n}{1 - r}$$ — Multiply out the bracket on the denominator.

$$= \frac{a - ar^{n-1} \times r}{1 - r}$$ — Use index laws to rewrite ar^n as $ar^{n-1} \times r$

$$= \frac{a - u_n \times r}{1 - r}$$ — The nth term of a geometric series is $u_n = ar^{n-1}$, so we can replace ar^{n-1} with u_n

$$S_n = \frac{a - u_n \times r}{1 - r}$$

$$= \frac{1024 - \left(1 \times \left(-\frac{1}{2}\right)\right)}{1 - \left(-\frac{1}{2}\right)}$$

$$= 683$$

Example 7.14

Find the smallest value of n such that the sum of $1 + 2 + 4 + 8 + \ldots$ to n terms would exceed 1 million.

$a = 1$

$r = \frac{u_2}{u_1} = \frac{2}{1} = 2$

$$S_n = \frac{a\left(1 - r^n\right)}{1 - r} = \frac{1\left(1 - 2^n\right)}{1 - 2} = \frac{1 - 2^n}{-1} = 2^n - 1$$ — Use the formula for calculating the sum of n terms of a geometric series and substitute in values of a and r.

(continued)

We want $S_n > 1\,000\,000$

$$2^n - 1 > 1\,000\,000$$
$$2^n > 1\,000\,001$$
$$n \log_e 2 > \log_e 1\,000\,001$$
$$n > 19.9315\ldots$$

> Take logs of both sides and use logarithmic laws to bring the power of 2 in front of $\log_e 2$.

Hence 20 terms are required for the sum to exceed 1 million.

> Check your working by calculating the sums of the first 19 and 20 terms and seeing if the sum of the first 20 terms is the first one to exceed one million. Using the calculator, $S_{19} = 524\,287$ and $S_{20} = 1\,048\,575$.

Example 7.15

The sum of the first three terms of a geometric series is 23·75. If the first term is 5, find the possible values of r.

$a = 5$ and $S_3 = 23.75$

$$S_n = \frac{a\left(1 - r^n\right)}{1 - r}$$
$$23.75 = \frac{5\left(1 - r^3\right)}{1 - r}$$
$$23.75(1 - r) = 5(1 - r^3)$$
$$95 - 95r = 20 - 20r^3$$

> Multiply both sides by 4 (to eliminate decimals) and multiply out brackets.

$$20r^3 - 95r + 75 = 0$$
$$4r^3 - 19r + 15 = 0$$

> Factorise the cubic equation.

$$(r - 1)(2r + 5)(2r - 3) = 0$$

so $r = 1, -\dfrac{5}{2}, \dfrac{3}{2}$

> Solve for r. We know from earlier that $r \neq 1$ because if $r = 1$, the first three terms are all 5 and $5 + 5 + 5 = 15 \neq 23.75$

Possible values of r are $-\dfrac{5}{2}$ and $\dfrac{3}{2}$

> It is good practice to check S_3 for each r value i.e. when $r = -\dfrac{5}{2}$, $S_3 = 23.75$ and when $r = \dfrac{3}{2}$, $S_3 = 23.75$.

Exercise 7D

★ 1   Find the sum for each geometric series.

a $1 + 3 + 9 + 27 + \ldots$ (10 terms)

b $1024 + 512 + 256 + 128 + \ldots$ (15 terms)

c $1 - 4 + 16 - 64 + \ldots$ (8 terms)

d $3125 - 625 + 125 - 25 + \ldots$ (12 terms)

e $1 + 3 + 9 + 27 + \ldots + 1594323$

f $1024 + 512 + 256 + \ldots + \dfrac{1}{8}$

g $1 - 4 + 16 - 64 + \ldots - 262144$

h $3125 - 625 + 125 + \ldots - \dfrac{1}{15625}$

2   How many terms are required to give the total stated in each series?

   a $5 + 10 + 20 + \ldots = 5115$

   b $2 + 6 + 18 + \ldots = 6560$

   c $1000 + 500 + 250 + \ldots = \dfrac{31875}{16}$

   d $\dfrac{1}{4} - \dfrac{1}{2} + 1 - 2 + \ldots = -\dfrac{1365}{4}$

★ 3   Calculate the 1st term in each geometric series.

   a $r = 2$ and $S_{15} = 98301$

   b $r = \dfrac{1}{2}$ and $S_{10} = \dfrac{25575}{64}$

   c $r = \dfrac{3}{2}$ and $S_6 = \dfrac{665}{32}$

   d $r = 3$ and $S_9 = 19682$

4   The sum of the first three terms of a geometric series is 76. The 1st term is 16. Find the possible values of r.

5   The 3rd term of a geometric series is 4·5 and the 5th term is 40·5, $r > 1$. Find the sum of the first 10 terms of the series.

6   Find the smallest value of n such that the sum of $2 + 4 + 8 + 16 + \ldots$ to n terms would exceed 1 million.

7   Find the smallest value of n such that the sum of $3 + 6 + 12 + 24 + \ldots$ to n terms would exceed 2 million.

8   A line 315 cm long is divided into six parts, such that the lengths of the parts form a geometric series. Given the length of the longest part is 32 times longer than the shortest part, find the length of the shortest part of the line.

9   The inventor of the game of chess was asked to name his reward for his invention. He asked for 1 grain of wheat to be placed on the first square of the chessboard, 2 grains of wheat on the second, 4 grains of wheat on the third and so on until all 64 squares were covered. His reward was all the grains of wheat on the chessboard. How many grains of wheat did he claim as his reward?

10 A ball is dropped from a height of 10 metres. It bounces to a height of 6 metres and continues to bounce. The heights that it bounces to each time follows a geometric sequence. Find the total distance travelled after it hits the ground for the fifth time.

Sum to infinity of a geometric series

Consider the series $S = 2 + 1 + 0.5 + 0.25 + \dots$

If you keep adding on more terms, the geometric series will never exceed a certain number. This is called the **limit** of the geometric series or the **sum to infinity** of the geometric series.

We can calculate the limit using the formula for the sum of n terms of a geometric series and the fact that $a = 2$ and $r = \frac{1}{2}$:

$$S_n = \frac{a\left(1 - r^n\right)}{1 - r}$$

$$= \frac{2\left(1 - \left(\frac{1}{2}\right)^n\right)}{1 - \frac{1}{2}}$$

$$= 4\left(1 - \left(\frac{1}{2}\right)^n\right)$$

and if we replace n with certain values to find the sum, we find that

when $n = 3$, $S_3 = 3.5$

$n = 5$, $S_5 = 3.875$

$n = 10$, $S_{10} = 3.99609\dots$

$n = 20$, $S_{20} = 3.999996\dots$

so when n gets larger, S_n becomes closer and closer to 4.

We say that the infinite series is **convergent** and has a sum to infinity of 4. (Convergent means that the series tends towards a specific value as more terms are added.)

The series $2 + 1 + 0.5 + 0.25 + \dots$ converges because $-1 < r < 1$

Not all series converge. Some series will diverge, such as $1 + 2 + 4 + 8 + 16 + \dots$

The sum to infinity of a geometric series exists only if $|r| < 1$, i.e. $-1 < r < 1$.

The sum to n terms of a geometric series is $S_n = \dfrac{a\left(1 - r^n\right)}{1 - r}$

If $-1 < r < 1$, $r^n \to 0$ as $n \to \infty$, so $S_\infty = \dfrac{a(1 - 0)}{1 - r}$

The sum to infinity of a geometric series is $S_\infty = \dfrac{a}{1 - r}$ if $-1 < r < 1$ i.e. $|r| < 1$

In the above example, the sum to infinity is:

$$S_\infty = \frac{2}{1 - \frac{1}{2}} = 4$$

The formula to find the sum to infinity of a geometric series, $S_\infty = \frac{a}{1 - r}$, will find the limit straight away, **if it exists**, without the need to find successive terms and add them. The limit in the example above was fairly obvious after a few summations, but not all series converge so quickly.

Example 7.16

In a geometric series, the sum of the first four terms is $\frac{650}{27}$ and the sum to infinity of the series is 30.

Find the possible values of r and, given the terms are all positive, find the 1st term a in the series.

$$S_n = \frac{a\left(1 - r^n\right)}{1 - r}$$

$$S_4 = \frac{650}{27}$$

$$\frac{a\left(1 - r^4\right)}{1 - r} = \frac{650}{27} \quad \text{①}$$

$$S_\infty = \frac{a}{1 - r}$$

$$S_\infty = 30$$

$$\frac{a}{1 - r} = 30 \quad \text{②}$$

$$30\left(1 - r^4\right) = \frac{650}{27}$$

Replace $\frac{a}{1 - r}$ in ① with 30 ②.

$$\left(1 - r^4\right) = \frac{65}{81}$$

$$r^4 = \frac{16}{81}$$

$$r = \pm\frac{2}{3}$$

The two possible values of r are $\pm\frac{2}{3}$

All the terms are positive in the geometric series so $r = \frac{2}{3}$, hence

$$\frac{a}{1 - \frac{2}{3}} = 30$$

$$a = 10$$

Example 7.17

Find the sum to infinity of each series, if it exists.

a $\quad 3 + 9 + 27 + 81 + 243 + \dots$ b $\quad 64 + 16 + 4 + 1 + \dfrac{1}{4} + \dots$

a $\quad a = 3, \quad r = \dfrac{u_2}{u_1} = \dfrac{9}{3} = 3$

$\quad r = 3$

$\quad S_\infty$ does not exist. •———— For the sum to infinity of a geometric series to exist, $-1 < r < 1$.

b $\quad a = 64, \; r = \dfrac{u_2}{u_1} = \dfrac{1}{4}$

$\quad -1 < \dfrac{1}{4} < 1$ so S_∞ exists •———— A statement about the S_∞ existing because of the value of r must be made.

$\quad S_\infty = \dfrac{a}{1 - r} = \dfrac{64}{1 - \dfrac{1}{4}} = \dfrac{256}{3}$

Example 7.18

Find the fraction equal to the recurring decimal fraction $0 \cdot 232\,323\,2323\dots$

$0 \cdot 232\,323\,2323 = 0 \cdot 23 + 0 \cdot 0023 + 0 \cdot 000\,023 + \dots$ •———— Rewrite $0 \cdot 232\,323\,2323\dots$ as a sum of terms to create an infinite geometric series. Use this to find a and r.

$a = 0 \cdot 23$

$r = 0 \cdot 01$

since $-1 < 0 \cdot 01 < 1$, then S_∞ exists

$S_\infty = \dfrac{a}{1 - r}$ •———— Formula for the sum to infinity of a geometric series.

$\quad = \dfrac{0 \cdot 23}{1 - 0 \cdot 01}$ •———— Using $a = 0 \cdot 23$ and $r = 0 \cdot 01$.

$\quad = \dfrac{0 \cdot 23}{0 \cdot 99}$

$\quad = \dfrac{23}{99}$

Example 7.19

Determine the values of x for which the geometric series $1 + 2x + 4x^2 + 8x^3 + \dots$ converges.

$-1 < 2x < 1$ if and only if $-0.5 < x < 0.5$ •———— For a geometric series to converge we need $|r| < 1$.

Hence the series will converge if only if $-0.5 < x < 0.5$.

Exercise 7E

1 Find the sum to infinity, if it exists, of each series.

    a $3 + 6 + 12 + 24 + \dots$ **b** $8 + 4 + 2 + 1 + \dots$

    c $243 - 81 + 27 - 9 + \dots$ **d** $\dfrac{1}{2} + 2 + 8 + 32 + \dots$

    e $20 + 12 + 7{\cdot}2 + \dots$ **f** $1 + \dfrac{11}{10} + \dfrac{121}{100} + \dots$

★ 2 Find the sum to infinity of each geometric series with:

    a   a 2nd term of 40 and a 5th term of 5

    b   a 2nd term of 8·1 and a 4th term of 6·561 (where $r > 0$)

    c   a 3rd term of 2·5 and a 6th term of 0·3125

    d   a 2nd term of 750 and a 5th term of –6

★ 3   **a**   The sum to infinity of a geometric series is $\dfrac{160}{3}$ and the common ratio is $\dfrac{1}{4}$. Find the 1st term of the series.

    b   The sum to infinity of a geometric series is 54 and the 1st term of the series is 36. Find the common ratio.

    c   The sum to infinity of a geometric series is 480 and the 1st term is 120. Find the common ratio and calculate S_6.

    d   The sum to infinity of a geometric series is 4 and the 1st term is 2. Find the common ratio and u_6.

    e   The sum to infinity of a geometric series is 25 and the common ratio is $\dfrac{2}{5}$. Find u_4.

4 The sum of the first three terms of a geometric series is 3·5 and the sum to infinity is 4. Find a and r.

5 The sum of the first two terms of a geometric series is 15 and the sum to infinity is 27. Find two possible geometric series.

★ 6 Express these recurring decimal fractions as a geometric series and as fractions in their simplest form.

    a   0·454 545...   **b**   0·272 727...   **c**   0·123 123 123...

    d   0·019 191 919...   **e**   0·419 1919...   **f**   1·738 3838...

7 Determine the values of x for which the geometric series

$3x + 9x^2 + 27x^3 + 81x^4 + \dots$ converges.

For Q6e, write 0·419 1919... as 0·4 + 0·019 1919...

Zeno's paradoxes

Zeno of Elea (490-430 BC) was a Greek philosopher who posed a series of problems which came to be known as Zeno's Paradoxes. One of them is the **Dichotomy argument** or the **Dichotomy paradox**. The Dichotomy paradox is framed as follows:

- Suppose an object or a person has to travel a given distance, d.

- Before they travel that distance, they must travel half that distance, i.e. $\dfrac{d}{2}$

- Then in order to travel $\dfrac{d}{2}$, first they must travel half that distance, i.e. $\dfrac{d}{4}$

and so on.

Given that this can go on forever, the distance d can never be travelled. For example, if Usain Bolt has to run 100 metres, first he must travel 50 metres; and in order to travel 50 metres, he must first travel 25 metres etc. Usain Bolt does indeed run 100 metres and so clearly the distance can be travelled with the resolution coming from

the fact that the infinite geometric series $\displaystyle\sum_{r=1}^{\infty}\left(\frac{1}{2}\right)^r$ converges to 1. Another of Zeno's

Paradox, titled Achilles and the Tortoise, involves a similar resolution.

Expanding $(1 - x)^{-1}$ and related functions

For all $n \in \mathbb{N}$, the binomial theorem states that:

$$(x + y)^n = \binom{n}{0}x^n + \binom{n}{1}x^{n-1}y^1 + \binom{n}{2}x^{n-2}y^2 + \cdots + \binom{n}{r}x^{n-r}y^r + \cdots \binom{n}{n}y^n$$

> ⚠️ See Chapter 1 for more about the binomial theorem.

This can also be written as

$$(x + y)^n = x^n + \frac{n}{1!}x^{n-1}y + \frac{n(n-1)}{2!}x^{n-2}y^2 + \frac{n(n-1)(n-2)}{3!}x^{n-3}y^3 + \ldots + y^n$$

Written in this way, the result can be extended to include **all** values of n which are rational, such as $n = \dfrac{2}{3}$, $n = -2$ or $n = -\dfrac{2}{5}$.

For a positive integer value of n, the series has at most $(n + 1)$ terms. However, if n is **not** a positive integer, then the series is infinite and is only valid for $|x| < y$.

So for $|x| < y$

$$(x + y)^n = x^n + \frac{n}{1!}x^{n-1}y + \frac{n(n-1)}{2!}x^{n-2}y^2$$
$$+ \frac{n(n-1)(n-2)}{3!}x^{n-3}y^3$$
$$+ \ldots + \frac{n(n-1)\ldots(n-r+1)}{r!}x^{n-r}y^r + \ldots$$

Consider $\dfrac{1}{1-r}$

then $\dfrac{1}{1-r} = \dfrac{1}{1+(-r)} = \left(1+(-r)\right)^{-1}$

Using the expansion above with $x = 1$, $y = -r$ and $n = -1$ gives:

$\left(1+(-r)\right)^{-1} = 1 + \dfrac{(-1)}{1!}(-r) + \dfrac{(-1)(-2)}{2!}(-r)^2 + \dfrac{(-1)(-2)(-3)}{3!}(-r)^3 + \ldots$

$\dfrac{1}{1-r} = 1 + r + r^2 + r^3 + \ldots$

With $|r| < 1$, this can be interpreted as the sum to infinity of a geometric series with a 1st term of 1 and a common ratio of r.

Example 7.20

Expand $\dfrac{1}{1-3x}$, $|x| < \dfrac{1}{3}$, in ascending powers of x giving the first four terms.

$\dfrac{1}{1-3x} = 1 + (3x)^1 + (3x)^2 + (3x)^3 + \ldots$

$\qquad = 1 + 3x + 9x^2 + 27x^3 + \ldots$

Use $\dfrac{1}{1-r} = 1 + r + r^2 + r^3 + \ldots$ with $r = 3x$.

Example 7.21

Expand $\dfrac{1}{1+2x}$, $|x| < \dfrac{1}{2}$, in ascending powers of x giving the first four terms.

$\dfrac{1}{1+2x} = \dfrac{1}{1-(-2x)}$

$\qquad = 1 + (-2x)^1 + (-2x)^2 + (-2x)^3 + \ldots$

$\qquad = 1 - 2x + 4x^2 - 8x^3 + \ldots$

Use $\dfrac{1}{1-r} = 1 + r + r^2 + r^3 + \ldots$ with $r = -2x$.

Example 7.22

Expand $\dfrac{1}{0.9}$ to four decimal places.

$$\dfrac{1}{0.9} = \dfrac{1}{1 - 0.1}$$
$$= 1 + (0.1)^1 + (0.1)^2 + (0.1)^3 + (0.1)^4 \dots$$
$$= 1 + 0.1 + 0.01 + 0.001 + 0.0001 + \dots$$
$$= 1.1111 (4\text{dp})$$

Use $\dfrac{1}{1-r} = 1 + r + r^2 + r^3 + \dots$ with $r = 0.1$.

Example 7.23

Expand $\dfrac{1}{2 + 3x}$, $|x| < \dfrac{2}{3}$, in ascending powers of x giving the first four terms.

$$\dfrac{1}{2 + 3x} = \dfrac{1}{2\left(1 + \dfrac{3x}{2}\right)}$$
$$= \dfrac{1}{2}\left(\dfrac{1}{\left(1 - \left(-\dfrac{3x}{2}\right)\right)}\right)$$
$$= \dfrac{1}{2}\left(1 + \left(-\dfrac{3x}{2}\right)^1 + \left(-\dfrac{3x}{2}\right)^2 + \left(-\dfrac{3x}{2}\right)^3 + \dots\right)$$
$$= \dfrac{1}{2}\left(1 - \dfrac{3x}{2} + \dfrac{9x^2}{4} - \dfrac{27x^3}{8} + \dots\right)$$
$$= \dfrac{1}{2} - \dfrac{3x}{4} + \dfrac{9x^2}{8} - \dfrac{27x^3}{16} + \dots$$

Put the expression into required form to use $\dfrac{1}{1-r} = 1 + r + r^2 + r^3 + \dots$

Use $\dfrac{1}{1-r} = 1 + r + r^2 + r^3 + \dots$ with $r = -\dfrac{3x}{2}$

Example 7.24

Find the expansion of $-\dfrac{2}{(x + 5)(x + 3)}$ up to and including the term in x^3.

$$-\dfrac{2}{(x + 5)(x + 3)} = \dfrac{-2}{(x + 5)(x + 3)} = \dfrac{A}{x + 5} + \dfrac{B}{x + 3}$$
$$= \dfrac{A(x + 3) + B(x + 5)}{(x + 5)(x + 3)}$$

First express the function as a sum of partial fractions. See pages 12–21 for a reminder of the process for finding partial fractions.

$A(x + 3) + B(x + 5) = -2$
Let $x = -3$
$2B = -2$
$B = -1$

Let $x = -5$

$-2A = -2$

$A = 1$

So $-\dfrac{2}{(x + 5)(x + 3)} = \dfrac{1}{x + 5} - \dfrac{1}{x + 3}$

> To expand $\dfrac{1}{x + 5} - \dfrac{1}{x + 3}$ find the expansion for each fraction separately.

$\dfrac{1}{x + 5} = \dfrac{1}{5\left(1 - \left(-\dfrac{x}{5}\right)\right)}$

> Note that this expansion is valid for $|x| < 5$. Express in the form $\dfrac{1}{1 - r}$

$= \dfrac{1}{5}\left(1 + \left(-\dfrac{x}{5}\right) + \left(-\dfrac{x}{5}\right)^2 + \left(-\dfrac{x}{5}\right)^3 + \ldots\right)$

$= \dfrac{1}{5} - \dfrac{x}{25} + \dfrac{x^2}{125} - \dfrac{x^3}{625} \ldots$

$\dfrac{1}{x + 3} = \dfrac{1}{3\left(1 - \left(-\dfrac{x}{3}\right)\right)}$

> Note that this expansion is valid for $|x| < 3$.

$= \dfrac{1}{3}\left(1 + \left(-\dfrac{x}{3}\right) + \left(-\dfrac{x}{3}\right)^2 + \left(-\dfrac{x}{3}\right)^3\right) \ldots$

$= \dfrac{1}{3} - \dfrac{x}{9} + \dfrac{x^2}{27} - \dfrac{x^3}{81} \ldots$

> The expansion for $\dfrac{1}{x + 5} - \dfrac{1}{x + 3}$ is found by summing the expansions together.

$\dfrac{1}{x + 5} - \dfrac{1}{x + 3} = \left(\dfrac{1}{5} - \dfrac{x}{25} + \dfrac{x^2}{125} - \dfrac{x^3}{625} + \ldots\right) - \left(\dfrac{1}{3} - \dfrac{x}{9} + \dfrac{x^2}{27} - \dfrac{x^3}{81}\right)$

$= -\dfrac{2}{15} + \dfrac{16x}{225} - \dfrac{98x^2}{3375} + \dfrac{544x^3}{50625} \ldots$

> This expansion is valid where the values for each separate expansion overlap, i.e. $|x| < 3$.

Exercise 7F

1 Expand each fraction in ascending powers of x, stating the first four terms of the expansion.

a $\dfrac{1}{1 - 2x}$, $|x| < \dfrac{1}{2}$

b $\dfrac{1}{1 + 4x}$, $|x| < \dfrac{1}{4}$

c $\dfrac{1}{1 + 10x}$, $|x| < \dfrac{1}{10}$

d $\dfrac{1}{1 + \dfrac{1}{3}x}$, $|x| < 3$

2 Expand each fraction in ascending powers of x, stating the first four terms of the expansion.

a $\dfrac{1}{3 + x}$

b $\dfrac{1}{3 - 2x}$

c $\dfrac{1}{2 + 4x}$

d $\dfrac{1}{5 + 2x}$

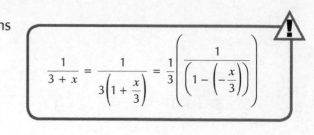

$$\frac{1}{3 + x} = \frac{1}{3\left(1 + \dfrac{x}{3}\right)} = \frac{1}{3}\left(\frac{1}{\left(1 - \left(-\dfrac{x}{3}\right)\right)}\right)$$

3 Given that the expansion $\dfrac{1}{1 - r} = 1 + r + r^2 + r^3 + \dots$ is only valid for $|r| < 1$, state what values of x the expansions in Question 2 are valid for.

4 Expand these in ascending powers of x, stating the first four terms of the expansion.

a $\dfrac{5}{1 - 2x}$

b $\dfrac{x}{1 + 3x}$

c $\dfrac{2x - 1}{1 + 5x}$

> For Q4a, expand $\dfrac{1}{1 - 2x}$ then multiply each term in the expansion by 5.

★ **5** Express these in partial fractions, then find the expansion of the function in ascending powers of x, up to and including the term in x^3.

State the set of values for which the expansion is valid:

> For Q4, $\dfrac{1}{1 + 5x}$ then multiply each term in the expansion by $2x-1$.

a $\dfrac{x}{(3 - 2x)(2 - x)}$

b $\dfrac{2}{x^2 + 2x - 8}$

6 Use the binomial expansion to find the first four terms of $\sqrt{(1 - x)}$ in ascending powers of x. Hence estimate the value of $\sqrt{0.9}$ to 3 decimal places.

7 Expand these fractions and state the range of validity.

a $\dfrac{1}{\sin x - \cos x}$

b $\dfrac{1}{\cos 2x}$

> First try to rewrite the fraction with denominator $1 - \tan x$.

> Rewrite the denominator using a trigonometric identity.

Maclaurin series

A series of the form

$$a_0 + a_1x + a_2x^2 + a_3x^3 + \ldots + a_rx^r + \ldots$$

is referred to as a **power series** when a_0, a_1, a_2, ..., a_r are real constants and x is regarded as a real variable.

> The symbol Σ means 'the sum of'. The terms that are added together are generated when $r = 0$, $r = 1$, $r = 2$,..., $r = n - 1$, $r = n$ are substituted into the general term (the expression after the Σ symbol).

If such a series has an infinite number of terms, it can also be written as $\displaystyle\sum_{r=0}^{\infty} a_r x^r$

In general, whether or not the series converges depends on x. In many cases, the sum diverges. In other cases, it converges to a particular limit.

In this section, we will be considering if a real function can be represented by a power series. This method is used by calculators and computers to accurately calculate values for trigonometric, exponential and logarithmic functions.

The Scottish mathematician Colin Maclaurin developed a method to approximate any function that is close to the origin. It allows any function that is differentiable to be expressed as a power series. This can be helpful in simplifying problems involving functions such as e^x and $\cos x$.

There are two conditions on the function:

- firstly, the function, f, must be differentiable up to the nth derivative

- secondly, each derivative can be evaluated for $x = 0$.

Some series converge to $f(x)$ for all values of x and others converge to $f(x)$ for a **limited** range of values of x.

Using the Maclaurin series

Consider a real function, $f(x)$, such that $f(0)$, $f'(0)$, $f''(0)$,... , $f^{(r)}(0)$, ... are defined, then we can form the infinite series:

$$f(x) = f(0) + f'(0)x + \frac{f''(x)x^2}{2!} + \frac{f'''(x)x^3}{3!} + \cdots$$

> This is provided in the exam formulae list.

When this series converges, it is known as the **Maclaurin series** (or **Maclaurin expansion**) of $f(x)$. For a given function, the series may converge for **all** values of x. However, more often the expansion only holds for a restricted range of values of x.

Also, if a power series tends to a limit:

- the power series can be differentiated and the differentiated series tends to the derivative of the limit

- the power series can be integrated and the integrated series tends to the integral of the limit.

That is, an expansion for $f'(x)$ can be found by differentiating the expansion of $f(x)$ and an expansion for $\int f(x)\,dx$ can be found by integrating the expansion of $f(x)$.

Example 7.25

Use Maclaurin's expansion to expand $f(x) = e^x$ to the term in x^3.

$$f(x) = e^x \qquad f(0) = 1$$
$$f'(x) = e^x \qquad f'(0) = 1$$
$$f''(x) = e^x \qquad f''(0) = 1$$
$$f'''(x) = e^x \qquad f'''(0) = 1$$

> Calculate the derivatives, up to $f'''(x)$ and then substitute $x = 0$ into each of the derivatives.

$$f(x) = f(0) + \frac{x}{1!}f'(0) + \frac{x^2}{2!}f''(0) + \dots + \frac{x^r}{r!}f^{(r)}(0) + \dots$$

then

$$e^x = 1 + x + \frac{1}{2!}x^2 + \frac{1}{3!}x^3 + \dots$$

$$= 1 + x + \frac{x^2}{2} + \frac{x^3}{6} + \dots, \text{ for all } x \in \mathbb{R}$$

Example 7.26

Use Maclaurin's expansion to expand $f(x) = \sin x$ to the term in x^4, x is measured in radians.

$$f(x) = \sin x \qquad f(0) = 0$$
$$f'(x) = \cos x \qquad f'(0) = 1$$
$$f''(x) = -\sin x \qquad f''(0) = 0$$
$$f'''(x) = -\cos x \qquad f'''(0) = -1$$
$$f^4(x) = \sin x \qquad f^4(0) = 0$$

> Calculate the derivatives, up to $f^4(x)$ and then substitute $x = 0$ into each of the derivatives.

so using

$$f(x) = f(0) + xf'(0) + \frac{x^2}{2!}f''(0) + \dots + \frac{x^r}{r!}f^{(r)}(0) + \dots$$

$$\sin x = 0 + x + 0 - \frac{x^3}{3!} + \dots$$

$$\sin x = x - \frac{x^3}{6} + \dots$$

Standard results and their range of validity

The following expansions are standard results. They can be quoted and used in an exam. You are also expected to know how to derive them. The range of values for which each series is valid for is also given.

$$e^x = 1 + \frac{x}{1!} + \frac{x^2}{2!} + \frac{x^3}{3!} + \dots \quad \text{for all } x \in \mathbb{R}$$

$$\ln(1 + x) = x - \frac{x^2}{2} + \frac{x^3}{3} - \frac{x^4}{4} + \dots \quad \text{for } -1 < x \leq 1$$

$$\sin x = \frac{x}{1!} - \frac{x^3}{3!} + \frac{x^5}{5!} - \dots \quad \text{for all } x \in \mathbb{R}$$

$$\cos x = 1 - \frac{x^2}{2!} + \frac{x^4}{4!} - \frac{x^6}{6!} + \dots \quad \text{for all } x \in \mathbb{R}$$

$$\tan^{-1} x = x - \frac{x^3}{3} + \frac{x^5}{5} - \frac{x^7}{7} + \dots \quad \text{for } -1 < x < 1$$

$$(1 + x)^n = 1 + \frac{n}{1!}x + \frac{n(n-1)}{2!}x^2 + \frac{n(n-1)(n-2)}{3!}x^3 + \dots \quad \text{for } -1 < x \leq 1$$

Example 7.27

Use Maclaurin's expansion to expand $f(x) = x\cos x$ to the term in x^4, where x is measured in radians.

> The product rule needs to be used in this example.

$$f(x) = x\cos x \qquad\qquad f(0) = 0$$
$$f'(x) = -x\sin x + \cos x \qquad f'(0) = 1$$

$$f''(x) = -x\cos x - \sin x - \sin x$$
$$\quad = -x\cos x - 2\sin x \qquad f''(0) = 0$$

$$f'''(x) = x\sin x - \cos x - 2\cos x$$
$$\quad = x\sin x - 3\cos x \qquad f'''(0) = -3$$

$$f^4(x) = x\cos x + \sin x + 3\sin x$$
$$\quad = x\cos x + 4\sin x \qquad f^4(0) = 0$$

> Calculate the derivatives, up to $f^4(x)$ and then substitute $x = 0$ into each of the derivatives.

$$f(x) = f(0) + xf'(0) + \frac{x^2}{2!}f''(0) + \dots + \frac{x^r}{r!}f^{(r)}(0) + \dots$$

$$x\cos x = 0 + x + 0 - \frac{3x^3}{3!} + 0 \dots$$

$$x\cos x = x - \frac{x^3}{2} + \dots$$

Example 7.28

Use Maclaurin's expansion to expand $f(x) = \cos 2x$ to the term in x^4, where x is measured in radians.

$$\cos x = 1 - \frac{x^2}{2!} + \frac{x^4}{4!} - \dots$$

> You can use the method in Example 7.24, or you can use the expansion for $f(x) = \cos x$ If using standard results, you must quote them in your solution.

then

$$\cos 2x = 1 - \frac{(2x)^2}{2!} + \frac{(2x)^4}{4!} - \dots$$

$$= 1 - \frac{4x^2}{2} + \frac{16x^4}{24} - \dots$$

> Substitute $2x$ for x in the standard result.

$$= 1 - 2x^2 + \frac{2x^4}{3} - \dots$$

Exercise 7G

★ **1** Use the Maclaurin series to derive a series for each function in ascending powers. State the first four terms in the expansion.

a $f(x) = e^x$ **b** $f(x) = \sin x$ **c** $f(x) = \sqrt{1-x}$

d $f(x) = \cos x$ **e** $f(x) = \ln(1-x)$ **f** $f(x) = \dfrac{1}{1-x}$

2 Find the first four non-zero terms in the Maclaurin expansion of each function.

> You may have to use the chain rule or product rule when differentiating.

a $f(x) = e^{3x}$ **b** $f(x) = (1+x)^{\frac{1}{2}}$ **c** $f(x) = \sin 2x$

d $f(x) = \cos 3x$ **e** $f(x) = e^x \cos x$ **f** $f(x) = x \sin x$

g $f(x) = e^x \sin 2x$

3 a Find the first four non-zero terms in the Maclaurin expansion of $f(x) = \sin^2 x$

b Hence obtain a series for $\cos^2 x$

> You can simplify working in Q3a by remembering that $2\sin x \cos x = \sin 2x$. For Q3b, remember that $\cos^2 x + \sin^2 x = 1$.

Using more than one Maclaurin expansion

Often questions will ask for an expansion of a composite function. This can be done by working out $f(0)$, $f'(0)$, $f''(0)$, ..., $f^{(r)}(0)$, ... and using

$$f(x) = f(0) + xf'(0) + \frac{x^2}{2!}f''(0) + \ldots + \frac{x^r}{r!}f^{(r)}(0) + \ldots$$

(although this can become cumbersome) or by using the standard results (which is probably the most efficient way of doing it). If a standard result is used, it must be quoted and any working to get the final solution shown.

Example 7.29

Use Maclaurin's expansion to expand $f(x) = e^{-3x}\sin 2x$ to the term in x^4, where x is measured in radians.

$$e^x = 1 + \frac{x}{1!} + \frac{x^2}{2!} + \frac{x^3}{3!} + \frac{x^4}{4!} + \ldots$$

$$= 1 + x + \frac{x^2}{2} + \frac{x^3}{6} + \frac{x^4}{24} + \ldots \quad \text{●} \longrightarrow \boxed{\text{Quote any standard results used.}}$$

$$\text{so } e^{-3x} = 1 + (-3x) + \frac{(-3x)^2}{2} + \frac{(-3x)^3}{6} + \frac{(-3x)^4}{24} + \ldots$$

$$= 1 - 3x + \frac{9x^2}{2} - \frac{27x^3}{6} + \frac{81x^4}{24} + \ldots$$

$$= 1 - 3x + \frac{9x^2}{2} - \frac{9x^3}{2} + \frac{27x^4}{8} + \ldots$$

$$\sin x = \frac{x}{1!} - \frac{x^3}{3!} + \frac{x^5}{5!} - \ldots$$

$$= x - \frac{x^3}{6} + \frac{x^5}{120} - \ldots \quad \text{●} \longrightarrow \boxed{\text{Quote any standard results used.}}$$

$$\text{so } \sin 2x = (2x) - \frac{(2x)^3}{6} + \ldots \quad \text{●} \longrightarrow \boxed{\text{Only need to expand up to the term } x^4.}$$

$$= 2x - \frac{8x^3}{6} + \ldots$$

$$= 2x - \frac{4x^3}{3} + \ldots$$

And so, $e^{-3x}\sin 2x = \left(1 - 3x + \frac{9x^2}{2} - \frac{9x^3}{2} + \frac{27x^4}{8} + \ldots\right)\left(2x - \frac{8x^3}{6} + \ldots\right)$

$$= 2x - \frac{8x^3}{6} - 6x^2 + 4x^4 + 9x^3 - 9x^4 + \ldots$$

$$= 2x - 6x^2 + \frac{23}{3}x^3 - 5x^4 + \ldots$$

Example 7.30

Use Maclaurin's expansion to expand $f(x) = e^{\sin x}$ to the term in x^3, where x is measured in radians.

$$e^x = 1 + \frac{x}{1!} + \frac{x^2}{2!} + \frac{x^3}{3!} + \frac{x^4}{4!} + \ldots$$

$$= 1 + x + \frac{x^2}{2} + \frac{x^3}{6} + \frac{x^4}{24} + \ldots \quad \bullet \longrightarrow \boxed{\text{Quote any standard results used.}}$$

so

$$e^{\sin x} = 1 + (\sin x) + \frac{(\sin x)^2}{2} + \frac{(\sin x)^3}{6} + \ldots$$

$$\sin x = \frac{x}{1!} - \frac{x^3}{3!} + \frac{x^5}{5!} - \ldots$$

$$= x - \frac{x^3}{6} + \frac{x^5}{120} - \ldots \quad \bullet \longrightarrow \boxed{\text{Quote any standard results used.}}$$

so

$$e^{\sin x} = 1 + (\sin x) + \frac{(\sin x)^2}{2} + \frac{(\sin x)^3}{6} + \ldots$$

$$= 1 + \left(x - \frac{x^3}{6}\right) + \frac{1}{2}\left(x - \frac{x^3}{6}\right)^2 + \frac{1}{6}\left(x - \frac{x^3}{6}\right)^3 + \ldots$$

$$= 1 + x - \frac{x^3}{6} + \frac{1}{2}(x^2) + \frac{1}{6}(x^3) + \ldots \quad \bullet \longrightarrow \boxed{\text{Only need to expand up to the term } x^3.}$$

$$= 1 + x + \frac{1}{2}x^2 + \frac{1}{6}x^3 \ldots$$

Exercise 7H

In this exercise, obtain the expansion for the first four non-zero terms.

★ 1 Use standard results to state the Maclaurin series for each function.

  a $f(x) = e^{3x}$ b $f(x) = e^{-2x}$ c $f(x) = e^{x^2}$

  d $f(x) = \sin 3x$ e $f(x) = \sin\frac{1}{2}x$ f $f(x) = \cos x^2$

2 If $f(x) = \cos^2 x$, obtain a Maclaurin expansion for $f(x)$ by:

  a calculating $f(0),\ f'(0),\ f''(0),\ \ldots,\ f^{(r)}(0),\ \ldots$ and using

   $$f(x) = f(0) + xf'(0) + \frac{x^2}{2!}f''(0) + \ldots + \frac{x^r}{r!}f^{(r)}(0) + \ldots$$

  b using standard results and $f(x) = \cos^2 x = \cos x \cos x$

  c using $f(x) = \cos^2 x = \frac{1}{2}(\cos 2x + 1)$

3 Find a Maclaurin expansion for each function as far as x^4.

   a $f(x) = e^{\sin 2x}$ **b** $f(x) = e^{2x}\sin 2x$ **c** $f(x) = (1 + e^x)^2$

4 Find a Maclaurin expansion for each function as far as x^3.

   a $f(x) = \ln(1 + e^x)$ **b** $f(x) = \ln(1 + \cos x)$

   c $f(x) = \ln\left(\dfrac{1 + x}{1 - x}\right)^2$

> In Q3f, use log rules first before finding the expansion of the function.

Summation and sigma notation

Sigma notation uses the sigma symbol, Σ, to denote a summation of a series of terms. (See page 247 for more about sigma.) For example, using the notion u_n for the nth term,

$$\sum_{r=1}^{10} u_n = u_1 + u_2 + u_3 + u_4 + u_5 + u_6 + u_7 + u_8 + u_9 + u_{10}$$

so $\displaystyle\sum_{r=1}^{n} f(r) = f(1) + f(2) + f(3) + \ldots + f(n)$

$\displaystyle\sum_{r=1}^{n} f(r)$ is the series with 1st term $f(1)$, 2nd term $f(2)$, 3rd term $f(3)$, … and last term $f(n)$.

So, for example

$$\sum_{r=1}^{n} r = 1 + 2 + 3 + \ldots + n$$

> When $r = 1, 2r + 3 = 5$; $r = 2, 3r + 3 = 7$; $r = 3, 2r + 3 = 9$; … and $r = n, 2r + 3 = 2n + 3$.

and $\displaystyle\sum_{r=1}^{n}(2r + 3) = 5 + 7 + 9 + \ldots + (2n + 3)$

Example 7.31

Calculate $\displaystyle\sum_{r=1}^{10}(4r - 1)$

$$\sum_{r=1}^{10}(4r - 1) = 3 + 7 + 11 + 15 + 19 + 23 + 27 + 31 + 35 + 39 = 210$$

$a = 3$

$d = 4$

$n = 10$

> This is an arithmetic series, so we can find values for a, d and n.

$S_n = \dfrac{1}{2}n[2a + (n - 1)d]$ could be used to evaluate $\displaystyle\sum_{r=1}^{10}(4r - 1)$

> Remember $S_n = \dfrac{1}{2}n[2a + (n - 1)d]$ cannot be used for all series that use Σ notation as not all series are arithmetic series.

(continued)

$$S_n = \frac{1}{2} \times 10\left[(2 \times 3) + ((10 - 1) \times 4)\right]$$

$$S_n = 5[6 + 36]$$

$$= 210$$

Example 7.32

Write each series in the form $\displaystyle\sum_{r=1}^{n} f(r)$, (i.e. using sigma notation).

a $1^2 + 2^2 + 3^2 + \ldots + n^2$ b $-1 + 1 + 3 + \ldots + 27$

a General term is r^2

$$1^2 + 2^2 + 3^2 + \ldots + n^2 = \sum_{r=1}^{n} r^2$$

> Each term can be written in the form r^2 and there are n terms in the series so the series is from $r = 1$ to $r = n$.

b General term is $2r - 3$, and the last term is 27.

$$2n - 3 = 27$$

$$\Rightarrow n = 15$$

$$-1 + 1 + 3 + \ldots + 27 = \sum_{r=1}^{15}(2r - 3)$$

> Each term can be written in the form $2r - 3$ so the general term is $2r - 3$.

Exercise 7I

★ 1 Write out each series in full and calculate the total.

a $\displaystyle\sum_{r=1}^{8}(2r + 1)$ b $\displaystyle\sum_{r=1}^{6} 4r^3$ c $\displaystyle\sum_{r=1}^{7}(3r + 2)(r - 1)$

d $\displaystyle\sum_{r=1}^{5} \frac{1}{2r + 3}$ e $\displaystyle\sum_{r=1}^{6} \frac{1}{2^r}$ f $\displaystyle\sum_{r=1}^{8} 2(-1)^r r$

g $\displaystyle\sum_{r=9}^{12}(3r - 1)$ h $\displaystyle\sum_{r=2}^{6}(2r - 1)(3r - 2)$

★ 2 Write each series using sigma notation.

a $3 + 5 + 7 + \ldots + n$

b $3 \times 1^3 + 3 \times 2^3 + 3 \times 3^3 + \ldots + 3 \times n^3$

c $\frac{1}{2} + \frac{1}{4} + \frac{1}{8} + \ldots + \frac{1}{2^n}$

d $1 + 4 + 9 + \ldots + 400$

e $52 + 48 + 44 + \ldots + 0$

f $(1 \times 2) + (2 \times 3) + (3 \times 4) + \ldots + (24 \times 25)$

g $(1^2 \times 3) + (2^2 \times 4) + (3^2 \times 5) + \ldots + (18^2 \times 20)$

h $-3 + 6 - 9 + 12 - 15 + 18 - 21 + 24 - 27 + 30$

> To deal with alternating signs, remember $(-1)^2 = 1$ and $(-1)^3 = -1$.

3 Find the value of n in each series.

a $\displaystyle\sum_{r=1}^{n}(2r - 1) = 256$ b $\displaystyle\sum_{r=1}^{n}(4r - 1) = 171$

> For an arithmetic series,
> $$S_n = \frac{1}{2}n\left[2a + (n - 1)d\right]$$

c $\displaystyle\sum_{r=1}^{n}(7 - 3r) = -490$

Standard results

The following results can be used to obtain formulae in n for expressions written with sigma notation.

Result 1

$$\sum_{r=1}^{n}1 = 1 + 1 + 1 + \ldots + 1 = n$$

> 1 is added n times.

$$\therefore \sum_{r=1}^{n}1 = n$$

Result 2

$$\sum_{r=1}^{n}kf(r) = kf(1) + kf(2) + kf(3) + \ldots + kf(n)$$

$$= k\big(f(1) + f(2) + f(3) + \ldots f(n)\big)$$

$$= k\sum_{r=1}^{n}f(r)$$

$$\therefore \sum_{r=1}^{n}kf(r) = k\sum_{r=1}^{n}f(r)$$

Result 3

$$\sum_{r=1}^{n}r = 1 + 2 + 3 + \ldots + (n - 1) + n$$

> This is an arithmetic series with n terms and $a = 1$ and $d = 1$.

$$\sum_{r=1}^{n}r = S_n = \frac{n}{2}\big(2a + (n - 1)d\big)$$

$$= \frac{n}{2}\big(2 + (n - 1) \times 1\big)$$

$$= \frac{n}{2}(n + 1)$$

$$\therefore \sum_{r=1}^{n}r = \frac{n}{2}(n + 1)$$

Result 4

$$\sum_{r=1}^{n}(f(r) + g(r)) = f(1) + g(1) + f(2) + g(2) + \ldots + f(n) + g(n)$$

$$= f(1) + f(2) + \ldots + f(n) + g(1) + g(2) + \ldots g(n)$$

$$= \sum_{r=1}^{n}f(r) + \sum_{r=1}^{n}g(r)$$

$$\therefore \sum_{r=1}^{n}(f(r) + g(r)) = \sum_{r=1}^{n}f(r) + \sum_{r=1}^{n}g(r)$$

Result 5

$$\sum_{r=1}^{n}(ar + b) = (a + b) + (2a + b) + (3a + b) + \ldots + (an + b)$$

Where a and b are constants.

$$= (a + 2a + 3a + \ldots + an) + (b + b + b + \ldots + b)$$

$$= \sum_{r=1}^{n}ar + \sum_{r=1}^{n}b$$

Using earlier results

$$\sum_{r=1}^{n}ar = a\sum_{r=1}^{n}r \quad \text{and} \quad \sum_{r=1}^{n}b = b\sum_{r=1}^{n}1$$

since $\sum_{r=1}^{n}r = \frac{n}{2}(n - 1)$ and $\sum_{r=1}^{n}1 = n$

$$\therefore \sum_{r=1}^{n}ar + b = a\sum_{r=1}^{n}r + \sum_{r=1}^{n}b = \frac{1}{2}an(n + 1) + bn$$

Result 6

$$\sum_{r=1}^{n}f(r + 1) - \sum_{r=1}^{n}f(r) = (f(2) + f(3) + f(4) + \ldots + f(n + 1))$$

$$-(f(1) + f(2) + f(3) + \ldots + f(n))$$

$$= f(n + 1) - f(1)$$

$$\therefore \sum_{r=1}^{n}f(r + 1) - \sum_{r=1}^{n}f(r) = f(n + 1) - f(1)$$

Result 7

$$\sum_{r=a}^{n} f(r) = f(a) + f(a + 1) + \ldots + f(n)$$

$$= \big(f(1) + f(2) + f(3) + \ldots + f(a - 1) + f(a) + \ldots + f(n)\big)$$
$$- \big(f(1) + f(2) + f(3) + \ldots + f(a - 1)\big)$$

$$= \sum_{r=1}^{n} f(r) - \sum_{r=1}^{a-1} f(r)$$

$$\therefore \sum_{r=a}^{n} f(r) = \sum_{r=1}^{n} f(r) - \sum_{r=1}^{a-1} f(r)$$

Result 1	$\sum\limits_{r=1}^{n} 1 = n$
Result 2	$\sum\limits_{r=1}^{n} kf(r) = k \sum\limits_{r=1}^{n} f(r)$
Result 3	$\sum\limits_{r=1}^{n} r = \dfrac{n}{2}(n + 1)$ This is provided in the exam formulae list.
Result 4	$\sum\limits_{r=1}^{n} \big(f(r) + g(r)\big) = \sum\limits_{r=1}^{n} f(r) + \sum\limits_{r=1}^{n} g(r)$
Result 5	$\sum\limits_{r=1}^{n} (ar + b) = a \sum\limits_{r=1}^{n} r + \sum\limits_{r=1}^{n} b = \dfrac{1}{2} an(n + 1) + bn$
Result 6	$\sum\limits_{r=1}^{n} f(r + 1) - \sum\limits_{r=1}^{n} f(r) = f(n + 1) - f(1)$
Result 7	$\sum\limits_{r=a}^{n} f(r) = \sum\limits_{r=1}^{n} f(r) - \sum\limits_{r=1}^{a-1} f(r)$

Example 7.33

Evaluate $\displaystyle\sum_{r=6}^{10} (2r + 1)$

Method 1

$$\sum_{r=6}^{10} (2r + 1) = 13 + 15 + 17 + 19 + 21 = 85$$

(continued)

This is straightforward but we need a method for more complicated questions. It is easier to work with sums if the lower limit is 1. To do this, we could express the question as:

[sum of 6th, 7th, 8th, 9th and 10th terms] = [first 10 terms of the series] − [first 5 terms of the series]

or

$$\sum_{r=6}^{10} (2r + 1) = \sum_{r=1}^{10} (2r + 1) - \sum_{r=1}^{5} (2r + 1)$$

> The upper limit on the second summation is always one less than the desired start (lower limit) in the question.

$$= 2\sum_{r=1}^{10} r + \sum_{r=1}^{10} 1 - \left(2\sum_{r=1}^{5} r + \sum_{r=1}^{5} 1\right)$$

> Use Result 5

$$= 2\left(\frac{1}{2} \times 10 \times 11\right) + 10 - 2\left(\frac{1}{2} \times 5 \times 6\right) - 5$$

> Use Result 3

$$= 85$$

Method 2

$$\sum_{r=1}^{n} (2r + 1) = 2\sum_{r=1}^{n} r + \sum_{r=1}^{n} 1$$

> ⚠ Exam questions usually expect you to find a formula as part of a question similar to Method 2.

$$= 2\left(\frac{n}{2}(n + 1)\right) + n$$

> Find a formula for $\sum_{r=1}^{n} (2r + 1)$ first using Result 5.

$$= n^2 + 2n$$

> Use Result 3

so

$$\sum_{r=6}^{10} (2r + 1) = \sum_{r=1}^{10} (2r + 1) - \sum_{r=1}^{5} (2r + 1)$$

$$= 10^2 + 2 \times 10 - \left(5^2 + 2 \times 5\right)$$

> Use Result 7

$$= 85$$

> Use the formula for $\sum_{r=1}^{n} (2r + 1)$ found in first part of the solution and replace $n = 10$ and $n = 5$.

Example 7.34

Show $\displaystyle\sum_{k=1}^{n}(10 - 6k) = 7n - 3n^2$

Hence write down an expression for $\displaystyle\sum_{k=1}^{2p}(10 - 6k)$

Show that $\displaystyle\sum_{k=p+1}^{2p}(10 - 6k) = 7p - 9p^2$

$$\sum_{k=1}^{n}(10 - 6k) = 10\sum_{k=1}^{n}1 - 6\sum_{k=1}^{n}k \quad\text{———— Use Result 5}$$

$$= 10n - 6\left(\frac{n}{2}(n + 1)\right) \quad\text{———— Use Results 1 and 3}$$

$$= 10n - 3n(n + 1)$$

$$= 7n - 3n^2 \quad\text{———— Simplify}$$

So $\displaystyle\sum_{k=1}^{2p}(10 - 6k) = 7(2p) - 3(2p)^2 = 14p - 12p^2$ — Substitute $2p$ for n in formula calculated in first part of question.

$$\sum_{k=p+1}^{2p}(10 - 6k) = \sum_{k=1}^{2p}(10 - 6k) - \sum_{k=1}^{p}(10 - 6k) \quad\text{———— Use Result 7}$$

$$= 14p - 12p^2 - \left(7p - 3p^2\right) \quad\text{— Using formulae calculated in earlier parts of question.}$$

$$= 7p - 9p^2 \quad\text{———— Simplify}$$

Exercise 7J

★ 1 Express the following series as a function of n without sigma notation:

    a $\displaystyle\sum_{r=1}^{n}3r$ b $\displaystyle\sum_{r=1}^{n}2$ c $\displaystyle\sum_{r=1}^{n}(5r + 6)$

    d $\displaystyle\sum_{r=1}^{n}(3 - 4r)$ e $\displaystyle\sum_{r=1}^{n}\left(7 + \frac{1}{2}r\right)$

2 a Find a formula for $\displaystyle\sum_{r=1}^{n}(3r + 1)$ and then evaluate $\displaystyle\sum_{r=10}^{20}(3r + 1)$

    b Find a formula for $\displaystyle\sum_{r=1}^{n}(4r - 1)$ and then evaluate $\displaystyle\sum_{r=5}^{17}(4r - 1)$

    c Find a formula for $\displaystyle\sum_{r=1}^{n}(2r - 3)$ and then evaluate $\displaystyle\sum_{r=20}^{100}(2r - 3)$

d Find a formula for $\sum_{r=1}^{n}(6-r)$ and then evaluate $\sum_{r=9}^{15}(6-r)$

e Find a formula for $\sum_{r=5}^{p}(2r+1)$

> Note the lower limit in Q2e.

3 Express each sum as a function of n without using sigma notation.

a $\sum_{r=1}^{n}\left[(r+1)^3-r^3\right]$ **b** $\sum_{r=1}^{n}\left[(r+2)^2-(r+1)^2\right]$

c $\sum_{r=1}^{n}\left[\cos(r+1)-\cos r\right]$ **d** $\sum_{r=1}^{n}\left[(2r+2)^3-(2r)^3\right]$

e $\sum_{r=1}^{n}\left(\frac{1}{r+1}\right)-\left(\frac{1}{r}\right)$

4 Show that $(n+1)^2-n^2=2n+1$

Use this to show that $\sum_{r=1}^{n}(2r+1)=n(n+2)$

> Remember
> $$\sum_{r=1}^{n}f(r+1)-\sum_{r=1}^{n}f(r)$$
> $$=f(n+1)-f(1).$$

5 Express $\dfrac{1}{r(r+1)}$ in partial fractions.

Hence find an expression for $\sum_{r=1}^{n}\dfrac{1}{r(r+1)}$

6 Show that $r(r+1)-(r-1)r=2r$

Use this to show that $\sum_{r=1}^{n}2r=n(n+1)$

More standard results

We know that $\sum_{r=1}^{n}r=\frac{n}{2}(n+1)$. In this section, we will find similar results for

$\sum_{r=1}^{n}r^2$ and $\sum_{r=1}^{n}r^3$.

Finding a result for $\sum_{r=1}^{n}r^2$

We know that $(r+1)^3=r^3+3r^2+3r+1$

so

$$\sum_{r=1}^{n}(r+1)^3=\sum_{r=1}^{n}\left(r^3+3r^2+3r+1\right)$$

> Since $(r+1)^3=r^3+3r^2+3r+1$.

$$\sum_{r=1}^{n}(r+1)^3=\sum_{r=1}^{n}r^3+3\sum_{r=1}^{n}r^2+3\sum_{r=1}^{n}r+\sum_{r=1}^{n}1$$

> Use Results 2 and 4

$$\sum_{r=1}^{n}(r+1)^3 - \sum_{r=1}^{n}r^3 = 3\sum_{r=1}^{n}r^2 + 3\sum_{r=1}^{n}r + \sum_{r=1}^{n}1$$

Rearrange the previous line.

$$(n+1)^3 - 1^3 = 3\sum_{r=1}^{n}r^2 + \frac{3}{2}n(n+1) + n$$

Use Result 6

$$3\sum_{r=1}^{n}r^2 = (n+1)^3 - \frac{3}{2}n(n+1) - n - 1$$

Rearrange the previous line to make $3\sum_{r=1}^{n}r^2$ the subject.

$$3\sum_{r=1}^{n}r^2 = (n+1)^3 - \frac{3}{2}n(n+1) - (n+1)$$

$$3\sum_{r=1}^{n}r^2 = \frac{1}{2}(n+1)\left(2(n+1)^2 - 3n - 2\right)$$

$$3\sum_{r=1}^{n}r^2 = \frac{1}{2}(n+1)\left(2n^2 + n\right)$$

$$\sum_{r=1}^{n}r^2 = \frac{1}{6}n(n+1)(2n+1)$$

This is provided in the exam formulae list.

Finding a result for $\displaystyle\sum_{r=1}^{n}r^3$

We know that $(r+1)^4 - r^4 = 4r^3 + 6r^2 + 4r + 1$

Since $(r+1)^4 - r^4 = 4r^3 + 6r^2 + 4r + 1$

so

$$\sum_{r=1}^{n}(r+1)^4 - r^4 = 4\sum_{r=1}^{n}r^3 + 6\sum_{r=1}^{n}r^2 + 4\sum_{r=1}^{n}r + \sum_{r=1}^{n}1$$

$$(n+1)^4 - 1^4 = 4\sum_{r=1}^{n}r^3 + 6\sum_{r=1}^{n}r^2 + 4\sum_{r=1}^{n}r + \sum_{r=1}^{n}1$$

Use Result 6

$$n^4 + 4n^3 + 6n^2 + 4n = 4\sum_{r=1}^{n}r^3 + 6\left(\frac{n}{6}(n+1)(2n+1)\right) + 4\left(\frac{n}{2}(n+1)\right) + n$$

$$n^4 + 4n^3 + 6n^2 + 4n = 4\sum_{r=1}^{n}r^3 + 2n^3 + 3n^2 + n + 2n^2 + 2n + n$$

$$4\sum_{r=1}^{n}r^3 = n^4 + 2n^3 + n^2$$

Simplify the right-hand side and use the standard results for $\displaystyle\sum_{r=1}^{n}r$ and $\displaystyle\sum_{r=1}^{n}r^2$

$$4\sum_{r=1}^{n}r^3 = n^2\left(n^2 + 2n + 1\right)$$

Rearrange the previous line to make $4\displaystyle\sum_{r=1}^{n}r$ the subject.

$$\sum_{r=1}^{n}r^3 = \frac{n^2}{4}(n+1)^2$$

$$\sum_{r=1}^{n}r^3 = \frac{n^2(n+1)^2}{4}$$

Given that $\displaystyle\sum_{r=1}^{n}r = \frac{n}{2}(n+1)$, then it can be seen that $\displaystyle\sum_{r=1}^{n}r^3 = \left(\sum_{r=1}^{n}r\right)^2$

This is provided in the exam formulae list.

Example 7.35

Express $\displaystyle\sum_{r=1}^{n} 3r^2$ as a function of n that does not involve sigma notation.

Use this formula to evaluate $\displaystyle\sum_{r=1}^{8} 3r^2$

$$\sum_{r=1}^{n} 3r^2 = 3\sum_{r=1}^{n} r^2 \qquad \boxed{\text{Use Result 2}}$$

$$= 3\left(\frac{1}{6}n(n+1)(2n+1)\right) \qquad \boxed{\text{Use the standard result for } \sum_{r=1}^{n} r^2}$$

$$= \frac{1}{2}n(n+1)(2n+1)$$

Hence

$$\sum_{r=1}^{8} 3r^2 = \frac{1}{2} \times 8 \times (8+1) \times (2 \times 8 + 1) = 612$$

Exercise 7K

★ 1   Express as a function of n without using sigma notation.

a $\displaystyle\sum_{r=1}^{n} 5r^2$

b $\displaystyle\sum_{r=1}^{n} \left(3r^2 + 2\right)$

c $\displaystyle\sum_{r=1}^{n} \left(r^2 + 3r + 2\right)$

d $\displaystyle\sum_{r=1}^{n} \left(3r^2 - 2r - 1\right)$

e $\displaystyle\sum_{r=1}^{n} \left(2 - 3r - 5r^2\right)$

f $\displaystyle\sum_{r=1}^{n} (2r + 1)(r - 3)$

★ 2   Express as a function of n without using sigma notation.

a $\displaystyle\sum_{r=1}^{n} 2r^3$

b $\displaystyle\sum_{r=1}^{n} \left(2r^3 + 3\right)$

c $\displaystyle\sum_{r=1}^{n} \left(r^3 + 3r^2 + r\right)$

d $\displaystyle\sum_{r=1}^{n} r^2(r + 1)$

e $\displaystyle\sum_{r=1}^{n} (r + 2)^3$

3   Use the standard results for $\displaystyle\sum_{r=1}^{n} 1,\ \sum_{r=1}^{n} r,\ \sum_{r=1}^{n} r^2$ and $\displaystyle\sum_{r=1}^{n} r^3$ to evaluate:

a $\displaystyle\sum_{r=1}^{10} 3r^2$

b $\displaystyle\sum_{r=1}^{5} 2r^3$

c $\displaystyle\sum_{r=1}^{8} \left(4r^2 + r\right)$

d $\displaystyle\sum_{r=1}^{7} (r + 1)(r - 2)$

e $\displaystyle\sum_{r=1}^{10} r^2(r - 1)$

f $\displaystyle\sum_{r=1}^{15} r(r - 2)(r + 2)$

4   Use standard results for $\sum\limits_{r=1}^{n}1,\ \sum\limits_{r=1}^{n}r,\ \sum\limits_{r=1}^{n}r^2$ and $\sum\limits_{r=1}^{n}r^3$ to evaluate:

a $\sum\limits_{r=5}^{10}4r^2$

b $\sum\limits_{r=6}^{18}r(r+1)$

c $\sum\limits_{r=3}^{12}2r^3$

d $\sum\limits_{r=20}^{50}r(r-1)(r+1)$

5   Find an expression without using sigma notion for $\sum\limits_{r=1}^{n}r(r+1)$

Hence calculate $(1 \times 2) + (2 \times 3) + (3 \times 4) + \ldots + (28 \times 29)$

6   Find an expression without using sigma notion for $\sum\limits_{r=1}^{n}r(r+1)(r+2)$

Hence calculate $(1 \times 2 \times 3) + (2 \times 3 \times 4) + \ldots + (20 \times 21 \times 22)$

Chapter review

1   For the arithmetic sequence 4, 7, 10, 13, ..., find:
   a   a formula for u_n
   b   a value of the 20th term
   c   the value of n such that $u_n = 334$
   d   the first term to exceed 500.

2   a   Calculate the sum of the first 15 terms of the arithmetic series $6 + 10 + 14 + \ldots$
   b   Calculate the sum of the arithmetic series $3 + 10 + 17 + \ldots + 136$
   c   How many terms are in the arithmetic series that starts $3 + 7 + 11 + \ldots$ and whose sum is 210?
   d   The sum of the first 10 terms of an arithmetic series is 120. The sum of the first 20 terms is 840. Calculate a and d.

3   a   A geometric sequence has a 1st term of 4 and a 4th term of 32. Find the 8th term of the sequence.
   b   A geometric sequence has a 2nd term of 12 and a 6th term of 3072. Given that $r > 0$, find the 1st term of the sequence and the common ratio.

4   a   Calculate the sum of the first 15 terms of the geometric series $2 + 6 + 18 + 54 + \ldots$
   b   Calculate the sum of the geometric series $5 + 10 + 20 + \ldots + 2560$
   c   What is the smallest number of terms in the geometric series $3 + 6 + 12 + 24 + \ldots$ so that the sum of the series exceeds $100\,000$?

5 a State why the sum to infinity exists for series $32 + 16 + 8 + 4 + \ldots$
   Find the sum to infinity of this series.

b The common ratio for a geometric series is $0{\cdot}6$ and the sum to infinity is $312{\cdot}5$. Find the 1st term of the series.

6 Find the first four non-zero terms in the Maclaurin expansion for each function.

   a $f(x) = \cos x$ **b** $f(x) = e^{3x}$

   c $f(x) = e^{2x}\sin 3x$ **d** $f(x) = \ln\left(\dfrac{1-x}{1+x}\right)$

7 Calculate

   a $\displaystyle\sum_{r=1}^{5}(3r + 1)$ **b** $\displaystyle\sum_{r=1}^{15}\left(5r^2 + 2r + 1\right)$

   c $\displaystyle\sum_{r=5}^{16}2r^3 + 3r$ **d** $\displaystyle\sum_{r=1}^{8}\dfrac{1}{r(r+1)}$

8 Use the standard results for $\displaystyle\sum_{r=1}^{n}r,\ \sum_{r=1}^{n}r^2$ and $\displaystyle\sum_{r=1}^{n}r^3$ to find an expression (as a single fraction) in n for the following:

   a $\displaystyle\sum_{r=1}^{n}r(r + 1)$ **b** $\displaystyle\sum_{r=1}^{n}r(r + 1)(2r - 1)$

- I can find and use the general term of an arithmetic sequence.
  ★ Exercise 7A Q2, Q3

- I can find the sum to n terms of an arithmetic series. ★ Exercise 7B Q5

- I can find and use the general term of a geometric sequence. ★ Exercise 7C Q5, Q8

- I can find and use the sum to n terms of a geometric series. ★ Exercise 7D Q1, Q3

- I can find and use the sum to infinity of a geometric series. ★ Exercise 7E Q2, Q3, Q6

- I can use the Maclaurin expansion to find a power series. ★ Exercise 7G Q1 ★ Exercise 7H, Q1

- I can use sigma notation. ★ Exercise 7I Q1, Q2

- I can use standard summation formulae for $\displaystyle\sum_{r=1}^{n}1,\ \sum_{r=1}^{n}r,\ \sum_{r=1}^{n}r^2$ and $\displaystyle\sum_{r=1}^{n}r^3$

  ★ Exercise 7J Q1 ★ Exercise 7K Q1, Q2

264

8 Matrices

This chapter will show you how to:

- understand and use matrix algebra
- calculate the determinant of a 2×2 and a 3×3 matrix
- determine the inverse of a 2×2 and a 3×3 matrix
- perform geometric transformations
- use Gaussian elimination to solve a 3×3 system of linear equations.

You should already know:

- the correct order of operations
- how to do fraction calculations
- how to solve simultaneous equations with two unknowns.

Matrix algebra

The term **matrix** was first used by the English mathematician James Sylvester in 1850. Earlier, in 1841, his friend Arthur Cayley had been developing determinants and using them to solve equations. After discussions with Sylvester, Cayley went on to develop a formal algebra for matrix operations.

A **matrix** is an **array** of numbers, or other mathematical objects, arranged in **rows** and **columns**. Each number or object is referred to as an **element** or **entry** and is identified by its position in the matrix. a_{ij} is the element of matrix A in the ith row and jth column.

A matrix is often denoted by a capital letter.

The **order** or size of a matrix is determined by the number of rows and columns it contains. A matrix with m rows and n columns is described as $\boldsymbol{m \times n}$, where $\boldsymbol{m}$ and $\boldsymbol{n}$ are the dimensions of the matrix.

matrix	$A = \begin{pmatrix} 3 & 5 \\ -2 & 7 \\ 8 & 4 \\ 0 & 3 \end{pmatrix}$	$B = \begin{pmatrix} 4 & 0 \\ -1 & 2 \end{pmatrix}$	$C = \begin{pmatrix} 1 & 5 & -7 \end{pmatrix}$
order	4×2	2×2	1×3
element	$a_{32} = 4$	$b_{21} = -1$	$c_{13} = -7$

There are many applications of matrices (plural of matrix) in mathematics and other sciences. Some applications take advantage of the compact way in which they can be used to represent a set of numbers and properties. In mathematics they are most commonly used in linear transformations and in solving systems of linear equations.

Matrix definitions and operations

In general, a matrix A, of order $m \times n$, has m rows and n columns and can be represented as shown, where a_{ij} denotes the element in the ith row and jth column.

$$A = \begin{pmatrix} a_{11} & \cdots & a_{1n} \\ \vdots & \ddots & \vdots \\ a_{m1} & \cdots & a_{mn} \end{pmatrix}$$

Equal matrices: Two matrices are defined as equal only if they are of the same order and their corresponding elements are equal.

The **zero matrix** is a matrix where all the elements are zero.

Transpose of a matrix: A new matrix can be formed from matrix A by writing row 1 as column 1, row 2 as column 2, row 3 as column 3 and so on. This new matrix is called the transpose of A and is denoted by A' or A^T

A matrix M is said to be **symmetric** if $M' = M$, e.g. $\begin{pmatrix} 2 & 4 & 7 \\ 4 & -1 & 3 \\ 7 & 3 & 5 \end{pmatrix}$

A matrix M is said to be **skew-symmetric** if $M' = -M$ e.g. $\begin{pmatrix} 0 & -4 & 7 \\ 4 & 0 & -3 \\ -7 & 3 & 0 \end{pmatrix}$

Note that in a skew-symmetric matrix, the leading diagonal has to be all zeroes.

Matrix addition: Only matrices with the same order can be added together. Two matrices A and B can be added together by adding each element of A to the corresponding element of B.

$$A = \begin{pmatrix} a & b \\ c & d \end{pmatrix} \text{ and } B = \begin{pmatrix} p & q \\ r & s \end{pmatrix} \Rightarrow A + B = \begin{pmatrix} a+p & b+q \\ c+r & d+s \end{pmatrix}$$

$$B + A = \begin{pmatrix} p+a & q+b \\ r+c & s+d \end{pmatrix} = A + B$$

Matrix subtraction: Only matrices with the same order can be subtracted. Matrix B can be subtracted from matrix A by subtracting each element of B from the corresponding element of A.

$$A = \begin{pmatrix} a & b \\ c & d \end{pmatrix} \text{ and } B = \begin{pmatrix} p & q \\ r & s \end{pmatrix} \Rightarrow A - B = \begin{pmatrix} a-p & b-q \\ c-r & d-s \end{pmatrix}$$

Scalar multiplication: A matrix can be multiplied by a scalar (number) by multiplying each entry (element) by the scalar.

If $A = \begin{pmatrix} a & b \\ c & d \end{pmatrix}$ then $kA = \begin{pmatrix} ka & kb \\ kc & kd \end{pmatrix}$ where k is a scalar.

Example 8.1

Let $A = \begin{pmatrix} 5 & 2 \\ -3 & 1 \end{pmatrix}$ $B = \begin{pmatrix} 3 & 6 & -2 \\ 4 & -7 & 1 \end{pmatrix}$ $C = \begin{pmatrix} 1 \\ 3 \\ -4 \end{pmatrix}$ $D = \begin{pmatrix} 2 & -4 & 0 \\ 3 & 11 & 5 \\ 15 & 3 & 1 \end{pmatrix}$

a   How many rows has matrix B?

b   State the order of each matrix A, B and C.

c   Write the entry in the second row and first column of D.

d   State the value of

   i a_{12} ii d_{32}

e   Write down B'.

a B has 2 rows.

b A 2×2, B 2×3 and C 3×1 •——— | Order is denoted by *number of rows × number of columns*.

c   3

d   i   2   ii   3 •——— | a_{ij} represents the entry in row i and column j in matrix A.

e $B' = \begin{pmatrix} 3 & 4 \\ 6 & -7 \\ -2 & 1 \end{pmatrix}$ •——— | To find the transpose, the first row becomes the first column and the second row becomes the second column.

Example 8.2

Evaluate these matrix calculations.

a $\begin{pmatrix} 3 & 5 \\ 1 & 2 \end{pmatrix} + \begin{pmatrix} -2 & 7 \\ 3 & 9 \end{pmatrix}$

b $\begin{pmatrix} 1 & 6 \\ 8 & 2 \\ 5 & 3 \end{pmatrix} + \begin{pmatrix} 12 & 2 \\ 9 & 0 \\ 2 & -3 \end{pmatrix}$

c $\begin{pmatrix} 2 & 7 \\ 4 & 2 \end{pmatrix} - \begin{pmatrix} -3 & 7 \\ -1 & 5 \end{pmatrix}$

d $2\begin{pmatrix} 1 & 5 & -2 & 3 \\ 3 & 7 & 6 & 2 \end{pmatrix}$

e $3\begin{pmatrix} 3 & 5 \\ 1 & 2 \end{pmatrix} + 2\begin{pmatrix} -2 & 7 \\ 3 & 9 \end{pmatrix}$

(continued)

a $\begin{pmatrix} 3 & 5 \\ 1 & 2 \end{pmatrix} + \begin{pmatrix} -2 & 7 \\ 3 & 9 \end{pmatrix} = \begin{pmatrix} 1 & 12 \\ 4 & 11 \end{pmatrix}$

b $\begin{pmatrix} 1 & 6 \\ 8 & 2 \\ 5 & 3 \end{pmatrix} + \begin{pmatrix} 12 & 2 \\ 9 & 0 \\ 2 & -3 \end{pmatrix} = \begin{pmatrix} 13 & 8 \\ 17 & 2 \\ 7 & 0 \end{pmatrix}$

c $\begin{pmatrix} 2 & 7 \\ 4 & 2 \end{pmatrix} - \begin{pmatrix} -3 & 7 \\ -1 & 5 \end{pmatrix} = \begin{pmatrix} 5 & 0 \\ 5 & -3 \end{pmatrix}$

d $2\begin{pmatrix} 1 & 5 & -2 & 3 \\ 3 & 7 & 6 & 2 \end{pmatrix} = \begin{pmatrix} 2 & 10 & -4 & 6 \\ 6 & 14 & 12 & 4 \end{pmatrix}$

e $3\begin{pmatrix} 3 & 5 \\ 1 & 2 \end{pmatrix} + 2\begin{pmatrix} -2 & 7 \\ 3 & 9 \end{pmatrix} = \begin{pmatrix} 9 & 15 \\ 3 & 6 \end{pmatrix} + \begin{pmatrix} -4 & 14 \\ 6 & 18 \end{pmatrix} = \begin{pmatrix} 5 & 29 \\ 9 & 24 \end{pmatrix}$

Exercise 8A

1 State the order of these matrices.

a $\begin{pmatrix} 1 & 7 \\ -2 & -3 \\ 6 & 4 \\ 2 & 3 \end{pmatrix}$ 　　　 **b** $(2 \ 1 \ 0)$ 　　　 **c** $\begin{pmatrix} -2 & 4 & 5 \\ 4 & 0 & 9 \\ 1 & -3 & 2 \end{pmatrix}$ 　　　 **d** $\begin{pmatrix} 3 \\ -5 \\ 1 \end{pmatrix}$

2 Find the value of x and y in each to make these pairs of matrices equal.

a $\begin{pmatrix} 2x & 7 \\ 3 & 9 \end{pmatrix}, \begin{pmatrix} 8 & 7 \\ 3 & 3y \end{pmatrix}$ 　　　 **b** $\begin{pmatrix} x+2 & 3 \\ 4 & 2 \\ -3 & y \end{pmatrix}, \begin{pmatrix} 5 & 3 \\ x+1 & 2 \\ y-1 & -1 \end{pmatrix}$

c $\begin{pmatrix} 1 & 3x+2y \\ 4x-3y & 5 \end{pmatrix}, \begin{pmatrix} 1 & 0 \\ 17 & 5 \end{pmatrix}$ 　　　 **d** $\begin{pmatrix} 2 & 2x+3y \\ -7 & 3 \end{pmatrix}, \begin{pmatrix} 2 & 11 \\ x-y & 3 \end{pmatrix}$

3 Write down the transpose of each of these matrices.

a $\begin{pmatrix} -1 & 3 \\ 2 & 4 \end{pmatrix}$ 　　　 **b** $\begin{pmatrix} 3 & 6 \\ 2 & 7 \\ 5 & 1 \end{pmatrix}$ 　　　 **c** $\begin{pmatrix} 2 & 4 & 1 & -2 \\ -5 & 0 & 1 & -3 \end{pmatrix}$

★ **4** Given $A = \begin{pmatrix} -1 & 3 \\ 2 & 0 \end{pmatrix}$, $B = \begin{pmatrix} 9 & 2 \\ -5 & 3 \end{pmatrix}$ and $C = \begin{pmatrix} -3 & 8 \\ 1 & -2 \end{pmatrix}$, evaluate:

a $2A + 3B$ 　　　 **b** $4A - 2B + C$ 　　　 **c** $B - 2A + 3C$

★ 5 Using A, B and C from Question 4, evaluate:

   a i $A + B$ ii $B + A$

   b i $(A + B)'$ ii $A' + B'$

   c i $A - B$ ii $B - A$

   d $(A')'$

   e i $3A'$ ii $(3A)'$

   f i $(A + B) + C$ ii $A + (B + C)$ iii $A + B + C$

   g i $2(A + B)$ ii $2A + 2B$

   Comment on your results.

6 For $A = \begin{pmatrix} 2 & 3 \\ 1 & 5 \end{pmatrix}$ and $B = \begin{pmatrix} \sqrt{3} & \frac{1}{3} \\ -2 & 0 \end{pmatrix}$, find:

   a $3A + 2B$ b $\frac{1}{2}A - \frac{1}{3}B$ c $\frac{3}{5}B - 2A$

7 Given $P = \begin{pmatrix} \frac{1}{2} & \frac{2}{3} \\ -2 & \frac{3}{5} \end{pmatrix}$, $Q = \begin{pmatrix} \frac{3}{2} & \frac{4}{3} \\ \frac{1}{5} & 2 \end{pmatrix}$ and $R = \begin{pmatrix} 2 & -1 \\ 3 & 7 \end{pmatrix}$, show that:

   a $P + Q = Q + P$ b $Q - P \neq P - Q$

   c $3(P - R) = 3P - 3R$ d $(P + Q) + R = P + (Q + R)$

8 Solve for x if $\begin{pmatrix} 2x & -6 \\ 8 & -11 \end{pmatrix} - \begin{pmatrix} 4 & 7 \\ 3 & 2 \end{pmatrix} + 2\begin{pmatrix} x & 3 \\ -1 & 5 \end{pmatrix} = \begin{pmatrix} 4 & -7 \\ 3 & -3 \end{pmatrix}$

9 If $A = \begin{pmatrix} 2 & 3 & a \\ 3 & -1 & 4 \\ 5 & 2b & 3 \end{pmatrix}$ is symmetric, what are the values of a and b?

10 Obtain the values of s and t if $M = \begin{pmatrix} 0 & 2s & 3 \\ 4 & 0 & 2 \\ -3 & t & 0 \end{pmatrix}$ is skew-symmetric.

- $A + B = B + A$ but $A - B \neq B - A$ (commutative law holds for addition but not subtraction)
- $(A + B) + C = A + (B + C) = A + B + C$ (associative law holds for addition)
- $k(A + B) = kA + kB$ (scalar multiplication is distributive over matrix addition)
- $(A')' = A$
- $(A + B)' = A' + B'$
- $kA' = (kA')$
- A is symmetric $\Rightarrow A' = A$
- A is skew–symmetric $\Rightarrow A' = -A$

Matrix multiplication

If A is an $m \times n$ matrix and B is an $n \times p$ matrix, their **matrix product** AB is an $m \times p$ matrix, in which the entries of AB are obtained by multiplying each entry of a row in A by the corresponding entry in the column of B and the products added.

$$\text{For example:} \quad \overset{\textbf{M1}}{\begin{pmatrix} a & b & c \\ d & e & f \end{pmatrix}} \overset{\textbf{M2}}{\begin{pmatrix} p \\ q \\ r \end{pmatrix}} = \begin{pmatrix} ap + bq + cr \\ dp + eq + fr \end{pmatrix}$$

$$\quad 2 \times 3 \quad 3 \times 1 \qquad 2 \times 1$$

A **2 × 3** matrix multiplied by a **3 × 1** matrix results in a **2 × 1** matrix.

Matrices must be **conformable** for multiplication. This means the number of columns in the first matrix must be the same as the number of rows in the second matrix.

In the example above, **M1** has 3 columns and **M2** has 3 rows.

Example 8.3

Evaluate:

a $\begin{pmatrix} 4 & 3 \\ -2 & 7 \end{pmatrix} \begin{pmatrix} 8 \\ 9 \end{pmatrix}$

b $\begin{pmatrix} -1 & 8 & 1 \\ 5 & 0 & 6 \end{pmatrix} \begin{pmatrix} 2 & -4 \\ -3 & 5 \\ 1 & 3 \end{pmatrix}$

a $\begin{pmatrix} 4 & 3 \\ -2 & 7 \end{pmatrix} \begin{pmatrix} 8 \\ 9 \end{pmatrix} = \begin{pmatrix} 4 \times 8 + 3 \times 9 \\ -2 \times 8 + 7 \times 9 \end{pmatrix}$

$\quad 2 \times 2 \quad 2 \times 1 \qquad 2 \times 1$

$\qquad\qquad = \begin{pmatrix} 32 + 27 \\ -16 + 63 \end{pmatrix}$

$\qquad\qquad = \begin{pmatrix} 59 \\ 47 \end{pmatrix}$

> Matrix 1 is a 2 × 2 matrix. Matrix 2 is a 2 × 1 matrix. When multiplied the product will be a 2 × 1 matrix. Multiply each entry in the row of matrix 1 by the corresponding entry in the column of matrix 2 and add their products.

b $\begin{pmatrix} -1 & 8 & 1 \\ 5 & 0 & 6 \end{pmatrix} \begin{pmatrix} 2 & -4 \\ -3 & 5 \\ 1 & 3 \end{pmatrix} = \begin{pmatrix} -1 \times 2 + 8 \times -3 + 1 \times 1 & -1 \times -4 + 8 \times 5 + 1 \times 3 \\ 5 \times 2 + 0 \times -3 + 6 \times 1 & 5 \times -4 + 0 \times 5 + 6 \times 3 \end{pmatrix}$

$\quad 2 \times 3 \qquad 3 \times 2 \qquad\qquad\qquad\qquad\qquad 2 \times 2$

$\qquad\qquad = \begin{pmatrix} -2 + (-24) + 1 & 4 + 40 + 3 \\ 10 + 0 + 6 & -20 + 0 + 18 \end{pmatrix}$

$\qquad\qquad = \begin{pmatrix} -25 & 47 \\ 16 & -2 \end{pmatrix}$

Example 8.4

If $A = \begin{pmatrix} 3 & 2 \\ -4 & 1 \end{pmatrix}$ find the values of p and q given that $A^2 = pA + qI$ where

I is the matrix $\begin{pmatrix} 1 & 0 \\ 0 & 1 \end{pmatrix}$

$A^2 = \begin{pmatrix} 3 & 2 \\ -4 & 1 \end{pmatrix}\begin{pmatrix} 3 & 2 \\ -4 & 1 \end{pmatrix} = \begin{pmatrix} 9 + (-8) & 6 + 2 \\ (-12) + (-4) & (-8) + 1 \end{pmatrix} = \begin{pmatrix} 1 & 8 \\ -16 & -7 \end{pmatrix}$

and

$A^2 = p\begin{pmatrix} 3 & 2 \\ -4 & 1 \end{pmatrix} + q\begin{pmatrix} 1 & 0 \\ 0 & 1 \end{pmatrix} = \begin{pmatrix} 3p + q & 2p \\ -4p & p + q \end{pmatrix}$

$\begin{pmatrix} 1 & 8 \\ -16 & -7 \end{pmatrix} = \begin{pmatrix} 3p + q & 2p \\ -4p & p + q \end{pmatrix}$ ⟵ Set the matrices and entries equal to each other.

$2p = 8 \Rightarrow p = 4$

$3p + q = 1 \Rightarrow q = -11$

$-4p = -4 \times 4 = -16$ and $p + q = 4 + -11 = -7$ ✓ ⟵ Check with the other entries

- The 2×2 matrix $\begin{pmatrix} 1 & 0 \\ 0 & 1 \end{pmatrix}$ is called the **unit matrix**, or **identity matrix**, of order 2 and is denoted by I. If A is a 2×2 matrix, then $IA = AI = A$

- Similarly the 3×3 matrix $\begin{pmatrix} 1 & 0 & 0 \\ 0 & 1 & 0 \\ 0 & 0 & 1 \end{pmatrix}$ is the unit matrix of order 3 and is also denoted by I.

Exercise 8B

★ 1 Evaluate these matrix products.

a $(2\ 7)\begin{pmatrix} 1 \\ 5 \end{pmatrix}$

b $(2\ 1\ 3)\begin{pmatrix} -1 \\ 3 \\ 2 \end{pmatrix}$

c $(3\ 2\ 5)\begin{pmatrix} x \\ y \\ z \end{pmatrix}$

d $\begin{pmatrix} 1 & 3 \\ -1 & 2 \end{pmatrix}\begin{pmatrix} 4 \\ 2 \end{pmatrix}$

e $\begin{pmatrix} 3 & 2 \\ 1 & -2 \end{pmatrix}\begin{pmatrix} 3 \\ -1 \end{pmatrix}$

f $\begin{pmatrix} 2 & 3 \\ -1 & 1 \end{pmatrix}\begin{pmatrix} -2 & 4 \\ 3 & -1 \end{pmatrix}$

g $\begin{pmatrix} 1 & 0 & -2 \\ 3 & 2 & 5 \end{pmatrix}\begin{pmatrix} -3 \\ 2 \\ 1 \end{pmatrix}$

h $\begin{pmatrix} 2 & 3 & -4 \\ -1 & 0 & 2 \end{pmatrix}\begin{pmatrix} 3 & -1 \\ 2 & 5 \\ 2 & 1 \end{pmatrix}$

i $\begin{pmatrix} 1 & 2 & 3 \\ -2 & 4 & 4 \\ 1 & 3 & -1 \end{pmatrix}\begin{pmatrix} 1 \\ 0 \\ -2 \end{pmatrix}$

j $\begin{pmatrix} \sqrt{3} & \frac{2}{3} \\ -1 & \sqrt{5} \end{pmatrix}\begin{pmatrix} 4 & \sqrt{3} \\ 5 & 2 \\ 3 & 1 \end{pmatrix}$

k $\begin{pmatrix} \cos\theta & \sin\theta \\ \sin\theta & \cos\theta \end{pmatrix}\begin{pmatrix} \sin\theta \\ \cos\theta \end{pmatrix}$

l $\begin{pmatrix} \cos\theta & \sin\theta \\ \sin\theta & -\cos\theta \end{pmatrix}\begin{pmatrix} \cos\theta \\ \sin\theta \end{pmatrix}$

2 If $A = \begin{pmatrix} 2 & 1 \\ 3 & 1 \end{pmatrix}$, $B = \begin{pmatrix} 4 & 2 \\ 0 & 5 \end{pmatrix}$, $C = \begin{pmatrix} -2 \\ 3 \end{pmatrix}$, find:

a i AB **ii** BA and comment on the result

b i $A(BC)$ **ii** $(AB)C$ and comment on the result

c i $(AB)'$ **ii** $B'A'$ and comment on the result.

3 Given that $A = \begin{pmatrix} 3 & -1 \\ 2 & 1 \end{pmatrix}$, calculate:

a A^2 **b** A^3

4 Find the values of x and y in each of these matrix equations:

a $\begin{pmatrix} 2 & 3 \\ x & 1 \end{pmatrix}\begin{pmatrix} -2 & 1 \\ 3 & -2 \end{pmatrix} = \begin{pmatrix} 5 & y \\ -5 & 2 \end{pmatrix}$

b $\begin{pmatrix} 3 & 4 \\ -2 & 1 \end{pmatrix}\begin{pmatrix} x \\ y \end{pmatrix} = \begin{pmatrix} -5 \\ -4 \end{pmatrix}$

★ **5** Evaluate the following matrix products.

a $\begin{pmatrix} 3 & -2 \\ 5 & 1 \end{pmatrix}\begin{pmatrix} 1 & 0 \\ 0 & 1 \end{pmatrix}$

b $\begin{pmatrix} 1 & 0 \\ 0 & 1 \end{pmatrix}\begin{pmatrix} 3 & -2 \\ 5 & 1 \end{pmatrix}$

c $\begin{pmatrix} p & q \\ r & s \end{pmatrix}\begin{pmatrix} 1 & 0 \\ 0 & 1 \end{pmatrix}$

d $\begin{pmatrix} 1 & 0 \\ 0 & 1 \end{pmatrix}\begin{pmatrix} p & q \\ r & s \end{pmatrix}$

6 a Given that $A = \begin{pmatrix} 2 & -1 \\ 3 & 1 \end{pmatrix}$ and $I = \begin{pmatrix} 1 & 0 \\ 0 & 1 \end{pmatrix}$ show that $A^2 = 3A - 5I$

b Using the result from part **a**, express A^3 in the form $pA + qI$

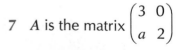
Multiply the equation through by A.

7 A is the matrix $\begin{pmatrix} 3 & 0 \\ a & 2 \end{pmatrix}$

a Show that A^2 can be expressed in the form $pA + qI$, stating the values of p and q.

b Obtain a similar expression for A^4.

8 If $A = \begin{pmatrix} 2 & 0 & 1 \\ -1 & 2 & 1 \\ 3 & 0 & 2 \end{pmatrix}$ and $B = \begin{pmatrix} -1 & 2 & 1 \\ 3 & 0 & 2 \\ 1 & 2 & 1 \end{pmatrix}$, find AB and BA.

Comment on the result.

9 Matrices A and B are defined by $A = \begin{pmatrix} 2 & x \\ 3 & 4 \end{pmatrix}$ $B = \begin{pmatrix} 4 & 6 \\ -2 & y \end{pmatrix}$

   a Find A^2.

   b Find the values of x and y if $B = 2A'$.

10 A matrix is said to be orthogonal if $A'A = I$.

   If $A = \begin{pmatrix} 0 & x \\ -1 & 0 \end{pmatrix}$, find the value(s) of x for A to be orthogonal.

11 Find the possible values x can take given that

   $A = \begin{pmatrix} x^2 & 2 \\ 1 & 2x \end{pmatrix}$, $B = \begin{pmatrix} 2 & 4 \\ 2 & x \end{pmatrix}$ and $AB = BA$.

- In general $AB \neq BA$
- $A(BC) = (AB)C = ABC$
- $A(B + C) = AB + AC$
- Unit matrix or identity matrix I, $AI = IA = A$
- $(AB)' = B'A'$
- A is orthogonal if $A'A = I = AA'$

Determinants and inverses of 2×2 and 3×3 matrices

The determinant of a matrix gives the matrix a value and is used when finding the inverse as well as having many other uses.

Determinant of a 2×2 matrix and a 3×3 matrix

The **determinant** of the matrix A is denoted by $\det(A)$ or $|A|$.

The determinant of a 2×2 matrix $A = \begin{pmatrix} a & b \\ c & d \end{pmatrix}$ is written as $\det(A) = \begin{vmatrix} a & b \\ c & d \end{vmatrix} = ad - bc$

The determinant of a 3×3 matrix $A = \begin{pmatrix} a & b & c \\ d & e & f \\ g & h & i \end{pmatrix}$ is written as $\det(A) = \begin{vmatrix} a & b & c \\ d & e & f \\ g & h & i \end{vmatrix}$

and is defined as:

$$\det(A) = a\begin{vmatrix} e & f \\ h & i \end{vmatrix} - b\begin{vmatrix} d & f \\ g & i \end{vmatrix} + c\begin{vmatrix} d & e \\ g & h \end{vmatrix}$$

$$= a(ei - fh) - b(di - fg) + c(dh - eg)$$

Note the minus sign in front of b in the middle of the calculation.

You can expand along any row or column using the table of signs $\begin{pmatrix} + & - & + \\ - & + & - \\ + & - & + \end{pmatrix}$

A matrix is said to be **singular** (**non-invertible**) if the determinant is zero and **non-singular** (**invertible**) if the determinant is not zero.

Example 8.5

1 Evaluate the determinants of these matrices.

a $A = \begin{pmatrix} 2 & -5 \\ -4 & 7 \end{pmatrix}$

b $B = \begin{pmatrix} 1 & 3 & -2 \\ 5 & 0 & 6 \\ 4 & -3 & 6 \end{pmatrix}$

a $\det(A) = \begin{vmatrix} 2 & -5 \\ -4 & 7 \end{vmatrix} = 2 \times 7 - (-5 \times -4) = -6$

b $\det(B) = \begin{vmatrix} 1 & 3 & -2 \\ 5 & 0 & 6 \\ 4 & -3 & 6 \end{vmatrix} = 1\begin{vmatrix} 0 & 6 \\ -3 & 6 \end{vmatrix} - 3\begin{vmatrix} 5 & 6 \\ 4 & 6 \end{vmatrix} + (-2)\begin{vmatrix} 5 & 0 \\ 4 & -3 \end{vmatrix}$

$= 1(0 \times 6 - 6 \times (-3)) - 3(5 \times 6 - 6 \times 4) + (-2)(5 \times (-3) - 0 \times 4)$

$= 18 - 18 + 30$

$= 30$

Exercise 8C

★ 1 Calculate the determinant of each matrix.

a $\begin{pmatrix} 1 & 4 \\ -2 & 3 \end{pmatrix}$

b $\begin{pmatrix} -2 & 1 \\ 2 & 4 \end{pmatrix}$

c $\begin{pmatrix} -3 & -2 \\ 5 & 4 \end{pmatrix}$

d $\begin{pmatrix} 3 & -6 \\ 1 & -2 \end{pmatrix}$

2 If $A = \begin{pmatrix} 2 & -1 \\ 3 & 4 \end{pmatrix}$ and $B = \begin{pmatrix} -1 & 3 \\ 2 & 1 \end{pmatrix}$, find:

    a AB and show that $\det(AB) = \det(A)\det(B)$

    b BA and show that $\det(BA) = \det(AB)$

    c A' and show that $\det(A') = \det(A)$

★ **3** Calculate the determinant of each matrix.

    a $\begin{pmatrix} 1 & 0 & 2 \\ -1 & 2 & 1 \\ 3 & 0 & 4 \end{pmatrix}$ **b** $\begin{pmatrix} 2 & 1 & 3 \\ 1 & -2 & 0 \\ -1 & 3 & 1 \end{pmatrix}$ **c** $\begin{pmatrix} 5 & 3 & -1 \\ 1 & 0 & 2 \\ -2 & 3 & 1 \end{pmatrix}$

4 $A = \begin{pmatrix} 1 & 2 & -1 \\ 0 & 3 & 4 \\ 2 & 1 & 3 \end{pmatrix}$ and $B = \begin{pmatrix} -1 & 3 & -2 \\ 2 & 1 & 2 \\ 1 & 0 & 4 \end{pmatrix}$

Find:

    a AB and show that $\det(AB) = \det(A)\det(B)$

    b BA and show that $\det(BA) = \det(AB)$

    c A' and show that $\det(A') = \det(A)$

- $\det(A) = \begin{vmatrix} a & b \\ c & d \end{vmatrix} = ad - bc$

- $\det(A) = \begin{vmatrix} a & b & c \\ d & e & f \\ g & h & i \end{vmatrix} = a\begin{vmatrix} e & f \\ h & i \end{vmatrix} - b\begin{vmatrix} d & f \\ g & i \end{vmatrix} + c\begin{vmatrix} d & e \\ g & h \end{vmatrix}$

$$= a(ei - fh) - b(di - fg) + c(dh - eg)$$

- A matrix is singular if the determinant is zero.
- A matrix is non-singular if the determinant is not zero.
- $\det(AB) = \det(A)\det(B)$
- $\det(A') = \det(A)$

Inverse of a 2 × 2 matrix

If $A = \begin{pmatrix} a & b \\ c & d \end{pmatrix}$ and $I = \begin{pmatrix} 1 & 0 \\ 0 & 1 \end{pmatrix}$

then $AI = \begin{pmatrix} a & b \\ c & d \end{pmatrix}\begin{pmatrix} 1 & 0 \\ 0 & 1 \end{pmatrix} = \begin{pmatrix} a & b \\ c & d \end{pmatrix}$

and $IA = \begin{pmatrix} 1 & 0 \\ 0 & 1 \end{pmatrix}\begin{pmatrix} a & b \\ c & d \end{pmatrix} = \begin{pmatrix} a & b \\ c & d \end{pmatrix}$

This shows that $AI = IA = A$. The 2×2 unit matrix is called the **identity matrix** for multiplication of 2×2 matrices.

Consider $A = \begin{pmatrix} 8 & 3 \\ 5 & 2 \end{pmatrix}$ and $B = \begin{pmatrix} 2 & -3 \\ -5 & 8 \end{pmatrix}$

$AB = \begin{pmatrix} 8 & 3 \\ 5 & 2 \end{pmatrix}\begin{pmatrix} 2 & -3 \\ -5 & 8 \end{pmatrix} = \begin{pmatrix} 1 & 0 \\ 0 & 1 \end{pmatrix}$ and $BA = \begin{pmatrix} 2 & -3 \\ -5 & 8 \end{pmatrix}\begin{pmatrix} 8 & 3 \\ 5 & 2 \end{pmatrix} = \begin{pmatrix} 1 & 0 \\ 0 & 1 \end{pmatrix}$

- Since $AB = BA = I$, B is called the **multiplicative inverse** of A.
- Similarly A can be called the **multiplicative inverse** of B.
- In general, the multiplicative inverse of a matrix A is denoted by A^{-1}, and $A A^{-1} = A^{-1}A = I$

In general the multiplicative inverse of a matrix is referred to as the **inverse** of a matrix. The additive inverse is more commonly known as its **negative**.

The inverse for $A = \begin{pmatrix} a & b \\ c & d \end{pmatrix}$ exists if and only if A **is non-singular**, i.e. $\det(A) \neq 0$

and is defined as $A^{-1} = \dfrac{1}{\det(A)}\begin{pmatrix} d & -b \\ -c & a \end{pmatrix} = \dfrac{1}{ad - bc}\begin{pmatrix} d & -b \\ -c & a \end{pmatrix}$

The matrix $\begin{pmatrix} d & -b \\ -c & a \end{pmatrix}$ is found by interchanging the entries in the main diagonal and changing the signs in the other diagonal. It is called the **adjoint** or **adjugate** of A, denoted by adj(A).

We can write $A^{-1} = \dfrac{\text{adj}(A)}{\det(A)}$

When $|A| = ad - bc = 0$ an inverse does not exist.

Example 8.6

Find the inverse of the matrix $A = \begin{pmatrix} 2 & 1 \\ 7 & 5 \end{pmatrix}$

$\det(A) = ad - bc = 2 \times 5 - 7 \times 1 = 3$, $\det(A) \neq 0$ ●——[$\det(A) \neq 0$, so an inverse does exist.]

$A^{-1} = \dfrac{1}{3}\begin{pmatrix} 5 & -1 \\ -7 & 2 \end{pmatrix} = \begin{pmatrix} \frac{5}{3} & \frac{-1}{3} \\ \frac{-7}{3} & \frac{2}{3} \end{pmatrix}$

Example 8.7

Solve this system of equations using matrices.

$$2x + y = 1$$
$$3x - 2y = 12$$

$$\begin{pmatrix} 2x + y \\ 3x - 2y \end{pmatrix} = \begin{pmatrix} 1 \\ 12 \end{pmatrix} \Rightarrow \begin{pmatrix} 2 & 1 \\ 3 & -2 \end{pmatrix}\begin{pmatrix} x \\ y \end{pmatrix} = \begin{pmatrix} 1 \\ 12 \end{pmatrix}$$

> Rewrite the system of equations in matrix form.

The inverse of $\begin{pmatrix} 2 & 1 \\ 3 & -2 \end{pmatrix}$ is $-\frac{1}{7}\begin{pmatrix} -2 & -1 \\ -3 & 2 \end{pmatrix}$

> Pre-multiply both sides of the matrix equation by the inverse $-\frac{1}{7}\begin{pmatrix} -2 & -1 \\ -3 & 2 \end{pmatrix}$
> Pre-multiplying A by B gives BA rather than AB; the order is important as in general $AB \neq BA$.

$$-\frac{1}{7}\begin{pmatrix} -2 & -1 \\ -3 & 2 \end{pmatrix}\begin{pmatrix} 2 & 1 \\ 3 & -2 \end{pmatrix}\begin{pmatrix} x \\ y \end{pmatrix} = -\frac{1}{7}\begin{pmatrix} -2 & -1 \\ -3 & 2 \end{pmatrix}\begin{pmatrix} 1 \\ 12 \end{pmatrix}$$

$$\begin{pmatrix} 1 & 0 \\ 0 & 1 \end{pmatrix}\begin{pmatrix} x \\ y \end{pmatrix} = -\frac{1}{7}\begin{pmatrix} -2 \times 1 + -1 \times 12 \\ -3 \times 1 + 2 \times 12 \end{pmatrix}$$

> $A^{-1}A = I$

$$\begin{pmatrix} x \\ y \end{pmatrix} = -\frac{1}{7}\begin{pmatrix} -14 \\ 21 \end{pmatrix} = \begin{pmatrix} 2 \\ -3 \end{pmatrix}$$

> $IA = A$

The solution to the system $\begin{matrix} 2x + y = 1 \\ 3x - 2y = 12 \end{matrix}$ is $x = 2$ and $y = -3$.

Example 8.8

a Find the inverse of the matrix $\begin{pmatrix} 3 & x \\ -2 & 4 \end{pmatrix}$

b For what value of x is this matrix singular?

a $\det(A) = \begin{vmatrix} 3 & x \\ -2 & 4 \end{vmatrix} = 3 \times 4 - (-2 \times x) = 12 + 2x$

Inverse $= \dfrac{1}{12 + 2x}\begin{pmatrix} 4 & -x \\ 2 & 3 \end{pmatrix}$

b $12 + 2x = 0 \Rightarrow x = -6$

> For a matrix to be singular $\det(A) = 0$.

Example 8.9

For a non-singular matrix B, where $B^2 = 2B - 3I$,

use matrix algebra to prove that:

a $B^3 = B - 6I$ b $B^{-1} = \frac{2}{3}I - \frac{1}{3}B$

a $B^2 = 2B - 3I$

$BB^2 = B(2B - 3I)$ ●────────── Pre-multiply both sides by B.

$B^3 = 2B^2 - 3BI$

$= 2(2B - 3I) - 3B$ ●────────── $BB = B^2$
Substitute $B^2 = 2B - 3I$ and $BI = B$

$= 4B - 6I - 3B$

$= B - 6I$

b **Method 1**

$B^2 = 2B - 3I$

$B^{-1}B^2 = B^{-1}(2B - 3I)$ ●────────── Pre-multiply both sides by B^{-1}.

$B^{-1}BB = 2B^{-1}B - 3B^{-1}I$

$IB = 2I - 3B^{-1}$ ●────────── Use the rules: $B^{-1}B = I$, $B^{-1}I = B^{-1}$

$3B^{-1} = 2I - B$

$B^{-1} = \frac{2}{3}I - \frac{1}{3}B$

Method 2

$B^2 = 2B - 3I$

$3I = 2B - B^2$

$= 2BI - BB$

$= B(2I - B)$

$I = B\frac{1}{3}(2I - B)$

Since $I = BB^{-1}$ ∴ $B^{-1} = \frac{1}{3}(2I - B) = \frac{2}{3}I - \frac{1}{3}B$

Exercise 8D

★ 1 Find the inverse of each of these 2×2 matrices, if it exists.

a $A = \begin{pmatrix} 2 & -4 \\ 1 & 3 \end{pmatrix}$ b $B = \begin{pmatrix} 8 & 2 \\ -3 & 7 \end{pmatrix}$ c $C = \begin{pmatrix} 11 & 2 \\ 3 & -1 \end{pmatrix}$

d $D = \begin{pmatrix} 3 & 2 \\ 6 & 4 \end{pmatrix}$ e $E = \begin{pmatrix} 5 & 4 \\ -2 & 7 \end{pmatrix}$ f $F = \begin{pmatrix} -3 & 1 \\ 9 & 8 \end{pmatrix}$

2   If $M = \begin{pmatrix} 2 & -1 \\ 3 & 4 \end{pmatrix}$ and $N = \begin{pmatrix} -1 & 3 \\ 2 & 1 \end{pmatrix}$, find:

   a M^{-1} b N^{-1} c $(MN)^{-1}$
   d $N^{-1}M^{-1}$ e $(NM)^{-1}$ f $M^{-1}N^{-1}$

3   a   Given that $A = \begin{pmatrix} -2 & 3 \\ 1 & 0 \end{pmatrix}$ and $B = \begin{pmatrix} 4 & 1 \\ 3 & -2 \end{pmatrix}$, show that:

   i $A^{-1}B^{-1} = (BA)^{-1}$ ii $(AB)^{-1} = B^{-1}A^{-1}$ iii $(A^{-1})^{-1} = A$

   iv $(A^{-1})' = (A')^{-1}$ v $(kA)^{-1} = \dfrac{1}{k}A^{-1}$

   b   Hence using matrix algebra show that $(AB)(AB)^{-1} = I$

4   Solve these systems of equations using matrices.
   a $x - y = 2$ b $2x + y = 1$ c $4x + 2y = 2$
       $x + y = 6$ $3x - y = -6$ $2x - 3y = -7$

★ 5   Find the inverse of these matrices and for what values of t are these matrices singular?

   a $\begin{pmatrix} t & 2 \\ -6 & 3 \end{pmatrix}$ b $\begin{pmatrix} 1 & 2t \\ 3 & 5 \end{pmatrix}$

   c $\begin{pmatrix} 2t & 4 \\ 8 & t \end{pmatrix}$ d $\begin{pmatrix} t+3 & 2 \\ 7 & t-2 \end{pmatrix}$

6   Let $A = \begin{pmatrix} 2 & 4 \\ 1 & 3 \end{pmatrix}$

   a   Show that $A^2 = 5A - 2I$

   b   Hence show, without evaluating A^3 or A^{-1}, that

      i $A^3 = 23A - 10I$ ii $A^{-1} = \frac{5}{2}I - \frac{1}{2}A$

   c   Find A^4

7   Let $A = \begin{pmatrix} x & 9 \\ 4 & x \end{pmatrix}$

   a   For what values of x is A singular?

   b   For $x = 6$, show that $A^2 = pA$, for some constant p, and determine the value of q such that $A^4 = qA$.

8 A is the matrix $\begin{pmatrix} 1 & 0 \\ k & -2 \end{pmatrix}$

    a Show that A^2 can be expressed in the form $mA + nI$, stating the values of m and n.

    b Obtain a similar expression for A^{-1}.

9 Let $A = \begin{pmatrix} p & 0 \\ 1-p & 1 \end{pmatrix}$

Find A^2 and A^3, expressing the entries of these matrices in terms of p in as simple a form as possible.

Hence suggest a formula for the matrix A^n where n is a positive integer.

- $AA^{-1} = A^{-1}A = I$
- $(AB)^{-1} = B^{-1}A^{-1}$
- $(A^{-1})' = (A')^{-1}$
- $(kA^{-1}) = \dfrac{1}{k}A^{-1}$

Inverse of a 3 × 3 matrix

If A is a 3×3 matrix, provided A^{-1} exists, then:

$$AA^{-1} = A^{-1}A = I \text{ where } I = \begin{pmatrix} 1 & 0 & 0 \\ 0 & 1 & 0 \\ 0 & 0 & 1 \end{pmatrix}$$

Just as in 2×2 matrices, a 3×3 matrix has an inverse, if and only if its determinant is non-zero. One method used to find the inverse of a 3×3 matrix is called **elementary row operations**. This method is also used in Gaussian elimination described later in the chapter.

There are three elementary row operations:

- any row R can be multiplied by a non-zero constant
- any row can be interchanged with another
- any row can be changed by adding it to any multiple of another.

To find the inverse using elementary row operations:

- Obtain the augmented matrix (This is a matrix obtained by appending the columns of two given matrices for the purpose of performing the same elementary row operations on each of the given matrices.)

- Reduce the augmented matrix $A \vdots I$ to the form $I \vdots B$ using elementary row operations. Then $B = A^{-1}$.

- Start with the bottom triangular half of A creating a triangle of zeros and then work on the top triangular half of A.

Example 8.10

Find the inverse of the 3×3 matrix $A = \begin{pmatrix} 1 & 2 & 1 \\ 1 & -1 & 2 \\ 3 & 0 & 2 \end{pmatrix}$

det $(A) = 9$ ●————

> det $(A) \neq 0$, so an inverse for A exists. See Example 8.5 part **b** for finding the determinant of a 3×3 matrix.

$\begin{pmatrix} 1 & 2 & 1 & \vdots & 1 & 0 & 0 \\ 1 & -1 & 2 & \vdots & 0 & 1 & 0 \\ 3 & 0 & 2 & \vdots & 0 & 0 & 1 \end{pmatrix}$

> Start with the augmented matrix.

$\begin{pmatrix} 1 & 2 & 1 & \vdots & 1 & 0 & 0 \\ 0 & -3 & 1 & \vdots & -1 & 1 & 0 \\ 0 & -6 & -1 & \vdots & -3 & 0 & 1 \end{pmatrix}$ $\quad \begin{matrix} R_2 \to R_2 - R_1 \\ R_3 \to R_3 - 3R_1 \end{matrix}$

$\begin{pmatrix} 1 & 2 & 1 & \vdots & 1 & 0 & 0 \\ 0 & -3 & 1 & \vdots & -1 & 1 & 0 \\ 0 & 0 & -3 & \vdots & -1 & -2 & 1 \end{pmatrix}$

$R_3 \to R_3 - 2R_2$

> You have created a triangle of zeros on the bottom half. Do the same on the top half.

$\begin{pmatrix} 3 & 6 & 0 & \vdots & 2 & -2 & 1 \\ 0 & -9 & 0 & \vdots & -4 & 1 & 1 \\ 0 & 0 & -3 & \vdots & -1 & -2 & 1 \end{pmatrix}$ $\quad \begin{matrix} R_1 \to 3R_1 + R_3 \\ R_2 \to 3R_2 + R_3 \end{matrix}$

$\begin{pmatrix} 9 & 0 & 0 & \vdots & -2 & -4 & 5 \\ 0 & -9 & 0 & \vdots & -4 & 1 & 1 \\ 0 & 0 & -3 & \vdots & -1 & -2 & 1 \end{pmatrix}$ $\quad R_1 \to 3R_1 + 2R_2$

> You have now created a triangle of zeros on the top half.
>
> To get the identity matrix, divide each row by the appropriate number. Divide row 1 by 9, row 2 by −9 and row 3 by −3.

$\begin{pmatrix} 1 & 0 & 0 & \vdots & \frac{-2}{9} & \frac{-4}{9} & \frac{5}{9} \\ 0 & 1 & 0 & \vdots & \frac{4}{9} & \frac{-1}{9} & \frac{-1}{9} \\ 0 & 0 & 1 & \vdots & \frac{1}{3} & \frac{2}{3} & \frac{-1}{3} \end{pmatrix}$

The inverse of $A = \begin{pmatrix} 1 & 2 & 1 \\ 1 & -1 & 2 \\ 3 & 0 & 2 \end{pmatrix}$ is $A^{-1} = \begin{pmatrix} \frac{-2}{9} & \frac{-4}{9} & \frac{5}{9} \\ \frac{4}{9} & \frac{-1}{9} & \frac{-1}{9} \\ \frac{1}{3} & \frac{2}{3} & \frac{-1}{3} \end{pmatrix} = \frac{1}{9} \begin{pmatrix} -2 & -4 & 5 \\ 4 & -1 & -1 \\ 3 & 6 & -3 \end{pmatrix}$

Check your answer using $AA^{-1} = I$ or $A^{-1}A = I$

Example 8.11

Let $A = \begin{pmatrix} 1 & -1 & 1 \\ 1 & 2 & 3 \\ 1 & -1 & 2 \end{pmatrix}$ and $B = \begin{pmatrix} 7 & 1 & -5 \\ 1 & 1 & -2 \\ -3 & 0 & 3 \end{pmatrix}$

a Show that $AB = kI$ for some constant k, where I is the 3×3 identity matrix and hence obtain the inverse matrix A^{-1}.

b Using the result from part **a** solve this system of equations:

$x - y + z = 6$

$x + 2y + 3z = 3$

$x - y + 2z = 2$

a $AB = \begin{pmatrix} 1 & -1 & 1 \\ 1 & 2 & 3 \\ 1 & -1 & 2 \end{pmatrix}\begin{pmatrix} 7 & 1 & -5 \\ 1 & 1 & -2 \\ -3 & 0 & 3 \end{pmatrix}$

> Multiply each entry in the row of the first matrix by the corresponding entry in the column of the second matrix and add their products

$= \begin{pmatrix} 7 + (-1) + (-3) & 1 + (-1) + 0 & (-5) + 2 + 3 \\ 7 + 2 + (-9) & 1 + 2 + 0 & (-5) + (-4) + 9 \\ 7 + (-1) + (-6) & 1 + (-1) + 0 & (-5) + 2 + 6 \end{pmatrix}$

$= \begin{pmatrix} 3 & 0 & 0 \\ 0 & 3 & 0 \\ 0 & 0 & 3 \end{pmatrix}$

$= 3I$ as required.

$AB = 3I$

$A^{-1}AB = 3A^{-1}I$ ← Pre-multiply by A^{-1}

$IB = 3A^{-1}$ ← Using $A^{-1}A = I$ and $IB = I$

$A^{-1} = \frac{1}{3}B$

$= \frac{1}{3}\begin{pmatrix} 7 & 1 & -5 \\ 1 & 1 & -2 \\ -3 & 0 & 3 \end{pmatrix}$

b $\begin{pmatrix} x - y + z \\ x + 2y + 3z \\ x - y + 2z \end{pmatrix} = \begin{pmatrix} 6 \\ 3 \\ 2 \end{pmatrix} \Rightarrow \begin{pmatrix} 1 & -1 & 1 \\ 1 & 2 & 3 \\ 1 & -1 & 2 \end{pmatrix}\begin{pmatrix} x \\ y \\ z \end{pmatrix} = \begin{pmatrix} 6 \\ 3 \\ 2 \end{pmatrix}$

> The system of equations can be written in matrix form

The inverse of $\begin{pmatrix} 1 & -1 & 1 \\ 1 & 2 & 3 \\ 1 & -1 & 2 \end{pmatrix}$ is $\frac{1}{3}\begin{pmatrix} 7 & 1 & -5 \\ 1 & 1 & -2 \\ -3 & 0 & 3 \end{pmatrix}$

$$\frac{1}{3}\begin{pmatrix} 7 & 1 & -5 \\ 1 & 1 & -2 \\ -3 & 0 & 3 \end{pmatrix}\begin{pmatrix} 1 & -1 & 1 \\ 1 & 2 & 3 \\ 1 & -1 & 2 \end{pmatrix}\begin{pmatrix} x \\ y \\ z \end{pmatrix} = \frac{1}{3}\begin{pmatrix} 7 & 1 & -5 \\ 1 & 1 & -2 \\ -3 & 0 & 3 \end{pmatrix}\begin{pmatrix} 6 \\ 3 \\ 2 \end{pmatrix}$$

Pre-multiply both sides by the inverse.

$$\Rightarrow \begin{pmatrix} x \\ y \\ z \end{pmatrix} = \frac{1}{3}\begin{pmatrix} 42 + 3 - 10 \\ 6 + 3 - 4 \\ -18 + 0 + 6 \end{pmatrix} = \frac{1}{3}\begin{pmatrix} 35 \\ 5 \\ -12 \end{pmatrix}$$

$$\Rightarrow x = \frac{35}{3}, y = \frac{5}{3}, z = -4$$

Exercise 8E

1 Find the inverse of the following 3×3 matrices, if they exist.

a $\begin{pmatrix} 1 & 0 & -2 \\ 2 & 3 & 2 \\ 1 & -2 & 1 \end{pmatrix}$
   b $\begin{pmatrix} 1 & 3 & 4 \\ 1 & -2 & 0 \\ 2 & 1 & 3 \end{pmatrix}$
   c $\begin{pmatrix} 1 & 1 & 1 \\ 3 & -1 & 2 \\ 2 & 3 & 1 \end{pmatrix}$

d $\begin{pmatrix} 1 & 0 & 1 \\ 4 & 3 & 1 \\ 2 & -1 & 3 \end{pmatrix}$
   e $\begin{pmatrix} 2 & 1 & 2 \\ 2 & 3 & 1 \\ -1 & 0 & 2 \end{pmatrix}$
   f $\begin{pmatrix} 1 & 1 & 2 \\ -1 & 1 & 3 \\ 2 & 1 & 1 \end{pmatrix}$

2 For what value of k does the matrix $\begin{pmatrix} 1 & 3 & -2 \\ 3 & -2 & 0 \\ 1 & k & 1 \end{pmatrix}$ not have an inverse?

★ **3** Let $A = \begin{pmatrix} 2 & 1 & 1 \\ 1 & 1 & 2 \\ -1 & -2 & 3 \end{pmatrix}$ and $B = \begin{pmatrix} 7 & -5 & 1 \\ -5 & 7 & -3 \\ -1 & 3 & 1 \end{pmatrix}$

a Show that $AB = kI$ for some constant k, where I is the 3×3 identity matrix and hence obtain the inverse matrix A^{-1}.

b Using the result from part **a** solve this system of equations:

$2x + y + z = 11$
$x + y + 2z = 9$
$-x - 2y + 3z = 8$

4 Find the inverse of the matrix $\begin{pmatrix} 2 & 3 & -1 \\ 1 & 2 & 1 \\ 3 & -1 & 2 \end{pmatrix}$ and hence solve this system of equations:

$2x + 3y - z = -7$
$x + 2y + z = 0$
$3x - y + 2z = 11$

Geometric transformations

A 2 × 2 matrix can be used to represent a transformation on the coordinate plane.

A point $P(x, y)$ on the coordinate plane can be transformed in a number of ways.

Reflection

Points can be reflected in the x-axis, y-axis, in the line $y = x$ and the line $y = -x$.

Reflection in the x-axis

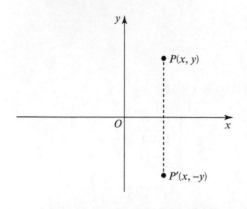

Reflection in the y-axis

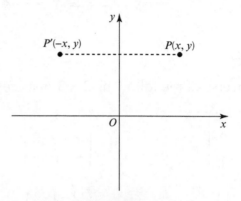

Reflection in the line $y = x$

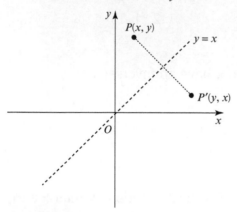

Reflection in the line $y = -x$

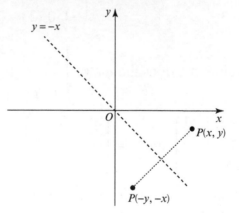

Rotation

A point can be rotated through $\theta°$ anti-clockwise about the origin.

Rotation anti-clockwise about the origin:

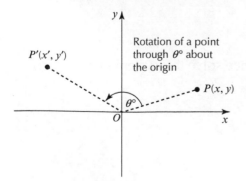

Rotation of a point through $\theta°$ about the origin

Dilation (enlargement or reduction)

2D shapes can be enlarged or reduced by a given scale factor.

Enlargement by scale factor 2:

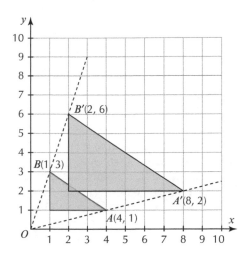

Transformation matrices

A linear transformation can be described as:

$(x, y) \rightarrow (ax + by, cx + dy)$ for a, b, c, and $d \in \mathbb{R}$

Let the point $P(x, y)$ under the transformation above have an image $P'(x', y')$ then $x' = ax + by$ and $y' = cx + dy$.

This can be represented using matrices:

$$\begin{pmatrix} x' \\ y' \end{pmatrix} = \begin{pmatrix} ax + by \\ cx + dy \end{pmatrix} \Rightarrow \begin{pmatrix} x' \\ y' \end{pmatrix} = \begin{pmatrix} a & b \\ c & d \end{pmatrix} \begin{pmatrix} x \\ y \end{pmatrix}$$

To obtain the image of point $P(x, y)$ under the transformation, pre-multiply the position vector of P, $\begin{pmatrix} x \\ y \end{pmatrix}$, by the matrix $\begin{pmatrix} a & b \\ c & d \end{pmatrix}$

To obtain the matrix associated with a particular transformation we need to find the images of (1, 0) and (0, 1) under that transformation.

$$\begin{pmatrix} a & b \\ c & d \end{pmatrix}\begin{pmatrix} 1 \\ 0 \end{pmatrix} = \begin{pmatrix} a \\ c \end{pmatrix} \text{ and } \begin{pmatrix} a & b \\ c & d \end{pmatrix}\begin{pmatrix} 0 \\ 1 \end{pmatrix} = \begin{pmatrix} b \\ d \end{pmatrix}$$

So under the transformation $\begin{pmatrix} a & b \\ c & d \end{pmatrix}$ the image of (1, 0) is (a, c) and the image of (0, 1) is (b, d).

Example 8.12

Find the following transformation matrices.

a R, the matrix associated with reflection in the x-axis.

b S, the matrix associated with reflection in the line $y = x$.

c D, the matrix associated with a dilation of scale factor 2 and centre of dilation the origin.

d M, the matrix associated with an anti-clockwise rotation about the origin of 60°.

a $R = \begin{pmatrix} a & b \\ c & d \end{pmatrix} = \begin{pmatrix} 1 & 0 \\ 0 & -1 \end{pmatrix}$

> Reflection in the x-axis maps the point (x, y) to $(x, -y)$. The mapping of $(1, 0)$ to $(1, 0)$ gives $(a, c) = (1, 0)$ and the mapping of $(0, 1)$ to $(0, -1)$ gives $(b, d) = (0, -1)$.

b $S = \begin{pmatrix} a & b \\ c & d \end{pmatrix} = \begin{pmatrix} 0 & 1 \\ 1 & 0 \end{pmatrix}$

> Reflection in the line $y = x$ maps the point (x, y) to (y, x). The mapping of $(1, 0)$ to $(0, 1)$ gives $(a, c) = (0, 1)$. The mapping of $(0, 1)$ to $(1, 0)$ gives $(b, d) = (1, 0)$.

c $D = \begin{pmatrix} a & b \\ c & d \end{pmatrix} = \begin{pmatrix} 2 & 0 \\ 0 & 2 \end{pmatrix}$

> Dilation of scale factor 2 maps the point (x, y) to $(2x, 2y)$. The mapping of $(1, 0)$ to $(2, 0)$ gives $(a, c) = (2, 0)$ and the mapping of $(0, 1)$ to $(0, 2)$ gives $(b, d) = (0, 2)$.

d $M = \begin{pmatrix} \cos 60° & -\sin 60° \\ \sin 60° & \cos 60° \end{pmatrix} = \begin{pmatrix} \frac{1}{2} & \frac{-\sqrt{3}}{2} \\ \frac{\sqrt{3}}{2} & \frac{1}{2} \end{pmatrix}$

> The transformation matrix associated with an anti-clockwise rotation about the origin through an angle of θ is $\begin{pmatrix} \cos\theta & -\sin\theta \\ \sin\theta & \cos\theta \end{pmatrix}$

> The formula to find M is given on the formulae list in your exam. If the rotation is 60° clockwise then use $\theta = -60°$.

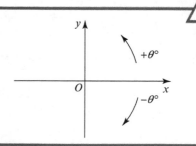

Example 8.13

Obtain the 2×2 transformation matrix, M, associated with a reduction of scale factor $\frac{1}{2}$ followed by an anti-clockwise rotation about the origin of 30°.

$A = \begin{pmatrix} \frac{1}{2} & 0 \\ 0 & \frac{1}{2} \end{pmatrix}$

> Let A be the transformation associated with the reduction of scale factor $\frac{1}{2}$

$$B = \begin{pmatrix} \cos 30° & -\sin 30° \\ \sin 30° & \cos 30° \end{pmatrix} = \begin{pmatrix} \dfrac{\sqrt{3}}{2} & \dfrac{-1}{2} \\ \dfrac{1}{2} & \dfrac{\sqrt{3}}{2} \end{pmatrix}$$

> Let B be the transformation matrix associated with the anti-clockwise rotation about the origin of 30°.

For the point $P(x, y)$, the image after transformation A is found by:

$$\begin{pmatrix} \dfrac{1}{2} & 0 \\ 0 & \dfrac{1}{2} \end{pmatrix} \begin{pmatrix} x \\ y \end{pmatrix}$$

> Apply transformation A to point P.

$$\begin{pmatrix} \dfrac{\sqrt{3}}{2} & \dfrac{-1}{2} \\ \dfrac{1}{2} & \dfrac{\sqrt{3}}{2} \end{pmatrix} \begin{pmatrix} \dfrac{1}{2} & 0 \\ 0 & \dfrac{1}{2} \end{pmatrix} \begin{pmatrix} x \\ y \end{pmatrix} = \begin{pmatrix} x' \\ y' \end{pmatrix}$$

> Apply transformation B. The order is important here. You must pre-multiply the first transformation by the second transformation.

$$M = BA$$

> Express M in terms of A and B.

$$\Rightarrow M = \begin{pmatrix} \dfrac{\sqrt{3}}{2} & \dfrac{-1}{2} \\ \dfrac{1}{2} & \dfrac{\sqrt{3}}{2} \end{pmatrix} \begin{pmatrix} \dfrac{1}{2} & 0 \\ 0 & \dfrac{1}{2} \end{pmatrix}$$

$$= \begin{pmatrix} \dfrac{\sqrt{3}}{4} & \dfrac{-1}{4} \\ \dfrac{1}{4} & \dfrac{\sqrt{3}}{4} \end{pmatrix}$$

Exercise 8F

★ 1 The points $A(3, 2)$, $B(2, 5)$, $C(-3, 4)$ and $D(5, -2)$ are transformed to

A', B', C' and D' by the transformation matrix $\begin{pmatrix} 2 & 3 \\ -1 & 0 \end{pmatrix}$

Find the coordinates of A', B', C' and D'.

★ 2 Find the 2×2 matrices corresponding to these linear transformations.

a   Reflection in the y-axis.

b   Reflection in the line $y = -x$.

c   Enlargement of scale factor 3 and centre of enlargement at the origin.

d   Anti-clockwise rotation about the origin of 135°.

3   Obtain the 2 × 2 matrices corresponding to the following linear transformations.

   a   Enlargement scale factor of 2 with centre at the origin followed by reflection in the x-axis.

   b   Reflection in the line $y = -x$ followed by a dilation of scale factor 2 with centre at the origin.

   c   Clockwise rotation of $\frac{\pi}{3}$ radians followed by enlargement of scale factor 4 with centre at the origin.

   d   Anti-clockwise rotation about the origin of $\frac{\pi}{2}$ followed by a reflection in the x-axis.

4   Show that a reflection in the y-axis followed by an anti-clockwise rotation of 270° about the origin is equivalent to a reflection in the line $y = x$.

5   Show that the image of a point (x, y) under a reflection in the line $y = x$ followed by a clockwise rotation of 60° about the origin is $\left(\dfrac{kx + y}{2}, \dfrac{x - ky}{2} \right)$, stating the value of k.

6   M is the matrix $\begin{pmatrix} \dfrac{1}{2} & \dfrac{-\sqrt{3}}{2} \\ \dfrac{\sqrt{3}}{2} & \dfrac{1}{2} \end{pmatrix}$ associated with a particular rotation about the origin O.

   A is a point with coordinates $\left(\dfrac{\sqrt{3}}{2} \quad \dfrac{1}{2} \right)$

   a   Calculate the length of OA and the angle OA makes with the positive direction of the x-axis.

   b   If A' is the image of A under the transformation associated with matrix M, find the coordinates of A'.

   c   With what angle of rotation about O is the matrix M associated?

   d   Find M^{-1}, the inverse matrix of M.

- Reflection in the x-axis is given by $\begin{pmatrix} 1 & 0 \\ 0 & -1 \end{pmatrix}$

- Reflection in the y-axis is given by $\begin{pmatrix} -1 & 0 \\ 0 & 1 \end{pmatrix}$

- Reflection in the line $y = x$ is given by $\begin{pmatrix} 0 & 1 \\ 1 & 0 \end{pmatrix}$

- Reflection in the line $y = -x$ is given by $\begin{pmatrix} 0 & -1 \\ -1 & 0 \end{pmatrix}$

- Enlargement of scale factor k with centre at the origin is given by $\begin{pmatrix} k & 0 \\ 0 & k \end{pmatrix}$

- Anti-clockwise rotation about the origin is given by $\begin{pmatrix} \cos\theta & -\sin\theta \\ \sin\theta & \cos\theta \end{pmatrix}$

Gaussian elimination to solve a 3×3 system of linear equations

A system of elementary row operations can be used to find the inverse of a 3×3 matrix. This method can be used to solve a 3×3 system of linear equations.

This method of applying an algorithm of elementary row operations to solve systems of linear equations was first explained in Europe by Sir Isaac Newton in the seventeenth century, although it was also widely used by Chinese mathematicians in 179 CE. It is named after the German mathematician Carl Friedrich Gauss.

In general a system of three linear equations:

$\begin{aligned} ax + by + cz &= p \\ dx + ey + fz &= q \\ gx + hy + iz &= r \end{aligned}$ can be expressed in matrix form as $\begin{pmatrix} a & b & c \\ d & e & f \\ g & h & i \end{pmatrix}\begin{pmatrix} x \\ y \\ z \end{pmatrix} = \begin{pmatrix} p \\ q \\ r \end{pmatrix}$

This can be reduced to the augmented matrix form:

$$\left(\begin{array}{ccc|c} a & b & c & p \\ d & e & f & q \\ g & h & i & r \end{array} \right)$$

Using elementary row operations to reduce this to upper triangular form (getting d, g and $h = 0$) would find the solution for z. Using back-substitution would then give the solutions to x and y.

If the matrix $\begin{pmatrix} a & b & c \\ d & e & f \\ g & h & i \end{pmatrix}$ is singular, the system of equations does not have a unique solution.

Example 8.14

Solve this system of equations:

$\begin{aligned} x + 2y + z &= 7 \\ x - y + 2z &= 5 \\ 3x \quad\quad + z &= 1 \end{aligned}$

$\left(\begin{array}{ccc|c} 1 & 2 & 1 & 7 \\ 1 & -1 & 2 & 5 \\ 3 & 0 & 1 & 1 \end{array} \right)$ — Rewrite as the augmented matrix.

$\left(\begin{array}{ccc|c} 1 & 2 & 1 & 7 \\ 1 & -1 & 2 & 5 \\ 0 & -6 & -2 & -20 \end{array} \right)$ $R_3 \to R_3 - 3R_1$ — Use elementary row operations to reduce this to upper triangular form.

(*continued*)

$$\begin{pmatrix} 1 & 2 & 1 & | & 7 \\ 0 & -3 & 1 & | & -2 \\ 0 & -6 & -2 & | & -20 \end{pmatrix} \qquad R_2 \rightarrow R_2 - R_1$$

$$\begin{pmatrix} 1 & 2 & 1 & | & 7 \\ 0 & -3 & 1 & | & -2 \\ 0 & 0 & -4 & | & -16 \end{pmatrix} \qquad R_3 \rightarrow R_3 - 2R_2$$

Row 3 gives: $-4z = -16 \Rightarrow z = 4$

Row 2 gives: $-3y + z = -2 \Rightarrow -3y + 4 = -2 \Rightarrow y = 2$

Row 1 gives: $x + 2y + z = 7 \Rightarrow x + 4 + 4 = 7 \Rightarrow x = -1$

The solution to the system of equations is $x = -1$, $y = 2$ and $z = 4$.

Inconsistent and redundant systems

There are three possibilities when solving a system of three linear equations for three unknowns:

- a unique solution

- infinitely many solutions, referred to as a **redundant system**

- no solutions, referred to as an **inconsistent system**.

These possibilities can be seen in context in Chapter 9 (Vectors) when dealing with the intersection of three planes.

Example 8.15

Solve this system of equations:

$x + y - 2z = 3$

$2x + 3y + z = -2$

$4x + 5y - 3z = 4$

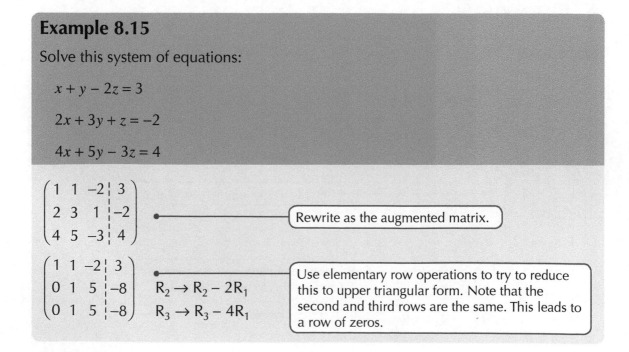

$$\begin{pmatrix} 1 & 1 & -2 & | & 3 \\ 2 & 3 & 1 & | & -2 \\ 4 & 5 & -3 & | & 4 \end{pmatrix}$$

Rewrite as the augmented matrix.

$$\begin{pmatrix} 1 & 1 & -2 & | & 3 \\ 0 & 1 & 5 & | & -8 \\ 0 & 1 & 5 & | & -8 \end{pmatrix} \qquad \begin{matrix} R_2 \rightarrow R_2 - 2R_1 \\ R_3 \rightarrow R_3 - 4R_1 \end{matrix}$$

Use elementary row operations to try to reduce this to upper triangular form. Note that the second and third rows are the same. This leads to a row of zeros.

$$\begin{pmatrix} 1 & 1 & -2 & | & 3 \\ 0 & 1 & 5 & | & -8 \\ 0 & 0 & 0 & | & 0 \end{pmatrix} \quad R_3 \rightarrow R_3 - R_2$$

> The row of zeros indicates that the system is redundant and there is no unique solution to the system.

There are an infinite number of solutions as z can have any value.

Given any z, then:

- from row 2: $\qquad y = -5z - 8$
- from row 1: $\qquad x = 7z + 11$

The general solution is then $x = 7z + 11$, $y = -5z - 8$, $z = z$.

In Example 8.15, a particular solution can be found by giving z a value.

For example, if $z = -1$, the solution would be $x = 4$, $y = -3$, $z = -1$.

Example 8.16

Solve this system of equations:

$x + y - 2z = 3$

$2x + 3y + z = -2$

$3x + 4y - z = 5$

$$\begin{pmatrix} 1 & 1 & -2 & | & 3 \\ 2 & 3 & 1 & | & -2 \\ 3 & 4 & -1 & | & 5 \end{pmatrix}$$

$$\begin{pmatrix} 1 & 1 & -2 & | & 3 \\ 0 & 1 & 5 & | & -8 \\ 0 & 1 & 5 & | & -4 \end{pmatrix} \quad \begin{array}{l} R_2 \rightarrow R_2 - 2R_1 \\ R_3 \rightarrow R_3 - 3R_1 \end{array}$$

$$\begin{pmatrix} 1 & 1 & -2 & | & 3 \\ 0 & 1 & 5 & | & -8 \\ 0 & 0 & 0 & | & 4 \end{pmatrix} \quad R_3 \rightarrow R_3 - R_2$$

> Row 3 reads as $0 = 4$, which is clearly impossible. This shows that the system is inconsistent and has no solutions.

Ill-conditioning

A system of equations is said to be **ill-conditioned** when a **small change** in the coefficients produces a relatively **large change** in the solution.

These small changes could occur when the data for the coefficients is a result of experiments which may have been rounded to either the nearest integer or one decimal place.

Consider the following:

The equations $2x + y = 2$ and $3x - y = 3$ have the solution $x = 1$ and $y = 0$.

By changing the coefficients of x slightly to 2·1 and 3·1 the equations become $2{·}1x + y = 2$ and $3{·}1x - y = 3$ and have the solution $x = 0{·}96$ and $y = -0{·}02$

A small change in the coefficients has led to a small change in the solutions and would be considered satisfactory in a real-life situation.

Now consider the following:

The equations $30x - 10y = -45$ and $31x - 10y = -15$ have the solution $x = 30$ and $y = 94{·}5$

Now let the equations be $29x - 10y = -45$ and $32x - 10y = -15$

The solution is now $x = 10$ and $y = 33{·}5$

Here a 3% change in the coefficient of x gives a 67% change in x and a 65% change in y.

This set of equations is said to be **ill-conditioned**.

In both of these examples we are looking for a point of intersection between two straight lines.

In the first example the gradients of the two lines are very different (their gradients are -2 and 3).

In the second example the lines are nearly parallel (their gradients are 3 and 3·1).

When the equations are nearly parallel a small change in any of the coefficients leads to a big change in the point of intersection.

Exercise 8G

1 Solve these systems of equations using Gaussian elimination.

a $x + y = 3$
$3y - z = 9$
$3x + y + z = 2$

b $x + y - z = -3$
$2x - 3y + 2z = -5$
$3x + 2y + z = 4$

c $x + z = 4$
$-x + 3y + 2z = -7$
$2x - 2y + 2z = 12$

d $2x + y + 2z = 6$
$x - 3y + 4z = -3$
$3x + y - z = 17$

e $2x + 4y + 2z = 3$
$4x + 2y - z = -5$
$x + 3y - 2z = -12$

f $x + y - 2z = -4$
$3x + y - z = 4$
$2x + 4y + z = 6$

2 A parabola, with equation $y = ax^2 + bx + c$, passes through the points (1, 4), (2, 7) and (−1, 10).

Use this information to set up a system of equations and solve it to find the equation of the parabola.

3 Find the general solution to this system of equations:

$x + 3y - 2z = 4$
$2x - y + z = -3$
$-3x - 2y + z = -1$

4 For what value of a does this system of equations have infinite solutions?

$x + 2y + 3z = 4$
$-x + 3y - 2z = -3$
$2x - 6y + az = 6$

5 For what value of a does this system of equations have no solutions?

$$x + 2y - z = -6$$
$$2x - y + z = 7$$
$$x - 3y + az = 4$$

6 Obtain the values of s and t for the system of equations below to have:

   i no solutions   ii infinite solutions   iii a unique solution.

$$x - 2y + z = 8$$
$$2x + y - z = -3$$
$$x + 3y + sz = t$$

7 The set of equations $3x + 4y = 11$, $10x + 13y = 29$ has solution $x = -27$, $y = 23$.
Find the solution to $3 \cdot 2x + 4y = 11$, $9 \cdot 9x + 13y = 29$
Comment on your result.

8 Identify which of these systems of equations demonstrates ill-conditioning.

   A $2x + y = 5$ B $20x + 5y = 15$

          $x - y = 7$ $21x + 5y = 16 \cdot 5$

   C $8x - 6y = 1$ D $4x - 2y = 20$

          $9x - 7y = 1 \cdot 1$ $x - 3y = 10$

Chapter review

Use these matrices throughout this chapter review:

$$A = \begin{pmatrix} 1 & 3 \\ 4 & 5 \end{pmatrix} \qquad B = \begin{pmatrix} 2 & 1 \\ -3 & 0 \end{pmatrix} \qquad C = \begin{pmatrix} \sqrt{2} & 1 \\ 4 & 3 \end{pmatrix}$$

$$D = \begin{pmatrix} 1 & 2 & -2 \\ 3 & 0 & 1 \\ 2 & -4 & 3 \end{pmatrix} \qquad E = \begin{pmatrix} 3 & 4 & 7 \\ 1 & -1 & 1 \\ 2 & 5 & 0 \end{pmatrix} \qquad F = \begin{pmatrix} -2 & 1 & 4 \\ 1 & 3 & 1 \\ 0 & 2 & 3 \end{pmatrix}$$

1 Evaluate:

   a $A + 3B - 2C$ b BC c C^2

   d D' e $\det(A)$ f $\det(E)$

   g C^{-1} h $2EF + DE$ i F^{-1}

2 Obtain the 2×2 matrix M such that $BM = C$.

3 Given that an inverse for P exists and $P^2 = 4P - 3I$:

   a show that: i $P^3 = 13P - 12I$ ii $P^{-1} = \frac{4}{3}I - \frac{1}{3}P$

   b find P^4 in the form $\alpha p + \beta I$

4 A is a 2×2 matrix such that $A^2 = 3A - 4I$ where I is the 2×2 identity matrix.

   a Find rational numbers p and q such that $A^3 = pA + qI$

   b If $B = xA + yI$, where x and y are rational numbers, find values of x and y for which $BA = I$

5 $M = \begin{pmatrix} 2 & -a \\ 4 & 6 \end{pmatrix}$

   a Find M^{-1}

   b For what value of a is M singular?

6 A linear transformation maps the point $(1, 0)$ to $(5, 2)$ and point $(0, 1)$ to $(-2, 1)$.
   Find the matrix associated with this transformation and find the image of $(3, 8)$.

7 Obtain the 2×2 matrices corresponding to these linear transformations.

   a Enlargement scale factor of 3 with centre at the origin followed by reflection in the y-axis.

   b Reflection in the line $y = -x$ followed by a rotation of $30°$ anti-clockwise about the origin.

8 Solve this system of equations for $\lambda \neq 3$.

   $x - 3y + z = 6$
   $2x + y - \lambda z = 2$
   $3x - 2y - 2z = 6$

   What happens when $\lambda = 3$?

9 Solve this system of equations for $t \neq -1$.

   $2x - y + 3z = 1$
   $4x + 3y + tz = 3$
   $2x + 4y - 4z = 2$

   Explain what happens when $t = -1$ and find the general solution.

10 By considering the system $(x + 0 \cdot 9y = 3$ and $x + 0 \cdot 99y = 2)$ show that the system $(x + y = 3$ and $x + 0 \cdot 99y = 2)$ is ill-conditioned.

- I understand and can use matrix algebra. ★ Exercise 8A Q4, Q5 ★ Exercise 8B Q1, Q5

- I can calculate the determinant of a 2×2 and a 3×3 matrix. ★ Exercise 8C Q1, Q3

- I can determine the inverse of a 2×2 and a 3×3 matrix. ★ Exercise 8D Q1, Q5 ★ Exercise 8E Q3

- I can use matrices to describe geometric transformations. ★ Exercise 8F Q1, Q2

- I can use Gaussian elimination to solve a 3×3 system of linear equations. ★ Exercise 8G Q1

9 Vectors

This chapter will show you how to:

- calculate a vector product
- work with lines in three dimensions
- work with planes.

You should already know:

- how to apply vector properties of addition, subtraction and magnitude
- how to determine position vectors and coordinates in three dimensions
- how to describe vectors defined in terms of unit vectors **i, j,** and **k**
- how to determine vector pathways
- how to work with the scalar product
- how to calculate the determinant of a 3×3 matrix.

Vectors

Vectors have both magnitude and direction as opposed to **scalars** which have magnitude only.

Geometrically, a vector can be described as a **directed line segment**, which illustrates both its magnitude and direction.

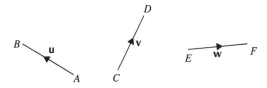

Definitions

A **position vector** is a vector that starts at the origin.

Parallel vectors have the same magnitude and are scalar multiples of each other.

If **a** is the vector $\begin{pmatrix} a_1 \\ a_2 \\ a_3 \end{pmatrix}$, the **length** (**magnitude**) of **a** is defined as $|\mathbf{a}| = \sqrt{a_1^2 + a_2^2 + a_3^2}$

A **unit vector** is a vector with magnitude 1. For example, the unit vector in the direction of **a** can be found by dividing the vector **a** by the magnitude of vector **a**.

Unit vector $= \dfrac{\mathbf{a}}{|\mathbf{a}|}$

The **standard basis vectors i**, **j** and **k** are unit vectors in the direction of the x, y and z axes respectively.

The position vector $\begin{pmatrix} a_1 \\ a_2 \\ a_3 \end{pmatrix}$ can be written as $a_1\mathbf{i} + a_2\mathbf{j} + a_3\mathbf{k}$

If P is the point $(1, -2, 3)$, then $\mathbf{p} = \mathbf{i} - 2\mathbf{j} + 3\mathbf{k}$ or $\mathbf{p} = \begin{pmatrix} x \\ y \\ z \end{pmatrix} = \begin{pmatrix} 1 \\ -2 \\ 3 \end{pmatrix}$

The **scalar** (or **dot**) **product** is defined as $\mathbf{a} \cdot \mathbf{b} = |\mathbf{a}||\mathbf{b}|\cos\theta$ and has magnitude only.

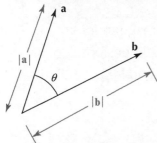

In component form: if $\mathbf{p} = \begin{pmatrix} a \\ b \\ c \end{pmatrix}$ and $\mathbf{q} = \begin{pmatrix} d \\ e \\ f \end{pmatrix}$, then $\mathbf{p} \cdot \mathbf{q} = ad + be + cf$

Properties of the scalar product

- $\mathbf{a} \cdot \mathbf{b} = \mathbf{b} \cdot \mathbf{a}$

- $\mathbf{a} \cdot \mathbf{a} = |\mathbf{a}|^2$

- $\mathbf{a} \cdot (\mathbf{b} + \mathbf{c}) = \mathbf{a} \cdot \mathbf{b} + \mathbf{a} \cdot \mathbf{c}$

- $\mathbf{a} \cdot \mathbf{b} = 0$ when $\mathbf{a}$ and $\mathbf{b}$ are perpendicular

---

Direction ratios and cosines

Given the vector $a = \begin{pmatrix} a_1 \\ a_2 \\ a_3 \end{pmatrix}$, the components uniquely determine the magnitude and the direction of **a**.

The ratio $a_1 : a_2 : a_3$ is referred to as the **direction ratio** of the vector.

If two vectors have equal direction ratios then they are parallel.

Vector **a** can be represented diagrammatically.

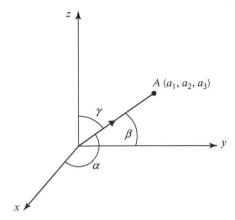

The vector **a** makes these angles with the axes:

- angle α with the x-axis
- angle β with the y-axis
- angle γ with the z-axis.

Therefore:

$$\cos\alpha = \frac{a_1}{|\mathbf{a}|}, \ \cos\beta = \frac{a_2}{|\mathbf{a}|} \text{ and } \cos\gamma = \frac{a_3}{|\mathbf{a}|}$$

The values $\frac{a_1}{|\mathbf{a}|}$, $\frac{a_2}{|\mathbf{a}|}$ and $\frac{a_3}{|\mathbf{a}|}$ are called the **direction cosines** of the vector **a**.

Example 9.1

Given that $\mathbf{a} = \begin{pmatrix} 2 \\ 1 \\ 3 \end{pmatrix}$, find the direction ratios and direction cosines of **a**.

The direction ratios are $2:1:3$.

$$|\mathbf{a}| = \sqrt{2^2 + 1^2 + 3^2} = \sqrt{14}$$

The direction cosines are $\frac{2}{\sqrt{14}}$, $\frac{1}{\sqrt{14}}$ and $\frac{3}{\sqrt{14}}$

Exercise 9A

1 Find the direction ratios of these vectors.

    a $2\mathbf{i} + 3\mathbf{j} - 4\mathbf{k}$

    b $\begin{pmatrix} -1 \\ 8 \\ -5 \end{pmatrix}$

    c   the vector joining points $A(2, 1, 0)$ and $B(-3, 2, 6)$.

2   Find the direction cosines of these vectors.

a $\mathbf{i} + \mathbf{j} + \mathbf{k}$

b $\begin{pmatrix} 3 \\ -5 \\ 2 \end{pmatrix}$

c $-3\mathbf{i} + 2\mathbf{j} + 7\mathbf{k}$

3   Show that the sum of the squares of the direction cosines is always equal to 1.

Vector product

A plane can be defined by any two non-parallel vectors **a** and **b**.

A normal, **n**, to the plane is a unit vector that is perpendicular to the plane in such a way that **a**, **b** and **n** form a right-handed system of vectors. You can model this with your right hand:

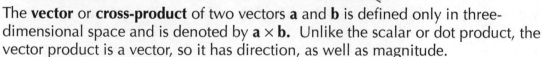

- **a** along the first finger
- **b** along the middle finger and
- **n** along the thumb

The **vector** or **cross-product** of two vectors **a** and **b** is defined only in three-dimensional space and is denoted by $\mathbf{a} \times \mathbf{b}$. Unlike the scalar or dot product, the vector product is a vector, so it has direction, as well as magnitude.

The vector product is defined as a vector that is perpendicular to both **a** and **b** and follows the right-hand rule.

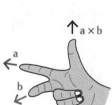

The vector product is defined by the formula:

$$\mathbf{a} \times \mathbf{b} = |\mathbf{a}||\mathbf{b}|\sin\theta\,\mathbf{n}$$

where:

n is the normal to the plane containing **a** and **b**

θ is the angle between the positive directions of **a** and **b**.

> **Properties of the vector product**
>
> - If either $\mathbf{a} = 0$ or $\mathbf{b} = 0$ then **n** is not defined and $\mathbf{a} \times \mathbf{b} = 0$.
>
> - $\mathbf{a} \times \mathbf{b}$ is a vector perpendicular to the plane in the same direction as **n**.
>
> - $|\mathbf{a} \times \mathbf{b}| = |\mathbf{a}||\mathbf{b}|\sin\theta$
>   This is the area of a parallelogram with sides determined by **a** and **b**.
>
>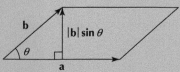
>
> - Parallel vectors have a vector product of zero: $|\mathbf{a} \times k\mathbf{a}| = |\mathbf{a}||k\mathbf{a}|\sin 0 = 0$.
>
> - $\mathbf{a} \times \mathbf{a} = 0$

- If $\mathbf{a} \neq 0$ and $\mathbf{b} \neq 0$ and $|\mathbf{a} \times \mathbf{b}| = 0$, then $\mathbf{a}$ and $\mathbf{b}$ are parallel vectors.

- $\mathbf{a} \times \mathbf{b} = -(\mathbf{b} \times \mathbf{a})$ (not commutative; see diagram on the right)

- $k\mathbf{a} \times \mathbf{b} = k(\mathbf{a} \times \mathbf{b})$ and $k\mathbf{a} \times l\mathbf{b} = kl\,(\mathbf{a} \times \mathbf{b})$

- The vector product is distributive over addition:

  $\mathbf{a} \times (\mathbf{b} + \mathbf{c}) = \mathbf{a} \times \mathbf{b} + \mathbf{a} \times \mathbf{c}$
  $(\mathbf{a} + \mathbf{b}) \times \mathbf{c} = \mathbf{a} \times \mathbf{c} + \mathbf{b} \times \mathbf{c}$

- $\mathbf{i}$, $\mathbf{j}$ and $\mathbf{k}$ form a right-hand system, therefore:

  $\mathbf{i} \times \mathbf{j} = \mathbf{k} \quad \mathbf{i} \times \mathbf{k} = -\mathbf{j} \quad \mathbf{j} \times \mathbf{i} = -\mathbf{k} \quad \mathbf{j} \times \mathbf{k} = \mathbf{i} \quad \mathbf{k} \times \mathbf{i} = \mathbf{j} \quad \mathbf{k} \times \mathbf{j} = -\mathbf{i}$

  These results can be remembered using this diagram. Clockwise gives a positive result and anti-clockwise gives a negative result.

- $\mathbf{i} \times \mathbf{i} = 0 \qquad \mathbf{j} \times \mathbf{j} = 0 \qquad \mathbf{k} \times \mathbf{k} = 0$

Vector product in component form

Let $\mathbf{a} = \begin{pmatrix} a_1 \\ a_2 \\ a_3 \end{pmatrix}$ and $\mathbf{b} = \begin{pmatrix} b_1 \\ b_2 \\ b_3 \end{pmatrix}$, then $\mathbf{a} \times \mathbf{b} = (a_1\mathbf{i} + a_2\mathbf{j} + a_3\mathbf{k}) \times (b_1\mathbf{i} + b_2\mathbf{j} + b_3\mathbf{k})$

By using the distributive law we get:

$$
\begin{aligned}
\mathbf{a} \times \mathbf{b} &= a_1b_1(\mathbf{i} \times \mathbf{i}) + a_1b_2(\mathbf{i} \times \mathbf{j}) + a_1b_3(\mathbf{i} \times \mathbf{k}) \\
&\quad + a_2b_1(\mathbf{j} \times \mathbf{i}) + a_2b_2(\mathbf{j} \times \mathbf{j}) + a_2b_3(\mathbf{j} \times \mathbf{k}) \\
&\quad + a_3b_1(\mathbf{k} \times \mathbf{i}) + a_3b_2(\mathbf{k} \times \mathbf{j}) + a_3b_3(\mathbf{k} \times \mathbf{k}) \\
&= 0 + a_1b_2\mathbf{k} + a_1b_3(-\mathbf{j}) + a_2b_1(-\mathbf{k}) + 0 + a_2b_3\mathbf{i} + a_3b_1\mathbf{j} + a_3b_2(-\mathbf{i}) + 0 \\
&= (a_2b_3 - a_3b_2)\mathbf{i} - (a_1b_3 - a_3b_1)\mathbf{j} + (a_1b_2 - a_2b_1)\mathbf{k}
\end{aligned}
$$

This is the determinant of the matrix $\begin{pmatrix} \mathbf{i} & \mathbf{j} & \mathbf{k} \\ a_1 & a_2 & a_3 \\ b_1 & b_2 & b_3 \end{pmatrix}$ from Chapter 8.

So $\mathbf{a} \times \mathbf{b} = \begin{vmatrix} \mathbf{i} & \mathbf{j} & \mathbf{k} \\ a_1 & a_2 & a_3 \\ b_1 & b_2 & b_3 \end{vmatrix} = (a_2b_3 - a_3b_2)\mathbf{i} - (a_1b_3 - a_3b_1)\mathbf{j} + (a_1b_2 - a_2b_1)\mathbf{k}$

Example 9.2

Given that $\mathbf{a} = \begin{pmatrix} 2 \\ 1 \\ 3 \end{pmatrix}$ and $\mathbf{b} = \begin{pmatrix} 1 \\ 5 \\ 2 \end{pmatrix}$, find $\mathbf{a} \times \mathbf{b}$.

$$\mathbf{a} \times \mathbf{b} = \begin{vmatrix} \mathbf{i} & \mathbf{j} & \mathbf{k} \\ 2 & 1 & 3 \\ 1 & 5 & 2 \end{vmatrix} = (2 - 15)\mathbf{i} - (4 - 3)\mathbf{j} + (10 - 1)\mathbf{k}$$

$$= -13\mathbf{i} - \mathbf{j} + 9\mathbf{k}$$

Example 9.3

Find the area of the triangle with vertices $A(2, 4, 5)$, $B(-3, 1, 2)$ and $C(3, -1, 6)$.

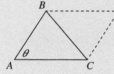

> The area of the triangle ABC is found by finding half the area of the parallelogram with sides $\overrightarrow{AB}$ and $\overrightarrow{AC}$.

$$\overrightarrow{AB} = \begin{pmatrix} -3 \\ 1 \\ 2 \end{pmatrix} - \begin{pmatrix} 2 \\ 4 \\ 5 \end{pmatrix} = \begin{pmatrix} -5 \\ -3 \\ -3 \end{pmatrix}$$

> You don't know the angle at A so use $|\mathbf{a} \times \mathbf{b}|$ to find the area.

$$\overrightarrow{AC} = \begin{pmatrix} 3 \\ -1 \\ 6 \end{pmatrix} - \begin{pmatrix} 2 \\ 4 \\ 5 \end{pmatrix} = \begin{pmatrix} 1 \\ -5 \\ 1 \end{pmatrix}$$

Area of $\triangle ABC = \frac{1}{2}\left|\overrightarrow{AB} \times \overrightarrow{AC}\right|$

$$= \frac{1}{2}\begin{vmatrix} \mathbf{i} & \mathbf{j} & \mathbf{k} \\ -5 & -3 & -3 \\ 1 & -5 & 1 \end{vmatrix} = \frac{1}{2}\left|((-3) - 15)\mathbf{i} - ((-5) - (-3))\mathbf{j} + (25 - (-3))\mathbf{k}\right|$$

$$= \frac{1}{2}\left|(-18\mathbf{i} + 2\mathbf{j} + 28\mathbf{k})\right|$$

$$= \frac{1}{2}\sqrt{(-18)^2 + 2^2 + 28^2}$$

$$= \frac{1}{2}\sqrt{1112}$$

$$= \frac{1}{2} \times 2\sqrt{278}$$

$$= \sqrt{278} \text{ square units}$$

Example 9.4

Find a unit vector **u** that is perpendicular to both $\mathbf{a} = \mathbf{i} + 2\mathbf{j} + 3\mathbf{k}$ and $\mathbf{b} = 2\mathbf{i} - \mathbf{j} + 2\mathbf{k}$.

$$\mathbf{u} = \frac{\mathbf{n}}{|\mathbf{n}|} = \frac{\mathbf{a} \times \mathbf{b}}{|\mathbf{a} \times \mathbf{b}|}$$

> The normal vector **n** is a vector perpendicular to both **a** and **b**. To find a unit vector **u** in the same direction as **n** you divide by $|\mathbf{n}|$.

$$\mathbf{a} \times \mathbf{b} = \begin{vmatrix} \mathbf{i} & \mathbf{j} & \mathbf{k} \\ 1 & 2 & 3 \\ 2 & -1 & 2 \end{vmatrix} = 7\mathbf{i} + 4\mathbf{j} - 5\mathbf{k}$$

$$|\mathbf{a} \times \mathbf{b}| = \sqrt{7^2 + 4^2 + (-5)^2} = \pm\sqrt{90} = \pm 3\sqrt{10}$$

$$\mathbf{u} = \pm\frac{1}{3\sqrt{10}}(7\mathbf{i} + 4\mathbf{j} - 5\mathbf{k})$$

Exercise 9B

★ 1 If $\mathbf{a} = \begin{pmatrix} 1 \\ 1 \\ 3 \end{pmatrix}$, $\mathbf{b} = \begin{pmatrix} 1 \\ -1 \\ 2 \end{pmatrix}$ and $\mathbf{c} = \begin{pmatrix} -1 \\ 3 \\ -2 \end{pmatrix}$, find:

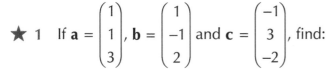

    a $\mathbf{a} \times \mathbf{b}$ b $\mathbf{b} \times \mathbf{a}$ c $\mathbf{a} \times \mathbf{a}$

    d $(\mathbf{a} \times \mathbf{b}) \times \mathbf{c}$ e $\mathbf{a} \times (\mathbf{b} \times \mathbf{c})$

2 Simplify the expression $(\mathbf{a} + 2\mathbf{b}) \times (3\mathbf{a} + 2\mathbf{b})$

3 Find the area of the triangle whose vertices are:

    a $(-1, -2, 3)$, $(-3, 1, 1)$ and $(1, 2, 2)$

    b $(2, 1, 1)$, $(2, 4, 0)$ and $(4, 3, 2)$

4 a Find a vector that is perpendicular to both $\mathbf{a} = -2\mathbf{i} + 5\mathbf{j} + 3\mathbf{k}$ and $\mathbf{b} = \mathbf{i} + 4\mathbf{j} - 2\mathbf{k}$

    b Find a unit vector that is perpendicular to both $\mathbf{a} = -2\mathbf{i} + 5\mathbf{j} + 3\mathbf{k}$ and $\mathbf{b} = \mathbf{i} + 4\mathbf{j} - 2\mathbf{k}$

Scalar triple product

To check that $\mathbf{a} \times \mathbf{b}$ is perpendicular to **a** and to **b** we can show that $\mathbf{a} \cdot (\mathbf{a} \times \mathbf{b})$ and $\mathbf{b} \cdot (\mathbf{a} \times \mathbf{b})$ are both zero.

With any three vectors **a**, **b** and **c**, we can find the scalar product of **a** and $\mathbf{b} \times \mathbf{c}$.

This is known as the **scalar triple product**: $\mathbf{a} \cdot (\mathbf{b} \times \mathbf{c})$.

The scalar triple product is a **number**, not a vector.

The scalar triple product can also be calculated using the determinant of a matrix in the same way as finding the vector product.

Let $\mathbf{a} = \begin{pmatrix} a_1 \\ a_2 \\ a_3 \end{pmatrix}$, $\mathbf{b} = \begin{pmatrix} b_1 \\ b_2 \\ b_3 \end{pmatrix}$ and $\mathbf{c} = \begin{pmatrix} c_1 \\ c_2 \\ c_3 \end{pmatrix}$,

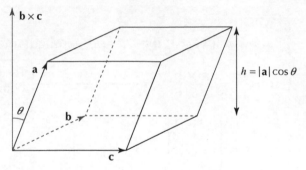

then $\mathbf{a} \cdot (\mathbf{b} \times \mathbf{c}) = \begin{vmatrix} a_1 & a_2 & a_3 \\ b_1 & b_2 & b_3 \\ c_1 & c_2 & c_3 \end{vmatrix}$

$$= a_1(b_2 c_3 - b_3 c_2) - a_2(b_1 c_3 - b_3 c_1) + a_3(b_1 c_2 - b_2 c_1)$$

The absolute value of the scalar triple product calculates the volume of a parallelepiped with sides given by vectors $\mathbf{a}$, $\mathbf{b}$ and $\mathbf{c}$.

The volume of a parallelepiped is found by calculating the area of base × height.

The base is a parallelogram with vectors $\mathbf{b}$ and $\mathbf{c}$ as its sides.

Area of base = $|\mathbf{b} \times \mathbf{c}|$

The height of the parallelepiped is given by $h = |\mathbf{a}| \cos\theta$ where θ is the angle between the plane $\mathbf{bc}$ and the vector $\mathbf{a}$.

Therefore $V = |\mathbf{b} \times \mathbf{c}| \cdot |\mathbf{a}| \cos\theta = \mathbf{a} \cdot (\mathbf{b} \times \mathbf{c})$

Any of the three pairs of parallel lines could have been used to find the volume so:

$V = \mathbf{a} \cdot (\mathbf{b} \times \mathbf{c}) = \mathbf{b} \cdot (\mathbf{c} \times \mathbf{a}) = \mathbf{c} \cdot (\mathbf{a} \times \mathbf{b})$

- $\mathbf{a} \cdot (\mathbf{b} \times \mathbf{c}) = (\mathbf{a} \times \mathbf{b}) \cdot \mathbf{c}$
- If any of $\mathbf{a}$, $\mathbf{b}$ or $\mathbf{c}$ are parallel then $\mathbf{a} \cdot (\mathbf{b} \times \mathbf{c})$ is zero.
- If any of $\mathbf{a}$, $\mathbf{b}$ or $\mathbf{c}$ are zero then $\mathbf{a} \cdot (\mathbf{b} \times \mathbf{c})$ is zero.

Example 9.5

Given $\mathbf{a} = 3\mathbf{i} + 2\mathbf{j} + \mathbf{k}$, $\mathbf{b} = \mathbf{i} - 3\mathbf{j} + 2\mathbf{k}$ and $\mathbf{c} = 3\mathbf{i} + \mathbf{j} + \mathbf{k}$, find $\mathbf{a} \cdot (\mathbf{b} \times \mathbf{c})$.

$\mathbf{a} \cdot \mathbf{b} \times \mathbf{c} = = \begin{vmatrix} 3 & 2 & 1 \\ 1 & -3 & 2 \\ 3 & 1 & 1 \end{vmatrix}$

$= 3((-3) - 2) - 2(1 - 6) + 1(1 - (-9))$

$= 5$

Example 9.6

Calculate the volume of the parallelepiped with edges $\mathbf{p} = \mathbf{i} - 2\mathbf{j} + \mathbf{k}$, $\mathbf{q} = 3\mathbf{i} + 3\mathbf{j}$ and $\mathbf{r} = 3\mathbf{i} + \mathbf{j} - 2\mathbf{k}$.

$\mathbf{p} \cdot (\mathbf{q} \times \mathbf{r}) = \begin{vmatrix} 1 & -2 & 1 \\ 3 & 3 & 0 \\ 3 & 1 & -2 \end{vmatrix}$

The absolute value is 24.

$= 1((-6) - 0) - (-2)((-6) - 0) + 1(3 - 9) = -24$

The volume of the parallelepiped is 24 cubic units.

Exercise 9C

⭐ **1** Given $\mathbf{p} = \begin{pmatrix} -2 \\ 1 \\ 4 \end{pmatrix}$, $\mathbf{q} = \begin{pmatrix} 1 \\ 1 \\ 2 \end{pmatrix}$, $\mathbf{r} = \begin{pmatrix} 5 \\ 3 \\ -2 \end{pmatrix}$ and $\mathbf{s} = \begin{pmatrix} 4 \\ -2 \\ -8 \end{pmatrix}$, find:

   a $\mathbf{p} \cdot (\mathbf{q} \times \mathbf{r})$ **b** $\mathbf{s} \cdot (\mathbf{p} \times \mathbf{r})$ **c** $\mathbf{p} \cdot (\mathbf{q} \times \mathbf{s})$

2 Calculate the volume of this parallelepiped:

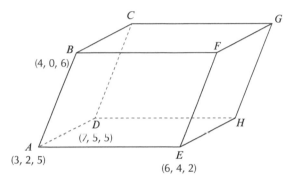

3 Given $\mathbf{a} = \mathbf{i} + 2\mathbf{j} + \mathbf{k}$, $\mathbf{b} = 2\mathbf{i} + \mathbf{k}$ and $\mathbf{c} = \mathbf{i} + \mathbf{j} - \mathbf{k}$, verify that
$\mathbf{a} \cdot (\mathbf{b} \times \mathbf{c}) = \mathbf{b} \cdot (\mathbf{c} \times \mathbf{a}) = \mathbf{c} \cdot (\mathbf{a} \times \mathbf{b})$

The equations of a straight line in three dimensions

We can determine the equation of any line in space if we know its direction and a point that lies on the line.

Let $\mathbf{b}$ be a vector that runs parallel to the line L, that is, the direction vector of line L and point $A(a_1, a_2, a_3)$ be a **known** point on the line.

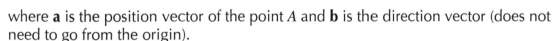

Let $R(x, y, z)$ be any point on the line.

Then $\overrightarrow{AR} = t\mathbf{b}$ for some scalar t. This gives $\mathbf{r} - \mathbf{a} = t\mathbf{b}$ and rearranging gives $\mathbf{r} = \mathbf{a} + t\mathbf{b}$ $(t \in \mathbb{R})$

where $\mathbf{a}$ is the position vector of the point A and $\mathbf{b}$ is the direction vector (does not need to go from the origin).

This is the **vector form of the equation of a straight line**.

By writing this in component form we get:

$$\begin{pmatrix} x \\ y \\ z \end{pmatrix} = \begin{pmatrix} a_1 \\ a_2 \\ a_3 \end{pmatrix} + t \begin{pmatrix} b_1 \\ b_2 \\ b_3 \end{pmatrix} = \begin{pmatrix} a_1 + tb_1 \\ a_2 + tb_2 \\ a_3 + tb_3 \end{pmatrix}$$

This gives the equation of a line in **parametric form** as:

$$x = a_1 + tb_1$$
$$y = a_2 + tb_2 \quad (t \in \mathbb{R})$$
$$z = a_3 + tb_3$$

By rearranging each equation to make t the subject we then get the **symmetric** form of the equation:

$$\frac{x - a_1}{b_1} = \frac{y - a_2}{b_2} = \frac{z - a_3}{b_3} \, (= t)$$

Note:

- The '$= t$' is omitted in the symmetric form but is needed to convert to the other forms.
- The symmetric form is also referred to as **standard** form, **Cartesian** form or **canonical** form.
- If any of the components of the direction vector is zero, then parts of the symmetric form will be undefined and it is better to use the parametric form.
- Each point on the line L is uniquely associated with a value of the parameter t.
- The equations of a particular line are not unique. The direction vector can be any multiple of **b**.

Example 9.7

a   Find the equation of the line passing through the point $(2, 1, 4)$ and parallel to $2\mathbf{i} + 3\mathbf{j} - 4\mathbf{k}$

b   Show that the point $(4, 4, 0)$ lies on this line.

a $2\mathbf{i} + 3\mathbf{j} - 4\mathbf{k}$ is parallel to the line so is a direction vector to the line.

$$\mathbf{a} = \begin{pmatrix} 2 \\ 1 \\ 4 \end{pmatrix} \qquad \mathbf{b} = \begin{pmatrix} 2 \\ 3 \\ -4 \end{pmatrix}$$

$$\frac{x - a_1}{b_1} = \frac{y - a_2}{b_2} = \frac{z - a_3}{b_3}$$

> Substitute the point and direction vector into the parametric form of the equation of a straight line.

$$\Rightarrow \frac{x - 2}{2} = \frac{y - 1}{3} = \frac{z - 4}{-4} \, (= t)$$

> The denominators are the coefficients of the direction vector.

b $\dfrac{4 - 2}{2} = 1, \quad \dfrac{4 - 1}{3} = 1, \quad \dfrac{0 - 4}{-4} = 1$

> Substitute the point $(4, 4, 0)$ into the line.

Because they are all the same answer (1), the point must lie on the line.

Example 9.8

Find the parametric equations of the line that passes through the points $A(-1, 2, -3)$ and $B(2, 3, 2)$.

$$\overrightarrow{AB} = \begin{pmatrix} 2 \\ 3 \\ 2 \end{pmatrix} - \begin{pmatrix} -1 \\ 2 \\ -3 \end{pmatrix} = \begin{pmatrix} 3 \\ 1 \\ 5 \end{pmatrix}$$

> To find the vector equation of the line you need a direction vector and a point on the line.
>
> The vector $\overrightarrow{AB}$ is parallel to the line. You could also use
> $$\overrightarrow{BA} = \begin{pmatrix} -3 \\ -1 \\ -5 \end{pmatrix}$$

The vector equation of the line with direction vector $3\mathbf{i}+\mathbf{j}+5\mathbf{k}$ and passing through the point $(2, 3, 2)$ is:

$\mathbf{r} = 2\mathbf{i}+3\mathbf{j}+2\mathbf{k}+t(3\mathbf{i}+\mathbf{j}+5\mathbf{k})$

The parametric equations are:

$$\begin{pmatrix} x \\ y \\ z \end{pmatrix} = \begin{pmatrix} 2 \\ 3 \\ 2 \end{pmatrix} + t\begin{pmatrix} 3 \\ 1 \\ 5 \end{pmatrix} \Rightarrow x = 2+3t,\ y = 3+t \text{ and } z = 2+5t$$

$\mathbf{r} = -\mathbf{i}+2\mathbf{j}-3\mathbf{k}+t(3\mathbf{i}+\mathbf{j}+5\mathbf{k})$ and
$x = -1+3t,\ y = 2+t$ and $z = -3+5t$

> You could have used point A instead of point B to give different equations. The equations of a particular line are not unique.

To see whether two lines are the same, check whether:

- a point on one line lies on the other
- their direction vectors are parallel.

Example 9.9

Find the parametric equations of the line that passes through the point $A(2, -1, 3)$ and is parallel to the line $\dfrac{x-1}{1} = \dfrac{y-3}{3} = \dfrac{z+2}{4}$

The direction vector of the line $\dfrac{x-1}{1} = \dfrac{y-3}{3} = \dfrac{z+2}{4}$ is $\begin{pmatrix} 1 \\ 3 \\ 4 \end{pmatrix}$

> This is the direction vector of the line you are looking for.

$\dfrac{x-2}{1} = \dfrac{y+1}{3} = \dfrac{z-3}{4}$

> This is the symmetric form of the line.

Exercise 9D

★ 1 a Find the symmetric equations of the line with parametric equations

$x = 3+2t$
$y = -2+t$
$z = 4-3t$

b Find the vector equation of the line with symmetric equation

$\dfrac{x-1}{1} = \dfrac{y-2}{-3} = \dfrac{z+3}{2}$

★ 2 Find the vector equation of the line through the point (4, 3, 7) in the

direction $\begin{pmatrix} 2 \\ -3 \\ 1 \end{pmatrix}$

★ 3 a State the symmetric equation of the line that passes through the point (−1, 3, 6) and is parallel to the vector $\mathbf{i} + 2\mathbf{j} + 5\mathbf{k}$

b Which of these points lie on this line?

    i (2, 1, 4)   ii (3, 11, 26)   iii (−2, 1, 1)

4 Find the vector equation of the line through the point (−2, 1, 0) parallel to the line with equation $\dfrac{x-1}{1} = \dfrac{y-2}{-3} = \dfrac{z+3}{2}$

5 Find the symmetric equations of the lines joining these pairs of points.

    a (1, 0, 2) and (3, 7, −2)   b (4, −1, 5) and (0, 2, 1)

    c (−2, 6, −1) and (−3, −2, −1)   d (0, 0, 4) and (2, 0, −3)

Intersection of two lines

Suppose that L_1 and L_2 are two lines in three-dimensional space.

There are three possibilities:

- the lines are **parallel** (their direction vectors are proportional)

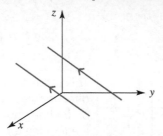

- the lines **intersect**

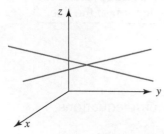

- the lines are **skew**, that is not parallel or intersecting. (Note that skew lines cannot occur in two-dimensional space.)

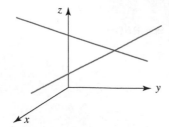

Example 9.10

Find the point of intersection of the lines $\dfrac{x-5}{3} = \dfrac{y-2}{1} = \dfrac{z-5}{1}$ and $\dfrac{x-2}{2} = \dfrac{y-1}{-1} = \dfrac{z-4}{3}$ if it exists.

Suppose the lines meet at the point (x, y, z).

> Write each equation in parametric form using parameters t_1 and t_2.

$x = 3t_1 + 5$, $y = t_1 + 2$, $z = t_1 + 5$ and $x = 2t_2 + 2$, $y = -t_2 + 1$, $z = 3t_2 + 4$

$$
\begin{aligned}
3t_1 + 5 &= 2t_2 + 2 & &\Rightarrow 3t_1 = 2t_2 - 3 & &①\\
t_1 + 2 &= -t_2 + 1 & &\Rightarrow\ t_1 = -t_2 - 1 & &②\\
t_1 + 5 &= 3t_2 + 4 & &\Rightarrow\ t_1 = 3t_2 - 1 & &③
\end{aligned}
$$

> Equate the corresponding coordinates.

$① - 3 \times ②$ gives $0 = 5t_2 \Rightarrow t_2 = 0$ and $t_1 = -1$

> Solve simultaneously to find t_1 and t_2.

$③$: $-1 = 3 \times 0 - 1 \Rightarrow -1 = -1$

> Substitute the values of t_1 and t_2 into $③$. This is consistent with both sets of parametric equations, so the lines intersect at the point.

The lines intersect at the point $x = 2t_2 + 2$, $y = -t_2 + 1$, $z = 3t_2 + 4$.

Point of intersection is $(2, 1, 4)$.

> Use the value of $t_2 = 0$ to find the point of intersection. This can be checked by substituting $t_1 = -1$ into $x = 3t_1 + 5$, $y = t_1 + 2$, $z = t_1 + 5$ and obtaining the same point.

Example 9.11

Let L_1 be the line with equation $\mathbf{r} = 3\mathbf{i} + 2\mathbf{j} - \mathbf{k} + t_1(\mathbf{i} - \mathbf{k})$ and L_2 be the line with equation $\mathbf{r} = \mathbf{i} - 2\mathbf{j} + \mathbf{k} + t_2(2\mathbf{i} + 2\mathbf{j} - \mathbf{k})$.

Show that L_1 and L_2 do not intersect.

$x = t_1 + 3$, $y = 2$, $z = -t_1 - 1$ and
$x = 2t_2 + 1$, $y = 2t_2 - 2$, $z = -t_2 + 1$

> Set up the parametric equations for each line.

$$
\begin{aligned}
t_1 + 3 &= 2t_2 + 1 & &\Rightarrow t_1 = 2t_2 - 2 & &①\\
2 &= 2t_2 - 2 & &\Rightarrow t_2 = 2 & &②\\
-t_1 - 1 &= -t_2 + 1 & &\Rightarrow t_1 = t_2 - 2 & &③
\end{aligned}
$$

> Equate the corresponding coordinates.

Substitute $t_2 = 2$ into $① \Rightarrow t_1 = 2$

> Substitute $t_2 = 2$ from $②$ into $①$.

Substitute $t_2 = 2$ and $t_1 = 2$ into $③ \Rightarrow 2 = 0$

> Substitute $t_2 = 2$ and $t_1 = 2$ into $③$.

The values do not satisfy $③$ so lines L_1 and L_2 do not intersect.

The angle between two lines
When two lines intersect we can find the angle between the two lines by finding the angle between their direction vectors.

Example 9.12
Find the acute angle between the lines $\frac{x-5}{3} = \frac{y-2}{1} = \frac{z-5}{1}$

and $\frac{x-2}{2} = \frac{y-1}{-1} = \frac{z-4}{3}$

Direction vectors for the lines are $\begin{pmatrix}3\\1\\1\end{pmatrix}$ and $\begin{pmatrix}2\\-1\\3\end{pmatrix}$ — Find the direction vectors.

$\mathbf{a} = \begin{pmatrix}3\\1\\1\end{pmatrix}$ $\mathbf{b} = \begin{pmatrix}2\\-1\\3\end{pmatrix}$

$\cos\theta = \frac{\mathbf{a}\cdot\mathbf{b}}{|\mathbf{a}||\mathbf{b}|}$ — Use the scalar product to find the angle between the two vectors.

$= \frac{3\times2+1\times(-1)+1\times3}{\sqrt{3^2+1^2+1^2}\times\sqrt{2^2+(-1)^2+3^2}} = \frac{8}{\sqrt{154}}$

$\theta = \cos^{-1}\left(\frac{8}{\sqrt{154}}\right)$

$= 49.86°$

Exercise 9E

1 Lines L_1 and L_2 have vector equations $\mathbf{r} = \mathbf{i} + 3\mathbf{j} - 2\mathbf{k} + t_1(4\mathbf{i} + \mathbf{k})$ and $\mathbf{r} = 5\mathbf{i} + 3\mathbf{j} + 8\mathbf{k} + t_2(-\mathbf{i} + 2\mathbf{k})$ respectively. Show that L_1 and L_2 intersect and find the point of intersection.

2 Find the acute angle between the lines whose equations are $x - 1 = y + 1 = z - 2$ and $x - 2 = \frac{y+2}{3} = \frac{z+1}{5}$, giving your answer to 3 s.f.

★ 3 For each pair of lines:
   i find the point of intersection if it exists
   ii find the acute angle between the lines if they intersect.

   a $\frac{x-5}{3} = \frac{y-16}{5} = \frac{z-9}{4}$ and $\frac{x-1}{1} = \frac{y-6}{1} = \frac{z-7}{2}$

   b $\mathbf{r} = \begin{pmatrix}1\\2\\-1\end{pmatrix} + t_1\begin{pmatrix}1\\0\\1\end{pmatrix}$ and $\mathbf{r} = \begin{pmatrix}2\\-1\\0\end{pmatrix} + t_2\begin{pmatrix}1\\1\\2\end{pmatrix}$

   c $x = t_1, y = 2t_1 - 2, z = -t_1 + 5$ and $x = -t_2 + 1, y = -3t_2 - 3, z = t_2 + 4$

4   The equations of three lines are:

$L_1: \dfrac{x-3}{-1} = \dfrac{y+2}{3} = \dfrac{z+1}{4}$

$L_2: \mathbf{r} = -2\mathbf{i} + 4\mathbf{j} + \mathbf{k} + t_1(-\mathbf{i} - 2\mathbf{k})$

$L_3: x = 2t_2 - 2,\ y = -3t_2 + 1,\ z = 3t_2$

a   Show that L_1 and L_2 intersect and find the position vector of the point of intersection.

b   Show that L_2 and L_3 intersect and find the position vector of the point of intersection.

c   Find the distance between these two points of intersection.

5   Find the point of intersection and the angle between the lines L_1 and L_2, when:

L_1 has equation $\dfrac{x-2}{1} = \dfrac{y-3}{3} = \dfrac{z-2}{-2}$

L_2 passes through $(3, 2, 4)$ and is parallel to $\mathbf{i} + 2\mathbf{j} - \mathbf{k}$

The equation of a plane

A plane, Π, can be uniquely defined if you know any of these:

- the normal to the plane and a point on the plane
- two non-parallel lines lying on the plane
- three points that lie in the plane.

Suppose that $\mathbf{n}$ is a vector perpendicular to the plane and A is a point on the plane having position vector $\mathbf{a}$.

Let R be a general point on the plane having position vector $\mathbf{r}$. $\overrightarrow{AR}$ will lie on the plane and is perpendicular to $\mathbf{n}$.

Therefore:

$\overrightarrow{AR} \cdot \mathbf{n} = 0$

$(\mathbf{r} - \mathbf{a}) \cdot \mathbf{n} = 0$

$\mathbf{r} \cdot \mathbf{n} - \mathbf{a} \cdot \mathbf{n} = 0$

$\mathbf{r} \cdot \mathbf{n} = \mathbf{a} \cdot \mathbf{n}$

This is the **scalar product form** of the equation of a plane. Since both $\mathbf{a}$ and $\mathbf{n}$ are known, this can be written as $\mathbf{r} \cdot \mathbf{n} = d$, where d is a constant.

By writing $\mathbf{r} = \begin{pmatrix} x \\ y \\ z \end{pmatrix}$ and $\mathbf{n} = \begin{pmatrix} n_1 \\ n_2 \\ n_3 \end{pmatrix}$ in component form we get the **Cartesian form** of

the equation of a plane: $n_1 x + n_2 y + n_3 z = d$, where $\mathbf{a} \cdot \mathbf{n} = d$

Given two non-parallel lines on a plane we will have two direction vectors that lie on the plane and hence can find a normal to the plane and two points that lie on the plane. If given three non-collinear points on the plane we can find two vectors on the plane and hence find a normal to the plane.

Example 9.13

Find the equation of the plane containing the point $(1, 3, 2)$ with normal $2\mathbf{i} + 3\mathbf{j} + \mathbf{k}$

The equation of a plane is given by $\mathbf{r} \cdot \mathbf{n} = \mathbf{a} \cdot \mathbf{n}$

Here $\mathbf{a} = \mathbf{i} + 3\mathbf{j} + 2\mathbf{k}$ and $\mathbf{n} = 2\mathbf{i} + 3\mathbf{j} + \mathbf{k}$

So $\mathbf{r} \cdot (2\mathbf{i} + 3\mathbf{j} + \mathbf{k}) = (\mathbf{i} + 3\mathbf{j} + 2\mathbf{k}).(2\mathbf{i} + 3\mathbf{j} + \mathbf{k}) = 2 + 9 + 2 = 13$

In Cartesian form the equation is:

$\quad 2x + 3y + z = 13$

Example 9.14

Find the equation of the plane that contains the lines $\dfrac{x+1}{2} = \dfrac{y-3}{3} = \dfrac{z+2}{1}$ and $\dfrac{x-1}{1} = \dfrac{y-3}{3} = \dfrac{z+2}{4}$

$$\mathbf{n} = \begin{vmatrix} \mathbf{i} & \mathbf{j} & \mathbf{k} \\ 2 & 3 & 1 \\ 1 & 3 & 4 \end{vmatrix}$$

> A normal to the plane can be found by finding the vector product of the two direction vectors of the straight lines.

$\quad = (12 - 3)\mathbf{i} - (8 - 1)\mathbf{j} + (6 - 3)\mathbf{k}$

$\quad = 9\mathbf{i} - 7\mathbf{j} + 3\mathbf{k}$

$\mathbf{a} = (-1, 3, -2)$ or $(1, 3, -2)$

> The position vector $\mathbf{a}$ of a point on the line could be either point from the two lines.

$\mathbf{r} \cdot \mathbf{n} = \mathbf{a} \cdot \mathbf{n}$

$\mathbf{r} \cdot (9\mathbf{i} - 7\mathbf{j} + 3\mathbf{k}) = (-\mathbf{i} + 3\mathbf{j} - 2\mathbf{k}) \cdot (9\mathbf{i} - 7\mathbf{j} + 3\mathbf{k})$

$\quad\quad\quad\quad\quad\quad = -9 - 21 - 6 = -36$

In Cartesian form the equation is:

$\quad 9x - 7y + 3z = -36$

> Always write out the equation.

Example 9.15

a   Find the equation of the plane that contains the points $A(2, 1, 3)$, $B(-1, 3, 1)$ and $C(1, 2, 7)$.

b   Show that the point $D(-2, 3, -9)$ also lies in the plane.

> First you need to find two vectors that are parallel to the plane.

a $\overrightarrow{AB} = \begin{pmatrix} -1 \\ 3 \\ 1 \end{pmatrix} - \begin{pmatrix} 2 \\ 1 \\ 3 \end{pmatrix} = \begin{pmatrix} -3 \\ 2 \\ -2 \end{pmatrix}$ and $\overrightarrow{AC} = \begin{pmatrix} 1 \\ 2 \\ 7 \end{pmatrix} - \begin{pmatrix} 2 \\ 1 \\ 3 \end{pmatrix} = \begin{pmatrix} -1 \\ 1 \\ 4 \end{pmatrix}$

$\mathbf{n} = \overrightarrow{AB} \times \overrightarrow{AC}$ — A normal to the plane is the vector product of these two vectors.

$$\mathbf{n} = \begin{vmatrix} \mathbf{i} & \mathbf{j} & \mathbf{k} \\ -3 & 2 & -2 \\ -1 & 1 & 4 \end{vmatrix} = (8 - (-2))\mathbf{i} - (-12 - 2)\mathbf{j} + (-3 - (-2))\mathbf{k}$$

$$= 10\mathbf{i} + 14\mathbf{j} - \mathbf{k}$$

$\mathbf{a} = 2\mathbf{i} + \mathbf{j} + 3\mathbf{k}$ — The position vector $\mathbf{a}$ can be any of the three points contained in the plane.

$\mathbf{r} \cdot \mathbf{n} = \mathbf{a} \cdot \mathbf{n}$

$\mathbf{r} \cdot (10\mathbf{i} + 14\mathbf{j} - \mathbf{k}) = (2\mathbf{i} + \mathbf{j} + 3\mathbf{k}) \cdot (10\mathbf{i} + 14\mathbf{j} - \mathbf{k}) = 20 + 14 - 3 = 31$

In Cartesian form the equation is:

$$10x + 14y - z = 31$$

Substitute the point $D(-2, 3, -9)$ into the equation.

b $(10 \times -2) + (14 \times 3) - (-9) = -20 + 42 + 9 = 31$

The point D satisfies the equation so D does lie in the plane.

The vector and parametric equations of a plane

If there are two non-parallel vectors $\mathbf{b}$ and $\mathbf{c}$ that are parallel to a particular plane and A is a point in the plane, then the plane can be uniquely defined.

To obtain the vector equation of the plane we consider the general point on the line R.

From the first diagram we can see that $\mathbf{r} = \mathbf{a} + \overrightarrow{AR}$

From the second diagram, we can get $\overrightarrow{AR}$ by combining a suitable multiple of vector $\mathbf{b}$ with a suitable multiple of vector $\mathbf{c}$.

So we can write $\overrightarrow{AR} = s\mathbf{b} + t\mathbf{c}$, where s and $t \in \mathbb{R}$.
Thus $\mathbf{r} = \mathbf{a} + s\mathbf{b} + t\mathbf{c}$

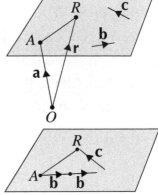

This is known as the **vector equation of the plane** that is parallel to the vectors $\mathbf{b}$ and $\mathbf{c}$ and contains the point with position vector $\mathbf{a}$.

If $\mathbf{a} = \begin{pmatrix} a_1 \\ a_2 \\ a_3 \end{pmatrix}$, $\mathbf{b} = \begin{pmatrix} b_1 \\ b_2 \\ b_3 \end{pmatrix}$ and $\mathbf{c} = \begin{pmatrix} c_1 \\ c_2 \\ c_3 \end{pmatrix}$, then the equation of the plane can be

written in parametric form, where $\mathbf{r} = \begin{pmatrix} x \\ y \\ z \end{pmatrix}$:

$$x = a_1 + sb_1 + tc_1$$
$$y = a_2 + sb_2 + tc_2 \quad (s, t \in \mathbb{R})$$
$$z = a_3 + sb_3 + tc_3$$

Example 9.16

Find the vector equation of the plane that contains three points A, B and C whose

position vectors are $\mathbf{a} = \begin{pmatrix} 2 \\ -1 \\ 3 \end{pmatrix}$, $\mathbf{b} = \begin{pmatrix} 3 \\ 1 \\ 1 \end{pmatrix}$ and $\mathbf{c} = \begin{pmatrix} 5 \\ -1 \\ 2 \end{pmatrix}$

$$\overrightarrow{AB} = \begin{pmatrix} 3 \\ 1 \\ 1 \end{pmatrix} - \begin{pmatrix} 2 \\ -1 \\ 3 \end{pmatrix} = \begin{pmatrix} 1 \\ 2 \\ -2 \end{pmatrix} \text{ and } \overrightarrow{AC} = \begin{pmatrix} 5 \\ -1 \\ 2 \end{pmatrix} - \begin{pmatrix} 2 \\ -1 \\ 3 \end{pmatrix} = \begin{pmatrix} 3 \\ 0 \\ -1 \end{pmatrix}$$

The vector equation of the plane is:

$\mathbf{r} = 2\mathbf{i} - \mathbf{j} + 3\mathbf{k} + s(\mathbf{i} + 2\mathbf{j} - 2\mathbf{k}) + t(3\mathbf{i} - \mathbf{k})$

> A, B and C lie in the plane so the vectors $\overrightarrow{AB}$ and $\overrightarrow{AC}$ will also lie in the plane.

This equation is not unique. It would be just as valid to say that the equation of the plane is:

$\mathbf{r} = 2\mathbf{i} - \mathbf{j} + 3\mathbf{k} + s\overrightarrow{BC} + t\overrightarrow{CA}$

or $\mathbf{r} = 3\mathbf{i} + \mathbf{j} + \mathbf{k} + s(\mathbf{i} + 2\mathbf{j} - 2\mathbf{k}) + t(3\mathbf{i} - \mathbf{k})$, for example.

Example 9.17

A plane Π has equation $3x - 2y + 3z = 12$. Find:

a the perpendicular distance from the plane Π to the origin

b the equation of the plane parallel to Π and containing the point $P(3, -2, 1)$

c the perpendicular distance from P to the plane Π.

$\mathbf{r} \cdot \mathbf{n} = d$

a $\hat{\mathbf{n}} = \dfrac{\mathbf{n}}{|\mathbf{n}|}$

> The perpendicular distance from a plane to the origin can be found by writing the equation of the plane in the form $\mathbf{r} \cdot \hat{\mathbf{n}} = d$, where $\hat{\mathbf{n}}$ is a unit vector normal to the plane, and d is the perpendicular distance from the plane to the origin.

The plane Π has equation $\mathbf{r} \cdot (3\mathbf{i} - 2\mathbf{j} + 3\mathbf{k}) = 12$ ── This can be rewritten.

$$\mathbf{r} \cdot \frac{3\mathbf{i} - 2\mathbf{j} + 3\mathbf{k}}{\sqrt{3^2 + (-2)^2 + 3^2}} = \frac{12}{\sqrt{3^2 + (-2)^2 + 3^2}}$$

$$\mathbf{r} \cdot \frac{3\mathbf{i} - 2\mathbf{j} + 3\mathbf{k}}{\sqrt{22}} = \frac{12}{\sqrt{22}}$$

The perpendicular distance of the plane from the origin is:

$\dfrac{12}{\sqrt{22}} \approx 2.56$ units.

b $\mathbf{n} = 3\mathbf{i} - 2\mathbf{j} + 3\mathbf{k}$

$\mathbf{p} = 3\mathbf{i} - 2\mathbf{j} + \mathbf{k}$ is the position vector of the point in the plane.

> The normal to plane Π is also a normal to any plane that is parallel to it.

The equation of the plane is $\mathbf{r} \cdot \mathbf{n} = \mathbf{p} \cdot \mathbf{n}$

$\mathbf{r} \cdot (3\mathbf{i} - 2\mathbf{j} + 3\mathbf{k}) = (3\mathbf{i} - 2\mathbf{j} + \mathbf{k}) \cdot (3\mathbf{i} - 2\mathbf{j} + 3\mathbf{k})$

$= 9 + 4 + 3 = 16$

In Cartesian form the equation is:

$3x - 2y + 3z = 16$

c   Rewrite $3x - 2y + 3z = 16$ as:

$\mathbf{r} \cdot \dfrac{3\mathbf{i} - 2\mathbf{j} + 3\mathbf{k}}{\sqrt{22}} = \dfrac{16}{\sqrt{22}}$

> To find the perpendicular distance of the point P from the plane Π you need to find the perpendicular distance from the plane containing P to the origin.

The perpendicular distance from P to the plane Π is:

$\dfrac{16}{\sqrt{22}} - \dfrac{12}{\sqrt{22}} = \dfrac{4}{\sqrt{22}}$ units.

Exercise 9F

★ 1   Find the Cartesian equation of the plane with normal $\begin{pmatrix} 1 \\ -2 \\ 2 \end{pmatrix}$ and containing the point $(4, 1, -3)$.

★ 2   a   Find the vector equation of the plane containing the points $A(2, 5, 1)$, $B(-1, 3, 5)$ and $C(4, -2, 3)$.

b   Express this equation in parametric form.

c   Show that the point $(-5, -8, 15)$ lies in the plane.

3   The plane Π has vector equation $\mathbf{r} = 2\mathbf{i} - \mathbf{j} + 3\mathbf{k} + s(\mathbf{i} + 2\mathbf{j} - 2\mathbf{k}) + t(3\mathbf{i} - \mathbf{k})$.
Show that the point $(1, 3, 0)$ lies in the plane Π.

4   The plane Π has equation $2x - 3y + z = 11$. Show that the point $(1, -2, 3)$ lies in the plane Π.

5   Find the Cartesian equation of the plane containing each sets of points:

a $(1, 0, 5)$, $(3, -2, 7)$ and $(-2, 1, 3)$

b $(2, 3, 5)$, $(1, -1, 0)$ and $(3, 1, 2)$.

6   Find the Cartesian equation of the plane containing the point $P(4, -3, 7)$ and the line $\dfrac{x + 2}{4} = \dfrac{y - 3}{1} = \dfrac{z - 3}{-3}$

7   Find the perpendicular distance from the plane $4x - 2y - 7z = -5$ to the origin.

8   A plane Π has equation $6x - 2y - 5z = -12$. Find:

a   the perpendicular distance from the plane Π to the origin

b   the equation of the plane parallel to Π and containing the point $P(1, 4, -1)$

c   the perpendicular distance from P to the plane Π.

Intersection of a line and a plane

Suppose L is a line and Π is a plane in three-dimensional space.

To find the intersection of the line and the plane we substitute the equation of the line in parametric form into the equation of the plane and solve for the parameter t.

There are three possibilities for the intersection of L and Π:

- there is one solution: L and Π have a unique point of intersection
- there are infinitely many solutions: the line L lies on the plane Π
- there are no solutions: the line L is parallel to the plane Π.

Example 9.18

Find the intersection of the plane $3x - 2y + z = 1$ and the line $\dfrac{x-5}{2} = \dfrac{y-1}{-1} = \dfrac{z-6}{1}$

The parametric equations of the line are:

$x = 2t + 5,\ y = -t + 1,\ z = t + 6$

$3(2t + 5) - 2(-t + 1) + t + 6 = 1$

$9t + 19 = 1$

> If there is a point of intersection then this point (x, y, z) will satisfy both the equation of the plane and the equation of the line. Substitute the parametric equations into the equation of the plane.

$t = -2$

$x = 1,\ y = 3$ and $z = 4$

> Substitute $t = -2$ into the parametric equations of the line.

The point of intersection of the line and the plane is $(1, 3, 4)$.

Example 9.19

Find the intersection of the plane $2x + y - 3z = 5$ and the line with parametric equations $x = 2t + 3,\ y = -t - 1,\ z = t$

$2(2t + 3) + (-t - 1) - 3t = 5 \Rightarrow 5 = 5$

> Substitute the parametric equations of the line into the equation of the plane.

This shows that t can be any value and the line must lie in the plane.

Example 9.20

Find the points of intersection between the line $\dfrac{x-1}{1} = \dfrac{y-9}{-2} = \dfrac{z-5}{1}$ and the coordinate planes.

The parametric equations of the line are:

$x = t + 1,\ y = -2t + 9,\ z = t + 5$

$z = t + 5 = 0 \Rightarrow t = -5$

> The line meets the xy plane when $z = 0$.

$x = t + 1 = -5 + 1 = -4$

$y = -2t + 9 = 10 + 9 = 19$

The point of intersection between the line and xy plane is $(-4, 19, 0)$.

$y = -2t + 9 = 0 \Rightarrow t = \frac{9}{2}$ ●————————| The line meets the xz plane when $y = 0$. |

$x = t + 1 = \frac{9}{2} + 1 = \frac{11}{2}$

$z = t + 5 = \frac{9}{2} + 5 = \frac{19}{2}$

The point of intersection between the line and xz plane is $(\frac{11}{2}, 0, \frac{19}{2})$.

$x = t + 1 = 0 \Rightarrow t = -1$ ●————————| The line meets the yz plane when $x = 0$. |

$y = -2t + 9 = 2 + 9 = 11$

$z = t + 5 = -1 + 5 = 4$

The point of intersection between the line and yz plane is $(0, 11, 4)$.

The angle between a line and a plane

If a line L and a plane Π do intersect, it is possible to find the angle of intersection between them. To calculate the angle of intersection θ, we first find the angle between the normal $\mathbf{n}$ to the plane and the direction vector $\mathbf{b}$ of the line x. The angle of intersection is the **complement** of the angle between the line and the normal.

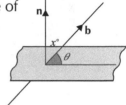

$\theta = 90 - x°$ or $\theta = \frac{\pi}{2} - x$

Example 9.21

Find the acute angle of intersection of the plane $3x - 2y + z = 1$ and the line

$\dfrac{x - 5}{2} = \dfrac{y - 1}{-1} = \dfrac{z - 6}{1}$

Method 1

Normal to the plane $= \begin{pmatrix} 3 \\ -2 \\ 1 \end{pmatrix}$ ●————| Express the normal and direction as vectors. |

Direction of the line $= \begin{pmatrix} 2 \\ -1 \\ 1 \end{pmatrix}$

The angle between the normal and the line:

$\cos\theta = \dfrac{3 \times 2 + (-2) \times (-1) + 1 \times 1}{\sqrt{3^2 + (-2)^2 + 1^2} \times \sqrt{2^2 + (-1)^2 + 1^2}} = \dfrac{9}{\sqrt{84}}$

$\theta = \cos^{-1}\left(\dfrac{9}{\sqrt{84}}\right)$

$= 10{\cdot}89°$

The acute angle of intersection is $90 - 10{\cdot}89 = 79{\cdot}11°$

Method 2

$$\sin \theta = \frac{\mathbf{b} \cdot \mathbf{n}}{|\mathbf{b}||\mathbf{n}|}$$

$$\boxed{\cos(90 - \theta) = \sin \theta}$$

$$\theta = \sin^{-1}\left(\frac{9}{\sqrt{84}}\right) = 79 \cdot 11°$$

Intersection of planes

Let Π_1 and Π_2 be two planes in three-dimensional space.

The two planes can:

- be parallel

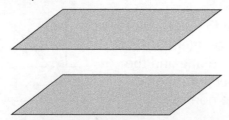

- intersect on a line

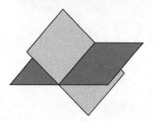

If the planes are parallel their normals are proportional.

Example 9.22

Show that the two planes Π_1 and Π_2 with equations $2x - 3y + z = 5$ and $-6x + 9y - 3z = 11$ are parallel.

Normal $\mathbf{n_1}$ of $\Pi_1 = \begin{pmatrix} 2 \\ -3 \\ 1 \end{pmatrix}$

Normal $\mathbf{n_2}$ of $\Pi_2 = \begin{pmatrix} -6 \\ 9 \\ -3 \end{pmatrix}$

$\mathbf{n_2} = -3\mathbf{n_1}$ so the planes are parallel.

If two planes intersect along a line, the equation of the line can be found using the following methods.

Example 9.23

Find the equation of the line of intersection between the two planes Π_1 and Π_2 with equations $4x - 2y + 2z = 2$ and $x + 2y + z = 8$

Method 1

$4x - 2y + 2z = 2$ ①

$x + 2y + z = 8$ ②

> For any point (x, y, z) that lies on the line of intersection, the point must satisfy each equation of the planes. First eliminate z. (Or you could choose to eliminate x or y.)

① − 2 × ② gives:

$2x - 6y = -14 \Rightarrow x = 3y - 7$

> There are infinitely many solutions to this equation. However, if you choose a value of x or y then the other value is fixed.

Suppose $y = t$, then $x = 3t - 7$

> Substitute these into the equation of plane Π_1. (Or you could choose to set $x = t$.)

$4(3t - 7) - 2t + 2z = 2 \Rightarrow z = 15 - 5t$

> Or you could substitute into plane Π_2.

$x = 3t - 8,\ y = t,\ z = 15 - 5t$

> These are the parametric equations of the line and can be rewritten in Cartesian form.

$$\frac{x + 8}{3} = \frac{y}{1} = \frac{z - 15}{-5}$$

> This line is not unique.

Method 2

Normal $\mathbf{n_1}$ of $\Pi_1 = \begin{pmatrix} 4 \\ -2 \\ 2 \end{pmatrix}$

> To find the equation of a line, you need the direction vector of the line and a point on the line. The direction vector of the line of intersection will be perpendicular to both normal vectors of the planes.

Normal $\mathbf{n_2}$ of $\Pi_2 = \begin{pmatrix} 1 \\ 2 \\ 1 \end{pmatrix}$

Direction vector $\mathbf{d}$ of the line is $\mathbf{n_1} \times \mathbf{n_2}$

$$\mathbf{d} = \begin{vmatrix} \mathbf{i} & \mathbf{j} & \mathbf{k} \\ 4 & -2 & 2 \\ 1 & 2 & 1 \end{vmatrix} = \mathbf{i}((-2) - 4) - \mathbf{j}(4 - 2) + \mathbf{k}(8 - (-2))$$

$$= -6\mathbf{i} - 2\mathbf{j} + 10\mathbf{k} = -2(3\mathbf{i} + \mathbf{j} - 5\mathbf{k})$$

$z = 0 \Rightarrow 4x - 2y = 2$

$ x + 2y = 8$

> To find a point on the line, set $z = 0$ in each equation of the plane if the line crosses the xy plane. (Similarly set $y = 0$ if the line crosses the xz plane.) Solve the resulting equations simultaneously.

$x = 2$ and $y = 3$

So $(2, 3, 0)$ is a point on the line.

> The coordinates are used to write the Cartesian equation of the line.

$$\frac{x - 2}{3} = \frac{y - 3}{1} = \frac{z}{-5}$$

> This line is not unique but does need to have the same direction as found in Method 1.

The angle between two planes

If two planes do intersect along a line then the angle between the planes can be found by finding the angle between their normal vectors.

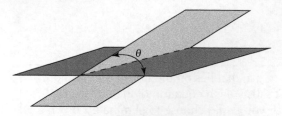

Example 9.24

Find the angle between the two planes Π_1 and Π_2 with equations $4x - 2y + 2z = 2$ and $x + 2y + z = 8$

Normal $\mathbf{n_1}$ of $\Pi_1 = \begin{pmatrix} 4 \\ -2 \\ 2 \end{pmatrix}$

Normal $\mathbf{n_2}$ of $\Pi_2 = \begin{pmatrix} 1 \\ 2 \\ 1 \end{pmatrix}$

The angle between the normals is:

$$\cos\theta = \frac{4 \times 1 + (-2) \times 2 + 2 \times 1}{\sqrt{4^2 + (-2)^2 + 2^2} \times \sqrt{1^2 + 2^2 + 1^2}} = \frac{2}{\sqrt{144}} = \frac{1}{6}$$

$$\theta = \cos^{-1}\left(\tfrac{1}{6}\right)$$

$$= 80 \cdot 41°$$

The intersection of three planes

Since a plane can be written in the form $ax + by + cz = k$, the intersection of three planes can be found by solving a 3×3 system of equations. We can use Gaussian elimination to solve this system for x, y and z (see Chapter 8).

The system of equations can have:

- a unique solution
- an infinite number of points
- no solution.

In a unique solution: the three planes meet at a single point. The three planes produce a 3×3 matrix which is non-singular.

An infinite number of points (**redundant system**) arises when:

- all three equations represent the same plane so any point in the plane will give a solution

- two of the planes are coincident and the third plane is not parallel to these two. The planes will intersect in a line and any point on the line will provide a solution

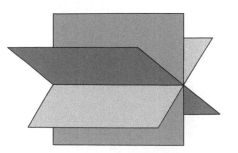

- three planes have a common line and any point on this line provides a solution.

No solution (**inconsistent system**) arises when:

- the equations are inconsistent with two or more of the planes parallel (but not coincident) so there is no common point

- the three planes are parallel

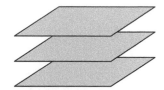

- two planes are parallel

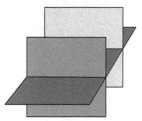

- each plane is parallel to the line of intersection of the other two planes so there is no common point.

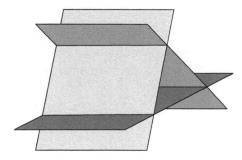

Example 9.25

Find the point of intersection of these three planes:

$$x + 2y + z = 7$$
$$x - y + 2z = 5$$
$$3x + z = 1$$

$$\begin{pmatrix} 1 & 2 & 1 & \vdots & 7 \\ 1 & -1 & 2 & \vdots & 5 \\ 3 & 0 & 1 & \vdots & 1 \end{pmatrix}$$

Rewrite as the augmented matrix.

$$\begin{pmatrix} 1 & 2 & 1 & | & 7 \\ 0 & -3 & 1 & | & -2 \\ 0 & 0 & -4 & | & -16 \end{pmatrix}$$

Carry out the elementary row operations.

The solution to the system of equations is:

$x = -1$, $y = 2$ and $z = 4$

This indicates a unique solution.

The planes intersect at the point $(-1, 2, 4)$.

$-1 + 2(2) + 4 = 7$ ✓

$-1 - 2 + 2(4) = 5$ ✓

$3(-1) + 4 = 1$ ✓

Check that $(-1, 2, 4)$ lies in all planes.

Remember to check your solution in all three equations.

Example 9.26

Find the line of intersection of these three planes:

$$x + y - 2z = 3$$
$$2x + 3y + z = -2$$
$$4x + 5y - 3z = 4$$

$$\begin{pmatrix} 1 & 1 & -2 & | & 3 \\ 2 & 3 & 1 & | & -2 \\ 4 & 5 & -3 & | & 4 \end{pmatrix}$$

Rewrite as the augmented matrix.

$$\begin{pmatrix} 1 & 1 & -2 & | & 3 \\ 0 & 1 & 5 & | & -8 \\ 0 & 0 & 0 & | & 0 \end{pmatrix}$$

Carry out the elementary row operations. This represents a redundant system, which means there are infinitely many solutions.

Let $z = t$, then from row 2 $y = -5t - 8$

from row 1 $x = 7t + 11$

The intersection of the planes is the line with parametric equations:

$x = 7t + 11$, $y = -5t - 8$, $z = t$

In symmetric form the equaton is:

Remember that the symmetric form is the same as the Cartesion form.

$$\frac{x - 11}{7} = \frac{y + 8}{-5} = \frac{z}{1}$$

Example 9.27

Find the line(s) of intersection of these three planes:

$$x + y - 2z = 3$$
$$2x + 3y + z = -2$$
$$3x + 4y - z = 5$$

$$\begin{pmatrix} 1 & 1 & -2 & | & 3 \\ 2 & 3 & 1 & | & -2 \\ 3 & 4 & -1 & | & 5 \end{pmatrix}$$

Notice that $\mathbf{n_3} = \mathbf{n_1} + \mathbf{n_2}$

$$\begin{pmatrix} 1 & 1 & -2 & | & 3 \\ 0 & 1 & 5 & | & -8 \\ 0 & 0 & 0 & | & 4 \end{pmatrix}$$

R_3 gives $0 = 4$ which is clearly impossible so this shows an inconsistent system and there are no solutions. Let $z = t$ and investigate the planes in pairs.

Planes 1 and 2

$x + y = 3 + 2t$ ①

$2x + 3y = -2 - t$ ②

② $- 2 \times$ ① $\Rightarrow y = -8 - 5t$ Substitute into ① to find x.

$x + (-8 - 5t) = 3 + 2t \Rightarrow x = 11 + 7t$

The intersection of planes 1 and 2 is the line with parametric equations:
$x = 7t + 11, \ y = -5t - 8, \ z = t$

In symmetric form the equation is: $\dfrac{x - 11}{7} = \dfrac{y + 8}{-5} = \dfrac{z}{1}$

Planes 1 and 3

$x + y = 3 + 2t$ ①

$3x + 4y = 5 + t$ ②

② $- 3 \times$ ① $\Rightarrow y = -4 - 5t$ Substitute into ① to find x.

$x + (-4 - 5t) = 3 + 2t \Rightarrow x = 7 + 7t$

The intersection of planes 1 and 3 is the line with parametric equations:
$x = 7 + 7t, \ y = -4 - 5t, \ z = t$

In symmetric form the equation is: $\dfrac{x - 7}{7} = \dfrac{y + 4}{-5} = \dfrac{z}{1}$

Planes 2 and 3

$2x + 3y = -2 - t$ ①

$3x + 4y = 5 + t$ ②

$2 \times$ ② $- 3 \times$ ① $\Rightarrow y = -16 - 5t$ Substitute into ① to find x.

$2x + 3(-16 - 5t) = 3 + 2t \Rightarrow x = 23 + 7t$

The intersection of planes 2 and 3 is the line with parametric equations:
$x = 23 + 7t, \ y = -16 - 5t, \ z = t$

In symmetric form the equation is: $\dfrac{x - 23}{7} = \dfrac{y + 16}{-5} = \dfrac{z}{1}$ Note that all three lines of intersection are parallel with direction vector $7\mathbf{i} - 5\mathbf{j} + \mathbf{k}$.

Exercise 9G

★ 1   For each line and plane find:

    i   the point of intersection

    ii   the angle between the line and the plane.

a $x = 2 + t, y = -3 + 2t, z = 2$ and $4x - y + z = 6$

b $x = 2 - 3t, y = 1 + t, z = 2 - 3t$ and $2x - 3y + z = 15$

c $\dfrac{x - 3}{2} = \dfrac{y - 6}{3} = \dfrac{z - 2}{1}$ and $x - 3y + 2z = 4$

d $\dfrac{x + 1}{-2} = \dfrac{y - 4}{1} = \dfrac{z}{1}$ and $3x + 2y - z = 5$

e $x = 3, y = 2t - 3, z = t + 1$ and $6x - y + 3z = 35$

f $x - 4 = \dfrac{y + 2}{3} = z - 1$ and $7x + 2y - 3z = 1$

2 Show that the line $\dfrac{x + 1}{-2} = y - 4 = z$ is parallel to the plane $2x + 4z = 3$

3 Show that the line $\dfrac{x - 1}{2} = \dfrac{y + 3}{4} = z - 2$ lies in the plane $3x - y - 2z = 2$

4 Show that the two planes with equations $x + 2y + 3z = 7$ and $-3x - 6y - 9z = 21$ are parallel.

★ **5** For each pair of planes find:

   i the line of intersection

   ii the angle between the planes.

   a $x + 3y - z = -2$ and $2x - y + 3z = 3$

   b $2x - 2y + z = -6$ and $x + y - 3z = -1$

   c $3x + 2y - z = 5$ and $x - y + 3z = 5$

★ **6** Find the point of intersection of the three planes with equations $x + 2y - z = 8$, $2x - y + 3z = 1$ and $3x + y - 4z = 15$

7 Find the line(s) of intersection of these sets of planes:

   a $\quad x + y - 3z = -1$ **b** $\quad x + 3y - z = -2$

       $2x - 2y + z = -6$ $2x - y + 3z = 3$

       $3x - y - 2z = -7$ $4x - 2y + 6z = 5$

Chapter review

1 Let A, B and C be the points $(3, -2, 5)$, $(1, 7, 3)$ and $(-2, 1, 6)$. Determine $\overrightarrow{AB} \times \overrightarrow{AC}$.

2 Evaluate $\mathbf{p} \cdot (\mathbf{q} \times \mathbf{r})$ given $\mathbf{p} = \begin{pmatrix} 1 \\ 2 \\ 7 \end{pmatrix}$, $\mathbf{q} = \begin{pmatrix} -3 \\ 5 \\ -2 \end{pmatrix}$ and $\mathbf{r} = \begin{pmatrix} 2 \\ 3 \\ -1 \end{pmatrix}$

3 Find the symmetric equation of the line through the point $(1, -2, 4)$ in the direction $\begin{pmatrix} 1 \\ 6 \\ -2 \end{pmatrix}$

4 Find the equation of the line that is perpendicular to the plane with equation $x + 2y + 3z = 7$ and passes through the point $(2, -5, 3)$.

5 Find the equation of the plane containing the point $A(1, -2, 9)$ and the line $\dfrac{x - 2}{2} = \dfrac{y + 7}{-3} = \dfrac{z}{5}$

6 Determine the equation of the plane that contains the points $(2, -5, 1)$, $(3, 0, 2)$ and $(-1, 2, 8)$.

7 **a** Obtain the parametric equation of the line L that passes through the point $(1, 2, 5)$ and is parallel to the line with equation $\dfrac{x-2}{2} = \dfrac{y+7}{-3} = \dfrac{z}{5}$

   b Plane Π contains the point $(-7, 1, 0)$ and is parallel to the plane $3x - y + z = 7$. Obtain the equation of the plane Π.

   c Find:

      i the point of intersection of L and Π

      ii the angle between L and Π.

8 Show that the line with parametric equations $x = t - 2$, $y = 2t + a$, $z = 1 - t$ $(a \neq 3)$ and the plane $x + 2y + 5z = 9$ do not intersect for all other values of a.

   What happens when $a = 3$?

9 **a** Let Π_1 be the plane containing the point $P(-1, 2, 4)$ and perpendicular to the line with equation $\dfrac{x-2}{1} = \dfrac{y-9}{-3} = \dfrac{z+2}{5}$. Determine the equation of Π_1.

   b Π_2 is another plane and has equation $2x - 5y + 4z = 12$. Find the equation of the line of intersection of Π_1 and Π_2.

   c Determine the acute angle between Π_1 and Π_2.

10 Lines L_1 and L_2 have equations $x - 5 = \dfrac{y+3}{-2} = \dfrac{z-3}{b}$ and $\dfrac{x-7}{2} = y - 8 = \dfrac{z+10}{-3}$ respectively.

   Find:

   a the value of b for which L_1 and L_2 intersect and the point of intersection

   b the acute angle between L_1 and L_2.

11 Find the point of intersection of the planes with these equations:

   $$x - 3y + 2z = 9$$
   $$3x + y - z = -1$$
   $$2x - y + 3z = 14$$

- I can calculate a vector product ★ Exercise 9B Q1
- I can calculate a scalar triple product ★ Exercise 9C Q1
- I can calculate equations of lines in three dimensions ★ Exercise 9D Q1 – Q3
- I can find the intersection of two lines in three dimensions ★ Exercise 9E Q3
- I can determine the equations of planes ★ Exercise 9F Q3
- I can find the intersection of lines with planes ★ 9G Q1
- I can find the intersection and angle between planes ★ Exercise 9G Q5 – Q6

10 Number theory

This chapter will show you how to:

- convert between different number bases
- use the Euclidean algorithm to find the greatest common divisor of two integers
- write the greatest common divisor of two integers a and b as a linear combination of a and b
- use the fundamental theorem of arithmetic and prime factorisations of natural numbers.

You should already know:

- what is meant by the absolute value of a real number
- the definitions of natural numbers, integers and rational numbers
- the definition of a prime number
- that the modern world uses 'base 10' counting system, which uses the digits 0 to 9.

Euclidean division

The foundation of many number theory results is a very simple idea:

- given two positive integers, we can calculate the **remainder** when the larger number is divided by the small number.

This idea is formally stated as the **division theorem**.

> ### The division theorem
> Given two integers a and b, with $b \neq 0$, there exist unique integers q and r such that:
>
> $a = qb + r$ and $0 \leq r < |b|$
>
> Here:
> - b is the **divisor**
> - q is the **quotient**
> - r is the **remainder**.

The process of finding the quotient and remainder is called **Euclidean division**.

There are some important points to note regarding Euclidean division.

- The word **unique** in the statement of the theorem means that the integers q and r are the **only** integers which satisfy the conditions of the theorem.

- Notice the condition on the integer r. It must be non-negative and **strictly less than the absolute value of** b.

- The theorem holds for negative integers a and b as well as positive. The theorem states that $r < |b|$, rather than just $r < b$.

Example 10.1

Use Euclidean division to find the quotient and remainder when 34 is divided by 5.

$34 = 6 \times 5 + 4$, and $0 \leqslant 4 < |5|$

> Writing in the form $a = qb + r$ gives $a = 34$, $q = 6$, $b = 5$, and $r = 4$.

The quotient q is 6.

The remainder r is 4.

> There are many integers q and r you could choose which would satisfy $34 = 5q + r$, for example:
> $34 = 2 \times 5 + 24$ (so $q = 2$ and $r = 24$), or
> $34 = 10 \times 5 - 16$ (so $q = 10$ and $r = -16$).
> However, when the condition $0 \leqslant r < |b|$ is imposed, there is only one pair q and r which you can choose.

Example 10.2

Use Euclidean division to find the quotient and remainder when –72 is divided by 7.

$-72 = -11 \times 7 + 5$, and $0 \leqslant 5 < |7|$

The quotient is –11.

The remainder is 5.

> Be careful when negative numbers are involved. The quotient can be negative, but the remainder must be **non-negative**. If you were to write $-72 = -10 \times 7 - 2$ then the remainder would be negative. This doesn't satisfy the conditions in the theorem.

Exercise 10A

1 Use Euclidean division to find the quotient and remainder when a is divided by b, in these cases.

a $a = 53, b = 4$ b $a = 159, b = 3$ c $a = 85, b = -6$

d $a = -126, b = 19$ e $a = -253, b = -64$ f $a = 1425, b = 57$

Number bases

We are familiar with expressing numbers as decimals – that is, with positions for units, tens, hundreds, thousands, and so on (along with positions after a decimal point, if necessary). This system is called **base 10** (or decimal), since the positions correspond to powers of 10. For example:

2479 = 2 thousands + 4 hundreds + 7 tens + 9 units

$$= 2 \times 10^3 + 4 \times 10^2 + 7 \times 10^1 + 9 \times 10^0$$

Numbers can be expressed in different bases. This may be useful in certain situations. For example, the natural language of computers is base 2, or **binary**. The **octal** base (base 8) and **hexadecimal** system (base 16) are also used in computer programming.

Other alternative number bases commonly used are associated with time. For example, there are 24 hours in one day and 60 minutes in one hour so different elements of time use base 24 and base 60.

When expressing numbers in different bases, it is very important to indicate the base being used. This is done by showing a subscript after the number itself. For example, **1324_5** means **1324 in base 5**.

Note that each digit in the number must be smaller than the base you are working in (just as in base 10).

Converting from other number bases to base 10

To convert a number to base 10 from another base, we multiply the digits in each position by the relevant power of the original base, then add together. To aid the process it is useful to know some small powers of common bases.

Base	Power of the base				
	4	3	2	1	0
base 2	$2^4 = 16$	$2^3 = 8$	$2^2 = 4$	$2^1 = 2$	$2^0 = 1$
base 3	$3^4 = 81$	$3^3 = 27$	$3^2 = 9$	$3^1 = 3$	$3^0 = 1$
base 5	$5^4 = 625$	$5^3 = 125$	$5^2 = 25$	$5^1 = 5$	$5^0 = 1$
base 8	$8^4 = 4096$	$8^3 = 512$	$8^2 = 64$	$8^1 = 8$	$8^0 = 1$

Example 10.3

Convert 212_3 to base 10.

$212_3 = (2 \times 3^2 + 1 \times 3^1 + 2 \times 3^0)_{10}$

> Multiply the digit in each position by the relevant power of 3.

$ = (18 + 3 + 2)_{10}$

> Clearly mark the base you are working in.

$ = 23_{10}$

Example 10.4

Convert the binary number 10011100 to decimal.

$10011100_2 = (1 \times 2^7 + 0 \times 2^6 + 0 \times 2^5 + 1 \times 2^4 + 1 \times 2^3 + 1 \times 2^2 + 0 \times 2^1 + 0 \times 2^0)_{10}$

$ = (128 + 16 + 8 + 4)_{10}$

$ = 156_{10}$

Example 10.5

Convert 7152 from the octal base to decimal.

$7152_8 = (7 \times 8^3 + 1 \times 8^2 + 5 \times 8^1 + 2 \times 8^0)_{10}$

$ = (3584 + 64 + 40 + 2)_{10}$

$ = 3690_{10}$

Exercise 10B

1 Convert these numbers from binary to decimal.

a 10_2	**b** 100_2	**c** 1010_2
d 1111_2	**e** 101100_2	**f** 10010010_2

2 Convert these numbers from octal to decimal.

a 37_8	**b** 73_8	**c** 266_8
d 13472_8	**e** 647271_8	**f** 1002742_8

3 Convert these numbers to base 10.

a 34_5	**b** 34_7	**c** 142_6
d 12211_3	**e** 8672_9	**f** 3310223_4

Converting to number bases other than base 10

The process for converting from base 10 to another base is slightly more involved than going the other way. There are three methods we can use, which all involve Euclidean division:

- **Method 1**   Use Euclidean division to find **quotients** when dividing by **powers** of the new base

- **Method 2**   Use Euclidean division to find **remainders** when dividing by the new base

- **Method 3**   Use Method 2 but with an alternative layout

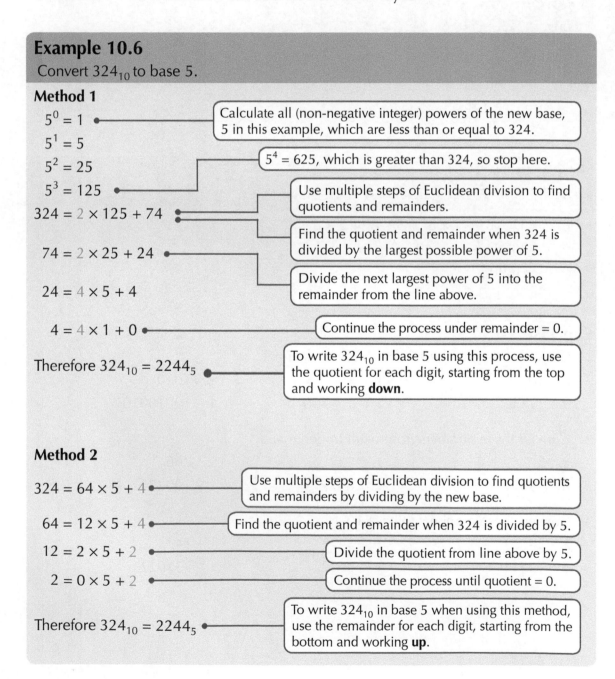

Example 10.6

Convert 324_{10} to base 5.

Method 1

$5^0 = 1$ ●——— Calculate all (non-negative integer) powers of the new base, 5 in this example, which are less than or equal to 324.

$5^1 = 5$

$5^2 = 25$ ——— $5^4 = 625$, which is greater than 324, so stop here.

$5^3 = 125$ ●

$324 = 2 \times 125 + 74$ ●——— Use multiple steps of Euclidean division to find quotients and remainders.

Find the quotient and remainder when 324 is divided by the largest possible power of 5.

$74 = 2 \times 25 + 24$ ●——— Divide the next largest power of 5 into the remainder from the line above.

$24 = 4 \times 5 + 4$

$4 = 4 \times 1 + 0$ ●——— Continue the process under remainder = 0.

Therefore $324_{10} = 2244_5$ ●——— To write 324_{10} in base 5 using this process, use the quotient for each digit, starting from the top and working **down**.

Method 2

$324 = 64 \times 5 + 4$ ●——— Use multiple steps of Euclidean division to find quotients and remainders by dividing by the new base.

$64 = 12 \times 5 + 4$ ●——— Find the quotient and remainder when 324 is divided by 5.

$12 = 2 \times 5 + 2$ ●——— Divide the quotient from line above by 5.

$2 = 0 \times 5 + 2$ ●——— Continue the process until quotient = 0.

Therefore $324_{10} = 2244_5$ ●——— To write 324_{10} in base 5 when using this method, use the remainder for each digit, starting from the bottom and working **up**.

Method 3

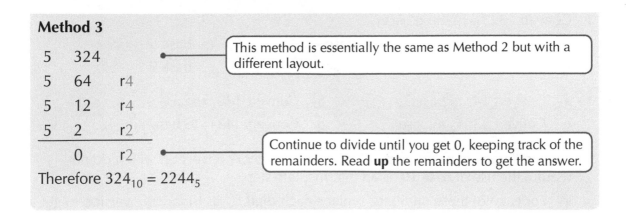

5 324

5 64   r4

5 12   r4

5 2   r2
_____

   0   r2

> This method is essentially the same as Method 2 but with a different layout.

> Continue to divide until you get 0, keeping track of the remainders. Read **up** the remainders to get the answer.

Therefore $324_{10} = 2244_5$

Converting between two number bases other than base 10

To convert between two number bases where neither base is 10, it is usually easiest to first convert to base 10 as an intermediate step.

Example 10.7

Convert 63_7 to base 5.

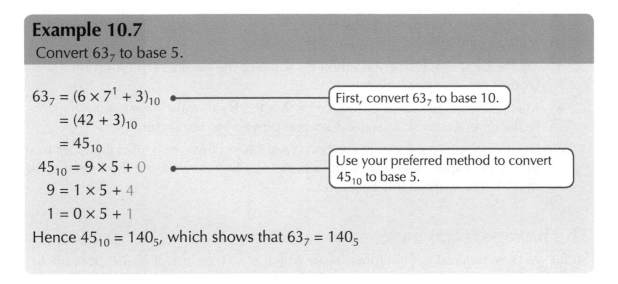

$63_7 = (6 \times 7^1 + 3)_{10}$

$\quad = (42 + 3)_{10}$

$\quad = 45_{10}$

$45_{10} = 9 \times 5 + 0$

$9 = 1 \times 5 + 4$

$1 = 0 \times 5 + 1$

> First, convert 63_7 to base 10.

> Use your preferred method to convert 45_{10} to base 5.

Hence $45_{10} = 140_5$, which shows that $63_7 = 140_5$

Exercise 10C

★ 1 Convert these numbers from base 10 to binary.

    a 2   **b** 7   **c** 26

    d 37   **e** 151   **f** 1376

★ 2 Convert these numbers from decimal to octal.

    a 18   **b** 96   **c** 105

    d 2378   **e** 5619   **f** 13001

★ 3   Convert 1323_{10} to these bases.

    a base 3   **b** base 4   **c** base 5

    d base 6   **e** base 7   **f** base 9

★ 4   **a** Convert 14_5 to base 3.   **b** Convert 132_6 to base 9.

    c Convert 2301_8 to base 7.   **d** Convert 4431_7 to base 8.

5   For some pairs of bases there are quick methods of conversion which do not require the use of base 10 as an intermediate step.

    a For each of these numbers, replace **each digit** by its binary conversion to get a binary number (use **two** digits for each replacement, so for example, replace 1_4 by 01_2).

        i 31_4 **ii** 132_4 **iii** 233_4

    b Convert each number from part **a** to binary, using base 10 as an intermediate step. How do your results compare to those you got when replacing each digit as in part **a**?

    c Try the same comparison when converting 637_9 to base 3. Does the digit replacement method give the correct answer?

    d Try the digit replacement method for some of the numbers in Question 4. Does it give the correct answer?

    e Make a conjecture about the relationship the bases must have for the digit replacement method to work. Can you prove your conjecture?

    f Does the digit replacement method work when converting from the octal base to binary? Experiment with some examples. How does the method need to be modified to work in this case?

The hexadecimal base

So far we have worked with number bases which are 10 or less. It is also possible to work with bases which are larger than 10. A common example of this is the **hexadecimal** base, which is base 16.

We have to be careful how we write numbers in base 16 (or any base larger than 10). We are used to working with ten symbols to express natural numbers:

    0, 1, 2, 3, 4, 5, 6, 7, 8, 9

In base 10, we express any natural numbers larger than 9 using a combination of these ten digits.

However, if we are working in base 16, we need **sixteen** symbols to work with. The hexadecimal system uses the letters A, B, C, D, E and F, in addition to the digits 0, 1, ..., 9, so that:

base 16	A_{16}	B_{16}	C_{16}	D_{16}	E_{16}	F_{16}
base 10	10_{10}	11_{10}	12_{10}	13_{10}	14_{10}	15_{10}

Example 10.8

Convert $87A3D_{16}$ from hexadecimal to decimal.

$$87A3D_{16} = (8 \times 16^4 + 7 \times 16^3 + 10 \times 16^2 + 3 \times 16^1 + 13)_{10}$$
$$= (524\,288 + 28\,672 + 2560 + 48 + 13)_{10}$$
$$= 555\,581_{10}$$

Use the usual method for conversion, remembering what the symbols A and D mean in base 10.

Example 10.9

Convert 7899_{10} to base 16.

Method 1

$7899 = 1 \times 4096 + 3803$

$3803 = 14 \times 256 + 219$

$219 = 13 \times 16 + 11$

$11 = 11 \times 1 + 0$

Therefore $7899_{10} = 1EDB_{16}$

Use Euclidean division, dividing by powers of 16: $16^0 = 1$, $16^1 = 16$, $16^2 = 256$ and $16^3 = 4096$.

Remember that $14_{10} = E_{16}$, $13_{10} = D_{16}$ and $11_{10} = B_{16}$.

Method 2

16	7899		
16	493	r11	B
16	30	r13	D
16	1	r14	E
	0	r1	1

Therefore $7899_{10} = 1EDB_{16}$

Keeping track of what each remainder is in hexadecimal makes it easier to read the final answer.

Remember that $14_{10} = E_{16}$, $13_{10} = D_{16}$ and $11_{10} = B_{16}$.

Remember to read **up** the remainders when using this method.

Exercise 10D

★ 1 Convert these numbers from hexadecimal to base 10.

a $1A_{16}$

b $C2_{16}$

c 82_{16}

d $1DE_{16}$

e ABC_{16}

f FFF_{16}

g $4B6E_{16}$

h $33AA_{16}$

i $B3A9F_{16}$

★ **2** Convert these numbers from decimal to hexadecimal.

a 53	**b** 109	**c** 228
d 791	**e** 1014	**f** 6555
g 10001	**h** 23910	**i** 77253

3 Convert these numbers to the bases indicated.

a $AB1_{16}$ to base 5 **b** $2F7_{16}$ to base 4

c 814_9 to base 16 **d** 3333_4 to base 16

e $BA7C_{16}$ to base 7 **f** AC_{16} to base 2

The Euclidean algorithm

It is often necessary to find the **greatest common divisor**, or **gcd**, of two integers. This is also referred to as the **highest common factor**. For example, when reducing fractions to their simplest form, the gcd of the numerator and denominator is cancelled as a common factor.

When the numbers involved are relatively small, it is often easy to spot the gcd (perhaps with a bit of trial and error). However, for larger numbers there can be more work involved. The **Euclidean algorithm** (or **Euclid's algorithm**) is a more formal method for finding the gcd of two integers.

This method was developed by the Greek mathematician Euclid, and dates back over 2000 years to around 300 BCE. Despite its age, the algorithm is still in use today.

Using the Euclidean algorithm

The Euclidean algorithm involves repeated use of Euclidean division. Suppose we wish to find the gcd of two integers a and b, and we assume that a is greater than b.

The steps in the Euclidean algorithm are:

1 find the remainder r when a is divided by b

2 find the remainder r_1 when b is divided by r

3 find the remainder r_2 when r is divided by r_1, and so on.

We continue in this way until we eventually arrive at a remainder which equals 0.

$$a = q \times b + r_1$$
$$b = q_1 \times r_1 + r_2$$
$$r_1 = q_2 \times r_2 + r_3$$
$$\vdots \qquad \vdots$$
$$r_{n-2} = q_{n-1} \times r_{n-1} + r_n$$
$$r_{n-1} = q_n \times r_n + 0$$

The last **non-zero** remainder is the gcd of a and b. Examples 10.10–10.12 show how to use the method.

Example 10.10

Use the Euclidean algorithm to find the greatest common divisor (gcd) of 123 and 33.

$123 = 3 \times 33 + 24$ — Use Euclidean division at each step.

$33 = 1 \times 24 + 9$ — The divisor from the line above appears on the left-hand side of the current line, and the remainder from the line above becomes the divisor in the current line.

$24 = 2 \times 9 + 6$

$9 = 1 \times 6 + 3$

$6 = 2 \times 3 + 0$ — The algorithm concludes when remainder = 0.

Therefore the greatest common divisor of 123 and 33 is 3. — The gcd is the last non-zero remainder.

Coprime integers

Two integers are said to be **coprime** (or **relatively prime**) if their gcd = 1.

Example 10.11

Show that the positive integers 189 and 1573 are coprime.

$1573 = 8 \times 189 + 61$ — Use the Euclidean algorithm to show that gcd(189, 1573) = 1.

$189 = 3 \times 61 + 6$

$61 = 10 \times 6 + 1$

$6 = 6 \times 1 + 0$

The last non-zero remainder = 1, so the gcd(189, 1573) = 1, and hence 189 and 1573 are coprime.

Example 10.12

Reduce the fraction $\dfrac{2739}{4107}$ to its simplest form.

$4107 = 1 \times 2739 + 1368$

$2739 = 2 \times 1368 + 3$

$1368 = 456 \times 3 + 0$ — The Euclidean algorithm shows that gcd(4107, 2739) = 3.

$\dfrac{2739}{4107} = \dfrac{913}{1369}$ — Divide the numerator and denominator by the gcd to simplify the fraction: $4107 \div 3 = 1369$ and $2739 \div 3 = 913$

Exercise 10E

This exercise should be completed without the use of a calculator.

★ 1 Use the Euclidean algorithm to determine the greatest common divisors of these pairs of positive integers.

    a   135 and 30   **b**   221 and 94   **c**   3673 and 1832

    d   11 707 and 40 601   **e**   34 512 and 2056   **f**   59 013 and 10 001

2 Find the highest common factor of these pairs of integers.

    a   32 and 248   **b**   69 and 143   **c**   251 and 699

    d   302 and 456   **e**   1017 and 1319   **f**   5604 and 9671

3 Determine which of these positive integers are coprime to 2695.

    a   1155   **b**   121   **c**   442

4 Use the Euclidean algorithm to find the largest integer which divides every number in the set {132, 603, 426, 150, 99}.

> Try to think of an efficient way to do this. You don't have to apply the Euclidean algorithm to every possible pair of numbers in the set.

5 Reduce these fractions to their simplest form.

    a $\dfrac{324}{684}$ **b** $\dfrac{4536}{6993}$ **c** $\dfrac{5103}{324}$

    d $\dfrac{64}{177}$ **e** $\dfrac{664}{308}$ **f** $\dfrac{22119}{30075}$

6 The **Fibonacci sequence** begins

     1, 1, 2, 3, 5, 8, 13, 21, 34, 55, 89, …

and is defined by the recurrence relation $u_{k+1} = u_k + u_{k-1}$ for all natural numbers $k \geqslant 2$.

    a   For $2 \leqslant k \leqslant 10$, calculate $\gcd(u_k, u_{k+1})$. What do you notice?

    b   Let a and b be integers. Prove that if another integer m divides a and m also divides $a + b$, then m must divide b.

    c   Use part **b** to prove that $\gcd(u_k, u_{k+1}) = 1$ for all integers $k \geqslant 2$.

    d   Calculate $\gcd(u_k, u_{k+2})$ for some examples, and formulate a conjecture. Can you prove it?

Proof of the Euclidean algorithm

The concept of mathematical proof is explored in detail in Chapter 11, but examining a proof of the Euclidean algorithm is good preparation for this. It also provides a deeper understanding of the working of the algorithm.

There are three steps involved in the proof:

1 show that the process concludes at some point

2 show that the algorithm produces a common divisor

3 show that the algorithm produces the greatest common divisor.

1 The process concludes at some point

This is very important from a practical point of view – it wouldn't be much use to keep applying Euclidean division if the process was never-ending.

At each step of the Euclidean algorithm, we apply the division theorem. This ensures that the remainder is strictly smaller than the divisor. When we move from one step to the next, the remainder from the previous step becomes the new divisor. This ensures that the remainders get smaller with each step.

However, the remainders must always be non-negative, so the smallest possible value is 0, which we must eventually reach. Therefore the process will conclude at some point.

2 The algorithm produces a common divisor

To prove that the positive integer produced by the Euclidean algorithm is a common divisor of a and b we use this **lemma** which concerns **integral linear combinations** of integers:

> Suppose that a, b and d are integers, and that d divides both a and b. Then d divides any integral linear combination of a and b. (An integral linear combination of two integers a and b is a number of the form $ax + by$, where x and y are integers.)

When we apply the Euclidean algorithm, we get these equations:

$$a = q \times b + r_1 \qquad ①$$
$$b = q_1 \times r_1 + r_2 \qquad ②$$
$$r_1 = q_2 \times r_2 + r_3 \qquad ③$$
$$\vdots \qquad \vdots$$
$$r_{n-2} = q_{n-1} \times r_{n-1} + r_n \qquad ⓝ$$
$$r_{n-1} = q_n \times r_n + 0 \qquad (n+1)$$

> A **lemma** is a preliminary or minor theorem, which is used to prove a more substantial theorem.

where $q, q_1, ..., q_n, r_1, ..., r_n \in \mathbb{Z}$

We must show that r_n divides both b and a.

Equation $(n+1)$ says that $r_{n-1} = q_n r_n$, which means that r_n divides r_{n-1}

Moving up one line, equation $ⓝ$ says that r_{n-2} is an integral linear combination of r_{n-1} and r_n

We have observed that r_n divides r_{n-1}, and clearly r_n divides r_n, so by the lemma r_n must divide r_{n-2}

Continuing in this way, we move up the equations and see that r_n divides $r_{n-3}, r_{n-4}, ...,$ r_2, r_1, and ultimately r_n must divide b and a.

3 The algorithm produces the greatest common divisor

To prove that r_n is the **greatest** integer which divides both a and b, we show that any integer s which divides a and b must also divide r_n

Rearranging equation ①, we see that $r_1 = a - q \times b$, so r_1 is an integral linear combination of a and b.

s divides both a and b, so by the lemma, s must also divide r_1.

Now move down one line and consider equation ②. Rearranging, we see that $r_2 = b - q_1 \times r_1$, so r_2 is an integral linear combination of b and r_1

Since s divides both b and r_1, by the lemma s must also divide r_2

Continuing in this way, we move down the equations to find that s must divide $r_3, r_4, ..., r_{n-1}$ and finally r_n

There was nothing special about our choice of s, so this shows that **any** divisor of a and b must also divide r_n

Therefore, r_n is the greatest common divisor of a and b.

Exercise 10F

1 Prove these results which, when combined, prove the lemma which was used in the proof of the Euclidean algorithm. Look at Chapter 11 if you are not sure how to get started. Assume that m, a and b are natural numbers.

   a Show that if m divides a, then m divides ka for *any* natural number k.

   b Show that if m divides a and m divides b, then m divides $a + b$.

   c Show that if m divides a and m divides b, then m divides $a - b$.

The extended Euclidean algorithm

The Euclidean algorithm allows us to find the greatest common divisor of two integers. The **extended Euclidean algorithm** is used to write $\gcd(a, b)$ as an integral linear combination of a and b.

There are two methods we can use when applying the extended Euclidean algorithm:

* **Method 1** Perform additional steps at each step of the Euclidean algorithm (Example 10.13)

* **Method 2** Apply the standard Euclidean algorithm first, before working backwards to find the required integral linear combination (Example 10.14).

Example 10.13

The greatest common divisor of 123 and 33 is 3 (from Example 10.10).
Use the extended Euclidean algorithm to write 3 as an integral linear combination of 123 and 33.

Euclidean algorithm steps	Additional steps in Extended Euclidean algorithm
$123 = 3 \times 33 + 24$	$24 = 123 - 3 \times 33$ Make the remainder the subject of the equation.
$33 = 1 \times 24 + 9$	$9 = 33 - 1 \times 24$ Make the remainder the subject of the equation. $9 = 33 - 1 \times (123 - 3 \times 33)$ Substitute the expression for 24 from previous step. $9 = 4 \times 33 - 1 \times 123$ Collect like terms to simplify as far as possible.
$24 = 2 \times 9 + 6$	$6 = 24 - 2 \times 9$ $6 = (123 - 3 \times 33) - 2 \times (4 \times 33 - 1 \times 123)$ Substitute the expressions for 24 and 9 from the previous two steps. $6 = 3 \times 123 - 11 \times 33$
$9 = 1 \times 6 + 3$	$3 = 9 - 1 \times 6$ $3 = (4 \times 33 - 1 \times 123) - 1 \times (3 \times 123 - 11 \times 33)$ $3 = 15 \times 33 - 4 \times 123$
$6 = 2 \times 3 + 0$	Remainder = 0 so no further steps necessary.

Therefore $3 = 15 \times 33 - 4 \times 123$ as required.

Example 10.14

Find the greatest common divisor of 315 and 55, and hence write gcd(315, 55) in the form $315a + 55b$ for some integers a and b.

$315 = 5 \times 55 + 40$ ①
Use the standard Euclidean algorithm to find the gcd.

$55 = 1 \times 40 + 15$ ②

$40 = 2 \times 15 + 10$ ③

$15 = 1 \times 10 + 5$ ④

$10 = 2 \times 5 + 0$

Hence gcd(315, 55) = 5.
Now work backwards through ① – ④ to find the required integral linear combination.

$5 = 1 \times 15 - (1 \times 10)$
Use ④ to write the remainder 5 as a linear combination of 10 and 15.

$\quad = 1 \times 15 - (1 \times (1 \times 40 - (2 \times 15)))$

$\quad = 3 \times 15 - (1 \times 40)$
Use ③ to replace 10 by a linear combination of 15 and 40, then simplify.

$\quad = 3 \times (55 - (1 \times 40)) - (1 \times 40)$

$\quad = 3 \times 55 - 4 \times 40$
Use ② to replace 15 by a linear combination of 40 and 55, then simplify.

$\quad = 3 \times 55 - 4 \times (315 - (5 \times 55))$

$\quad = 23 \times 55 - 4 \times 315$
Use ① to replace 40 by a linear combination of 55 and 315, then simplify.

So gcd(315, 55) = $315 \times (-4) + 55 \times 23$.

Expressing an integer multiple of d as a linear combination of a and b

Having expressed a number d as a linear combination of a and b, it is then straightforward to express any integer multiple of d as a linear combination of a and b.

Example 10.15

Given that $5 = 23 \times 55 - 4 \times 315$ (as shown in Example 10.14), express 30 as a linear combination of 55 and 315.

$30 = 6 \times 5$
Identify the correct multiple of 5.

$\quad = 6 \times (23 \times 55 - 4 \times 315)$

$\quad = (6 \times 23) \times 55 - (6 \times 4) \times 315$
Multiply the original equation by 6 and simplify. The required multiples of 55 and 315 are 6×23 and $6 \times (-4)$.

$\quad = 138 \times 55 - 24 \times 315$

Exercise 10G

This exercise should be completed without the use of a calculator.

1   Use the extended Euclidean algorithm to find integers p and q such that these equalities hold.

   a $9p + 21q = 3$ **b** $16p + 6q = 2$

   c $105p + 49q = 7$ **d** $17p - 4q = 1$

   e $650p - 110q = 10$ **f** $2079p + 399q = 21$

★ 2   For each pair of integers a and b, write $\gcd(a, b)$ as an integral linear combination of a and b.

   a $a = 135, b = 63$ **b** $a = 819, b = 1881$

   c $a = 228, b = 102$ **d** $a = 360, b = 220$

   e $a = 1573, b = 1859$ **f** $a = 13\,225, b = 18\,125$

3   A **linear Diophantine equation** is an equation of the form $ax + by = c$, where a, b and c are integers. Equations of this form may or may not have solutions x and y which are integers.

   For example, the equation $3x - 2y = 4$ has integer solutions, namely $x = 2$ and $y = 1$. However, the equation $4x + 2y = 3$ does **not** have any integer solutions.

   a   Suppose that the linear Diophantine equation $ax + by = c$ has integer solutions x and y, and let $d = \gcd(a, b)$. Show that d must divide c.

   b   Suppose that c is a multiple of d, where $d = \gcd(a, b)$. Show that the linear Diophantine equation $ax + by = c$ has integer solutions.

   c   Use the conclusions of parts **a** and **b** to make an 'if and only if' statement about the conditions needed for a linear Diophantine equation to have integer solutions.

   > ⚠ Look at Chapter 11 if you're unsure what to do in this question.

4   Find integer solutions x and y to these equations.

   a $255x + 399y = 6$

   b $275x + 1105y = 15$

   c $1656x + 348y = -24$

The fundamental theorem of arithmetic

We have previously used the fact that any positive integer can be written as a product of its factors, for example, when simplifying fractions. We have also seen that it is possible to factorise certain numbers in different ways. For example, we can factorise 30 as 3×10, but also as 6×5.

However, if we allow only **prime factors** (factors which are prime numbers), then there is essentially only one way to factorise a given natural number.

The **fundamental theorem of arithmetic** states that every positive integer greater than 1 can be expressed uniquely as a product of prime numbers. For example:

$$42 = 2 \times 3 \times 7$$
$$80 = 2^4 \times 5$$

This result is one of the foundational results of number theory.

> **The fundamental theorem of arithmetic**
>
> Every integer $n > 1$ can be represented in exactly one way as a product of prime powers.
>
> $$n = p_1^{\alpha_1} p_2^{\alpha_2} \ldots p_k^{\alpha_k} = \prod_{i=1}^{k} p_i^{\alpha_i}$$
>
> where $p_1 < p_2 < \ldots < p_k$ and the α_i are positive integers. This is the **prime factorisation** of n.

If the prime factorisations of two positive integers are known, it becomes straightforward to calculate the greatest common divisor and **least common multiple (lcm)** of the two positive integers.

If a and b are positive integers with

$$a = p_1^{\alpha_1} p_2^{\alpha_2} \ldots p_k^{\alpha_k}$$

and

$$b = p_1^{\beta_1} p_2^{\beta_2} \ldots p_k^{\beta_k}$$

where $p_1, \ldots, p_k$ are primes and the α_i and β_i are non-negative integers, then

$$\gcd(a, b) = p_1^{\min(\alpha_1, \beta_1)} p_2^{\min(\alpha_2, \beta_2)} \ldots p_k^{\min(\alpha_k, \beta_k)}$$

and the least common multiple of a and b, **lcm(a, b)**, is given by

$$\text{lcm}(a, b) = p_1^{\max(\alpha_1, \beta_1)} p_2^{\max(\alpha_2, \beta_2)} \ldots p_k^{\max(\alpha_k, \beta_k)}$$

Example 10.16

Use the prime factorisations of 204 and 414 to calculate gcd(204, 414) and lcm(204, 414).

$$204 = 2^2 \times 3^1 \times 17^1$$
$$414 = 2^1 \times 3^2 \times 23^1$$

Write the terms as products of prime powers.

Therefore:

$$\gcd(204, 414) = 2^{\min(1, 2)} \times 3^{\min(1, 2)} \times 17^{\min(0, 1)} \times 23^{\min(0, 1)}$$

If a specific prime does not explicitly appear in the prime factorisation of a number, its power is 0.

$$= 2^1 \times 3^1 \times 17^0 \times 23^0$$
$$= 6$$

Similarly:

$$\text{lcm}(204, 414) = 2^{\max(1, 2)} \times 3^{\max(1, 2)} \times 17^{\max(0, 1)} \times 23^{\max(0, 1)}$$
$$= 2^2 \times 3^2 \times 17^1 \times 23^1$$
$$= 14\,076$$

Example 10.17

The polynomial $f(x) = x^5 - 3x^4 - 5x^3 + 15x^2 + 4x - 12$ has roots $x = 1$, $x = -1$, $x = 2$, $x = -2$ and $x = 3$. Show that there does not exist an integer a such that $f(a) = 6$.

$f(x) = (x - 1)(x + 1)(x - 2)(x + 2)(x - 3)$ ●————— | You are given the five roots of f.

The roots of f are distinct, so for any integer a, $f(a)$ must be divisible by five distinct integers. By the fundamental theorem of arithmetic, 6 may only be written as a product of, at most, four distinct integers, for example, $6 = 1 \times (-1) \times (-2) \times 3$. ●—————

| If you use a theorem or other result in an answer, make sure you say that you are using it.

Therefore, there cannot exist an integer a such that $f(a) = 6$.

Exercise 10H

★ 1 By writing out their prime factorisations, find the greatest common divisor and least common multiple of each pair of integers.

   a 585 and 975   b 6776 and 3724   c 2880 and 2016

   d 2837 and 3757   e 500 and 80   f 675 and 5525

2 a Use the Euclidean algorithm to show:

      i gcd(72,15) = 3   ii gcd(72,14) = 2

   b **Without** using the Euclidean algorithm, but using part **a**, find the greatest common divisor of each pair of integers.

      i 72 and 30   ii 72 and 42   iii 72 and 210

3 The polynomial $f(x) = x^4 - 10x^3 + 25x^2 - 36$ has roots $x = -1$, $x = 2$, $x = 3$ and $x = 6$. Show that there is no integer a such that $f(a) = 5$.

4 a Let a be an integer, with $1 \le a \le 100$. For how many values of a does the equation $ax + 7y = 1$ have integer solutions?

   b Let b be an integer, with $1 \le b \le 100$. For how many values of b does the equation $bx + 6y = 1$ have integer solutions?

Chapter review

1 Convert 1327_{10} to these bases.

   a base 9   b base 5   c binary

   d octal   e hexadecimal

2 Convert these numbers to base 10.

   a 341_5 b 100101001_2

   c 7462_8 d $23AD5_{16}$

3 **a** Convert 643_7 to the octal base.

   b Convert $B33_{16}$ to base 3.

   c Convert 414_5 to base 9.

   d Convert 513_6 to binary.

4 Find the greatest common divisor of 8155 and 3110.

5 A recurrence relation is defined by $u_1 = 1$, $u_2 = 1$ and $u_k = u_{k-1} + 2u_{k-2}$ for all integers $k \geqslant 3$.

   a Write down the first 10 numbers in the sequence.

   b Prove by induction that u_k is odd for all $k \geqslant 1$.

   c Prove that $\gcd(u_{k+1}, u_k) = 1$ for all $k \geqslant 1$.

> For part **b**, see Chapter 11 for more on proof by induction. ⚠

   d Calculate $\gcd(u_{k+2}, u_k)$ for all $k \geqslant 1$. Formulate and prove a conjecture regarding $\gcd(u_{k+2}, u_k)$.

   e Is it the case that $\gcd(u_{k+3}, u_k) = 1$ for all $k \geqslant 1$? If yes, prove it; if no, provide a counterexample.

6 Reduce the fraction $\dfrac{12\,974}{15\,600}$ to its simplest form.

7 Write 17 as an integral linear combination of 323 and 629.

8 Express $\gcd(102, 796)$ as an integral linear combination of 102 and 796.

9 In each case write down the prime factorisations of a and b, and use these to calculate $\gcd(a, b)$ and $\operatorname{lcm}(a, b)$.

   a $a = 128, b = 396$ **b** $a = 735, b = 385$

   c $a = 108, b = 90$ **d** $a = 1573, b = 1287$

   e $a = 880, b = 1496$ **f** $a = 2125, b = 2375$

- I can convert numbers to base 10 from other bases. ★ Exercise 10B Q1–Q3 ★ Exercise 10D Q1

- I can convert numbers from base 10 to other bases. ★ Exercise 10C Q1–Q3 ★ Exercise 10D Q2

- I can convert between arbitrary number bases. ★ Exercise 10C Q4

- I can use the Euclidean algorithm to find the greatest common divisor of two integers. ★ Exercise 10E Q1

- I can use the extended Euclidean algorithm to express $\gcd(a, b)$ as an integral linear combination of a and b. ★ Exercise 10G Q2

- I can find prime factorisations of natural numbers, and use this to find greatest common divisors and least common multiples. ★ Exercise 10H Q1

11 Proof

This chapter will show you how to:

- write mathematical statements using logical notation
- disprove a conjecture by providing a counterexample
- use proof by induction
- use methods of direct proof
- use proof by contrapositive
- use proof by contradiction.

You should already know:

- the definitions of the sets $\mathbb{N}$, $\mathbb{W}$, $\mathbb{Z}$, $\mathbb{R}$, $\mathbb{Q}$ and $\mathbb{C}$
- how to manipulate algebraic expressions
- the definition of a matrix
- the definition of a function
- how to use summation notation.

Logic and mathematical notation

A mathematical proof uses logical reasoning to establish the truth of a mathematical statement. Although the terminology may not have been used, you will already have proved many statements on an informal basis when solving exercises throughout this book. This chapter will examine some proof techniques in more detail. To lay the groundwork, you must first become familiar with the mathematical language involved.

Notation

Throughout this chapter (and elsewhere in this book) we will be dealing with **sets** of mathematical objects. It is important to understand the notation used to describe these sets.

- $\mathbb{N} = \{1, 2, 3, 4, \ldots\}$ denotes the set of **natural numbers**.

- $\mathbb{W} = \{0, 1, 2, 3, \ldots\}$ denotes the set of **whole numbers**. $\mathbb{W}$ includes all the natural numbers and 0. Alternative notation which is sometimes used for the set of whole numbers is $\mathbb{N}_0$.

- $\mathbb{Z} = \{\ldots, -2, -1, 0, 1, 2, \ldots\}$ denotes the set of **integers**. $\mathbb{Z}$ includes all the whole numbers, along with their negatives.

- $\mathbb{Q}$ denotes the set of **rational** numbers. $\mathbb{Q}$ consists of all integers along with all positive or negative fractions. Any number which can be written as an exact decimal must be a rational number.

- $\mathbb{R}$ denotes the set of **real** numbers. $\mathbb{R}$ includes all rational numbers, and also all surds and other irrational numbers such as π and e.

- $\mathbb{C}$ denotes the set of **complex** numbers. $\mathbb{C}$ includes all real numbers, and also numbers involving $i = \sqrt{-1}$.

- $M_n(\mathbb{R})$ denotes the set of all $n \times n$ **matrices** with **real** entries.

We also use the following symbols in proofs:

- $\therefore$ means 'therefore'

- $\because$ means 'since' or 'because'.

Existential and universal statements

Many mathematical statements make reference to sets of numbers or other mathematical objects, such as the real numbers, the integers, 2×2 matrices, real-value functions and so on. We will use two statements:

- universal statements

- existential statements

and examine their use regarding sets, and the logical notation used to describe them.

Universal statements

A **universal statement** asserts that a property holds for **all** objects in a given set (for example, all real numbers, all injective functions, all 2×2 matrices). Such statements involve the **universal quantifier** $\forall$, which can be read as 'for all'.

Example 11.1

Write these universal statements using logical notation.

a   For all real numbers x, x^2 is greater than or equal to zero.

b   Every natural number is greater than or equal to 1.

a $\forall x \in \mathbb{R}, x^2 \geqslant 0$ ●——— When using logical notation to write a universal statement, the first part of the sentence should involve the universal quantifier, the variable, and the set. After the comma is the conclusion of the statement, written in terms of the variable.

b $\forall n \in \mathbb{N}, n \geqslant 1$ ●——— The statement does not give a name for the variable, but n is a sensible choice for a natural number.

Existential statements

An **existential statement** asserts that a property holds for **some** object in a given set. Such statements involve the **existential quantifier** $\exists$, which can be read as 'there exists'.

It is possible to have both universal and existential quantifiers in the same statement.

Example 11.2

Write these existential statements using logical notation.

a There exists a real number which equals 6 when multiplied by 2.

b Some integer is less than –43.

a $\exists\, r \in \mathbb{R}: 2r = 6$ ●————

> In an existential statement, the two parts of the sentence are separated with a colon, to be read as 'such that'.

b $\exists\, x \in \mathbb{Z}: x < -43$

Example 11.3

Write these statements using logical notation.

a For all natural numbers n, there exists a real number y such that $ny = 1$.

b There exists a real number x such that $xy = 0$ for all real numbers y.

a $\forall\, n \in \mathbb{N},\, \exists\, y \in \mathbb{R}: ny = 1$

b $\exists\, x \in \mathbb{R}: \forall\, y \in \mathbb{R},\, xy = 0$

Exercise 11A

1 Decide whether each statement is universal or existential.

   a There exists an integer a which equals 47.

   b For all real numbers x, $x - 2.5 < x$.

   c Every 2×2 matrix is invertible.

   d $\dfrac{3}{y} + 2 = 1$ for some complex number y.

   e Each element in the set $\{1, 2, 3, 4, 5\}$ is a natural number.

★ 2 Write these statements using logical notation.

   a There exists an integer a which is greater than 5.

   b Every integer a is greater than 5.

   c For all real numbers x, $x - 0.3 < x$.

   d There exists a 2×2 matrix with real entries whose determinant equals 3.

   e Some rational number q is equal to 3 when multiplied by 4.

Conditional statements

There are many examples of statements P and Q where P **implies** Q. (P and Q are often used as labels for mathematical statements.)

Suppose that P is the statement 'Jack lives in Edinburgh', and Q is the statement 'Jack lives in Scotland'.

Edinburgh is in Scotland, so clearly if Jack lives in Edinburgh, then he must also live in Scotland. Therefore P implies Q, and in logical notation we write $P \Rightarrow Q$.

Note that here it is **not** the case that $Q \Rightarrow P$. Jack could live in Scotland, but not live in Edinburgh.

Statements of the form $P \Rightarrow Q$ are called **conditional** statements. There are a number of ways of saying $P \Rightarrow Q$ in words, all of which you should be familiar with:

- P **implies** Q

- Q is **implied by** P

- **If** P **then** Q

- P **only if** Q

- P is **sufficient** for Q

- Q is **necessary** for P

Example 11.4

Write these statements using logical notation.

a $n < 3$ implies $n < 5$

b $\frac{x}{3} = 2$ is implied by $x = 6$

c If an integer x is positive, then x^2 is positive.

d For $z + 4$ to equal 0, it is necessary that $z = -7$.

e $f(x)$ equals $3 \cdot 6$ if $g(x)$ equals $3 \cdot 6$

f $f(x)$ equals $3 \cdot 6$ only if $g(x)$ equals $3 \cdot 6$

a $n < 3 \Rightarrow n < 5$ — Replace the word 'implies' with the correct mathematical symbol.

b $x = 6 \Rightarrow \frac{x}{3} = 2$ — Be careful with the direction of the implication.

c $x \in \mathbb{Z}, x > 0 \Rightarrow x^2 > 0$ — The two parts in the first statement (x being an integer, and x being positive) are separated by a comma.

d $z + 4 = 0 \Rightarrow z = -7$ — Notice that this statement is not actually true. However, it can still be written using logical notation.

e $g(x) = 3 \cdot 6 \Rightarrow f(x) = 3 \cdot 6$

f $f(x) = 3 \cdot 6 \Rightarrow g(x) = 3 \cdot 6$ — Note the difference between this and part **e**. If P holds **only** when Q holds, then P implies Q.

The **converse** of a conditional statement $P \Rightarrow Q$ is the statement $Q \Rightarrow P$, with the direction of implication reversed.

There are also examples of pairs of statements P and Q which are **equivalent**. This means that $P \Rightarrow Q$ and $Q \Rightarrow P$. In this situation we say P **if and only if** Q, or that P is **necessary and sufficient** for Q, and write $P \Leftrightarrow Q$. (You may sometimes see 'if and only if' abbreviated to 'iff'.)

Example 11.5

Express these statements using logical notation.

a $x + 5 = 2$ is equivalent to $x - 7 = -10$

b $x = 0$ if and only if $x^2 = 0$

c $y^3 = 64$ is necessary and sufficient for $y = 4$

a $x + 5 = 2 \Leftrightarrow x - 7 = -10$

b $x = 0 \Leftrightarrow x^2 = 0$

c $y^3 = 64 \Leftrightarrow y = 4$

Exercise 11B

★ 1 Write these statements using logical notation.

    a If x is greater than 15, then x is greater than 4.

    b $ab = 16$ if $a = 8$

    c n being a natural number implies n is an integer.

    d $f(x) = 4x^2$ implies $f'(x) = 8x$

    e $c = 4$ only if $c^2 = 16$

    f $y = -3$ if and only if $y^3 = -27$

    g $\cos \phi = 1$ is necessary for $\phi = \pi$

    h $\phi = 0$ is sufficient for $\tan \phi = 0$

2 For each pair of statements P and Q, decide if $P \Rightarrow Q$, $Q \Rightarrow P$, $P \Leftrightarrow Q$ or there are no implications at all.

    a $P: x = 4$ $Q: x < 6$

    b $P: y = 2$ $Q: y^2 = 4$

    c $P: 6a = 12$ $Q: 3a = 6$

    d $P: \dfrac{b}{4} = -2$ $Q: \dfrac{b}{3} = -1$

3 In each case, insert the correct symbol ($\Rightarrow$, $\Leftarrow$, or $\Leftrightarrow$) between statement P and statement Q.

    a P: Vera lives in the UK.

        Q: Vera lives in London.

    b P: The game is football.

        Q: The game involves a ball.

    c P: Anna sat the exam.

        Q: Anna achieved 80% in the exam.

    d P: Richard is Lianne's brother.

        Q: Lianne is Richard's sister.

    e P: Steven has £600 in his bank account.

        Q: Steven has over £500 in his bank account.

4   For each pair of statements P and Q, decide if $P \Rightarrow Q$, $Q \Rightarrow P$, $P \Leftrightarrow Q$ or there are no implications at all.

   a P: r divides 20 Q: r divides 30

   b P: q divides 15 Q: q divides 5

   c P: s divides 36 Q: s divides -36

   d P: t divides 40 Q: t divides 80

5   A function f is called **injective** (or **one-to-one**) if, for all x and y in the domain of f, $f(x) = f(y)$ if and only if $x = y$. For the statements P and Q, decide if $P \Rightarrow Q$, $Q \Rightarrow P$, $P \Leftrightarrow Q$ or there are no implications at all.

   P: $f(x)$ is injective Q: $f(x) = x - 3$ for all $x \in \mathbb{R}$

★ 6   Write down the converse of each statement.

   a   If $a = 5$, then a is an integer.

   b   For a 3×3 matrix A, $\det(A) \neq 0$ implies A is invertible.

   c   3 divides x if 3 divides y.

   d   2 divides x only if 2 divides y.

   e B being a square matrix is necessary for B^{-1} to exist.

   f $f(x) = 3$ is sufficient for $f'(x) = 0$

   g $ax = 0$ for all $x \in \mathbb{R}$ implies $a = 0$

   h   There exists an integer which divides 5 only if there exists a rational number which divides 5.

7   Write down the converse of each statement.

   a   David has travelled to Italy if David has travelled to Europe.

   b   The food being a banana is necessary for the food to be fruit.

   c   Laura plays a sport only if Laura plays badminton.

   d   Hameed being a father is sufficient for Hameed to have a son.

   e   Brian working for the local council implies Brian has a job.

The negation of a statement

Every mathematical statement you will meet (in Advanced Higher Maths) is either true or false. Therefore, for any statement P there is a corresponding statement 'not P', written $\neg P$. This is the statement that, if true, makes P false.

It is straightforward to write down the negation of many statements. For example, the negation of the statement $x = 4$ is simply $x \neq 4$, while the negation of the statement $y > 3$ is $y \leqslant 3$. However, we need to be careful when considering the negation of universal and existential statements.

For example, consider the universal statement 'for all x in $\mathbb{R}$, $2x > x$'. For this to be false, there needs to exist only **one** real number x for which $2x \leqslant x$. This illustrates that the negation of a universal statement is an existential statement.

Similarly, consider the existential statement 'there exists a natural number n that is greater than 10'. For this to be false, **all** natural numbers must be less than 10. This illustrates that the negation of an existential statement is a universal statement.

You should also be aware of the negations of statements involving **and** or **or** (as shown in Example 11.7).

Example 11.6

Write down the negations of these universal and existential statements.

a P: $\forall a \in \mathbb{N}, 4a > a$

b P: $\exists x \in \mathbb{R}: \frac{x}{3} = 9$

c P: $\forall x \in \mathbb{R}, \exists y \in \mathbb{R}: xy = 6$

d P: $\exists a \in \mathbb{Z}: \forall b \in \mathbb{Z}, a < b$

a $\neg P$: $\exists a \in \mathbb{N}: 4a \leq a$ — Remember to include the possibility that $4a = a$.

b $\neg P$: $\forall x \in \mathbb{R}, \frac{x}{3} \neq 9$

c $\neg P$: $\exists x \in \mathbb{R}: \forall y \in \mathbb{R}, xy \neq 6$

d $\neg P$: $\forall a \in \mathbb{Z}, \exists b \in \mathbb{Z}: a \geq b$

Example 11.7

Write down the negations of these statements.

a P: $x > 3$ and $x < 5$

b P: $x < -2$ or $x \geq 4$

a $\neg P$: $x \leq 3$ or $x \geq 5$ — The statement P consists of two parts: '$x > 3$' and '$x < 5$', both of which must be true for P to be true. Therefore, for P to be false, only one of '$x > 3$' and '$x < 5$' needs to be false.

b $\neg P$: $x \geq -2$ and $x < 4$ — In this case, only one of '$x < -2$' and '$x \geq 4$' needs to be true for P to be true. Therefore, for P to be false, both '$x < -2$' and '$x \geq 4$' must be false.

General rules

For mathematical statements P and Q:

- $\neg(P \text{ and } Q) = \neg P \text{ or } \neg Q$

- $\neg(P \text{ or } Q) = \neg P \text{ and } \neg Q$

Exercise 11C

★ 1 Write down the negations of each statement.

a $x < 3$

b $y \geqslant 2 \cdot 743$

c $x^2 - 2x + 4 \neq 6$

d $\forall z \in S, z < 3$

e $\exists r \in \mathbb{Q}: rt = \frac{1}{2}$

f $\forall A \in M_2(\mathbb{R}), A \neq 0$

g $\exists x \in \mathbb{C}: \forall y \in \mathbb{C}, \dfrac{y}{x} = y$

h $\forall p \in \mathbb{Z}, \exists q \in \mathbb{Z}: pq = 30$

★ 2 Write down the negations of each statement.

a $x < 4$ and $x > 2$

b $x < 3$ or $x > 5$

c $ab = 1$ or $a = 0$

d $\forall p \in \mathbb{Q}, pq = 0$ and $r > q$

e $x < 3 \cdot 165$ and $(x > 1 \cdot 212$ or $x = 0)$

f (y is irrational and $y > \pi$) or (y is irrational and $y < -\pi$)

Disproving a conjecture by providing a counterexample

You have seen examples of mathematical statements which are true, along with some ways to prove them. There are also countless examples of mathematical statements which are false – attempting to prove such statements would be a fruitless exercise. This section shows how to use **counterexamples** to disprove certain mathematical statements.

In mathematics, a **conjecture** is a statement which may be true but has not yet been proven. Some conjectures may be very easy to prove, others may be very easy to **disprove**, and there are some conjectures which are very difficult to either prove *or* disprove. For example, the Riemann hypothesis is a famous conjecture connected to prime numbers. It was proposed in 1859 and, despite the best efforts of numerous mathematicians, remains unproven over 150 years later.

In practice, when presented with a conjecture which is not obviously true, often the sensible first thing to do is to try and find a counterexample. Counterexamples are particularly useful when trying to disprove universal statements about large sets of numbers (for example, all natural numbers, all positive real numbers, etc.), as only **one** counterexample is needed to disprove the statement.

Example 11.8

By providing a counterexample, show that this statement is false:

If n is a natural number, then $3n$ is an odd number.

$n = 1$ is a natural number, and $3n = 3 \times 1 = 3$ which is odd.

So $n = 1$ is not a counterexample. — You need to provide one natural number n for which $3n$ is **not** odd. Start with some small examples and see what happens.

$n = 2$ is a natural number, but $3n = 3 \times 2 = 6$ which is **not** odd. — Only one counterexample is necessary to disprove the statement.

So $n = 2$ is a counterexample to the statement, and therefore the statement is false. — It is important to explicitly write down your counterexample.

Example 11.9

Find a counterexample to this statement:

For all real numbers a and b, $a - b > 0 \Rightarrow a^2 - b^2 > 0$

Let $a = 1$ and $b = -2$

Both 1 and −2 are real numbers, and $a - b = 1 - (-2) = 3$ which is greater than 0.

Hence taking $a = 1$ and $b = -2$ satisfies the first part of the statement.

However, $a^2 - b^2 = 1^2 - (-2)^2 = -3$ which is not greater than 0, so the second part of the statement is false.

Therefore $a = 1$ and $b = -2$ is a counterexample to the statement.

You are looking for real numbers a and b such that $a > b$ but $a^2 \leqslant b^2$. b must be a negative real number which has absolute value larger than the absolute value of a.

Exercise 11D

★ 1 By providing a counterexample, show that each statement about the real numbers is false.

   a   The square of every real number is positive.

   b   If x is a real number, then $x^2 \geqslant x$

   c   For every real number x, $2x \geqslant x$

   d   For every real number x, there exists another real number y such that $xy = 1$

2 Find counterexamples to these statements.

   a $n - n^2 + 100 > 0$ for all natural numbers n.

   b $x^3 - x^2 - 10x < 80$ for all real numbers x.

3   Disprove this statement:

Let $f(x) = ax^2 + bx + c$, where $a, b, c \in \mathbb{R}$ and $a \neq 0$. Then $f(x)$ has real roots.

4   Show that this statement is false:

Let $g(x)$ be a non-constant function with domain $\mathbb{R}$. If x and y are real numbers with $x \neq y$, then $g(x) \neq g(y)$.

★ 5   Find a counterexample to the following: every non-zero 2×2 matrix with real entries is invertible.

Proof by induction

Mathematical induction is a technique of proof which is often used when proving statements which are true for **all** natural numbers. An example of such a statement is:
$2^n \geq 2n$ for all $n \in \mathbb{N}$.

It is straightforward to prove this statement for any **particular** value of n. However, to prove the general statement we must deal with an infinite number of cases. Induction is the easiest way to prove this statement.

Proving a statement by induction involves two steps:

1   First, we prove the statement for the smallest value of n (often $n = 1$). This is known as the **base case**.

2   Secondly, we prove a general rule showing that if the statement is true for some arbitrary number $n = k$, then the statement must be true for the next value $n = k + 1$. This is known as the **inductive step**.

Taken together, these two steps prove the statement for **all** the natural numbers.

Think of proof by induction as climbing a ladder. The steps of the ladder correspond to the natural numbers: the first rung at the bottom corresponds to $n = 1$, the second rung to $n = 2$, and so on. Proving the base case allows us to step on to the first rung of the ladder, while the inductive step allows us to climb one rung higher from wherever we are standing. We can then climb up through all the natural numbers.

Example 11.10

Prove using mathematical induction that $2^n \geq 2n$ for all $n \in \mathbb{N}$.

Consider $n = 1$ •————

> The statement you are being asked to prove involves the set $\mathbb{N}$. Since 1 is the smallest natural number, this is the base case.

When $n = 1$ the LHS of the inequation is $2^1 = 2$, while the RHS is $2 \times 1 = 2$.

Since $2 = 2$, the statement holds when $n = 1$.

> Usually (but not always) the base case is quite easy to prove.

Assume true for $n = k$

Assume that the statement holds for some natural • number k. That is, assume $2^k \geq 2k$

> This is the inductive step. It is important that you assume nothing special about the number k.

Consider $n = k + 1$

Aim: to show that $2^{k+1} \geq 2(k+1)$

> It is a good idea to write down what you are trying to prove. You need to prove the statement for $n = k + 1$.

$2^{k+1} = 2^k \times 2$

> Break off 2^k so you can use the inductive step.

$\geq 2k \times 2$

> By the inductive step.

$= 4k$

$= 2k + 2k$

$\geq 2k + 2$ as $2k \geq 2$ for all $k \in \mathbb{N}$.

$= 2(k + 1)$

> Take out the common factor of 2.

Therefore, if the statement is true for $n = k$, it is true for $n = k + 1$.

Since it is true for $n = 1$, by induction it is true for all $n \in \mathbb{N}$.

> Make sure to clearly explain what you have shown.

Hence $2^n \geq 2n$ for all $n \in \mathbb{N}$.

Example 11.11

Prove using mathematical induction that $1 + 2 + 3 + \dots + n = \dfrac{n(n+1)}{2}$ for all $n \in \mathbb{N}$.

Consider $n = 1$

When $n = 1$, the sum on the LHS of the equation is simply 1, while the

RHS $= \dfrac{1 \times (1+1)}{2} = \dfrac{2}{2} = 1$. So the result holds for $n = 1$.

Assume true for $n = k$

Assume the statement holds for some natural number k, so assume:

$1 + 2 + 3 + \dots + k = \dfrac{k(k+1)}{2}$

Consider $n = k + 1$

Aim: to show that:

$1 + 2 + 3 + \dots + k + (k+1) = \dfrac{(k+1)(k+2)}{2}$

> You need to prove the statement for $n = k + 1$.

$= \dfrac{(k+1)((k+1)+1)}{2}$

$1 + 2 + 3 + \dots + k + (k+1) = (1 + 2 + 3 + \dots + k) + (k+1)$

> Break off the first k terms so you can use the inductive step.

$= \dfrac{k(k+1)}{2} + (k+1)$

> By the inductive step.

$= \dfrac{k(k+1)}{2} + \dfrac{2(k+1)}{2}$

> Write fractions over a common denominator.

$= \dfrac{(k+1)(k+2)}{2}$

> Take out the common factor of $k+1$ in the numerator.

$$= \frac{(k+1)((k+1)+1)}{2}$$

Therefore, if the result is true for $n = k$ it is true for $n = k + 1$. Since the result is true for $n = 1$, by induction

$$1 + 2 + 3 + \dots + n = \frac{n(n+1)}{2} \text{ for all } n \in \mathbb{N}.$$

Some statements are not true for all natural numbers n, but are true when n is larger than a particular value. We can still use induction in examples such as these, but the base case may not be $n = 1$.

Example 11.12

Prove that for $n \geqslant 5$, $4n < 2^n$.

Consider $n = 5$ — You can check that the statement is false for $n < 5$.

When $n = 5$, the LHS is $4 \times 5 = 20$, while the RHS is $2^5 = 32$.

So the result holds for $n = 5$.

Assume true for $n = k$

Assume the statement holds true for $n = k$, so $4k < 2^k$

Consider $n = k + 1$

Aim: to prove that $4(k + 1) < 2^{k+1}$

$4(k+1) = 4 + 4k$ — Break off the $4k$ so you can use the inductive step.

$\quad < 4 + 2^k$

$\quad < 2^k + 2^k$ (since $k \geqslant 5$ we have $2^k \geqslant 32$, so $4 < 2^k$)

$\quad = 2 \times 2^k$

$\quad = 2^{k+1}$

Therefore, if the result holds for $n = k$ it holds for $n = k + 1$. Since the result holds for $n = 5$, by induction $4n < 2^n$ for all $n \geqslant 5$.

Example 11.13

Prove this identity using mathematical induction:

$$\sum_{j=1}^{n} j^2 = \frac{n(n+1)(2n+1)}{6} \text{ for all } n \in \mathbb{N}.$$

Consider $n = 1$

When $n = 1$, the sum on the LHS is:

$$\sum_{j=1}^{1} j^2 = 1^2 = 1$$

The RHS is:

$$\frac{1 \times (1+1) \times (2 \times 1 + 1)}{6} = \frac{1 \times 2 \times 3}{6}$$

$$= \frac{6}{6} = 1$$

So the result holds for $n = 1$.

Assume true for $n = k$

Assume:

$$\sum_{j=1}^{k} j^2 = \frac{k(k+1)(2k+1)}{6}$$

Consider $n = k + 1$

Aim: to show that:

$$\sum_{j=1}^{k+1} j^2 = \frac{(k+1)((k+1)+1)(2(k+1)+1)}{6}$$

You need to prove the statement for $n = k + 1$. Take care when writing down what you need to prove – it is easy to make a mistake if you are not careful with brackets.

$$= \frac{(k+1)(k+2)(2k+2+1)}{6}$$

$$= \frac{(k+1)(k+2)(2k+3)}{6}$$

$$\sum_{j=1}^{k+1} j^2 = \sum_{j=1}^{k} j^2 + (k+1)^2$$

Removing the last term from the sum lets you use the inductive step with the sum from $j = 1$ to k.

$$= \frac{k(k+1)(2k+1)}{6} + (k+1)^2$$

By the inductive step.

$$= \frac{k(k+1)(2k+1)}{6} + \frac{6(k+1)^2}{6}$$

Simplify the fractions.

$$= \frac{k(k+1)(2k+1) + 6(k+1)^2}{6}$$

$$= \frac{(k+1)\left[(k(2k+1)+6(k+1))\right]}{6}$$

$$= \frac{(k+1)\left[(2k^2+k+6k+6)\right]}{6}$$

$$= \frac{(k+1)(2k^2+7k+6)}{6}$$

$$= \frac{(k+1)(k+2)(2k+3)}{6}$$

$$= \frac{(k+1)((k+1)+1)(2(k+1)+1)}{6}$$

Therefore, if the statement holds for $n = k$, it holds for $n = k + 1$. Since the statement holds for $n = 1$, by induction:

$$\sum_{j=1}^{n} j^2 = \frac{n(n+1)(2n+1)}{6} \text{ for all } n \in \mathbb{N}.$$

Example 11.14

Show that $3^{2n} + 7$ is divisible by 8 for all natural numbers n.

Consider $n = 1$

$3^{2 \times 1} + 7 = 9 + 7 = 16$, and since $16 = 8 \times 2$ it is divisible by 8. Thus the statement holds for $n = 1$.

Assume true for $n = k$

Assume $3^{2k} + 7$ is divisible by 8.

Consider $n = k + 1$

Aim: to show that $3^{2(k+1)} + 7$ is divisible by 8.

$$3^{2(k+1)} + 7 = 3^{2k+2} + 7$$
$$= 3^{2k} \times 3^2 + 7$$
$$= 9 \times 3^{2k} + 7$$
$$= (8 + 1) \times 3^{2k} + 7 \quad \longleftarrow \boxed{\text{Writing } 9 = 8 + 1 \text{ lets you extract } 3^{2k} + 7.}$$
$$= 8 \times 3^{2k} + (3^{2k} + 7)$$
$$= 8 \times 3^{2k} + 8p \text{ (for some integer } p) \quad \longleftarrow \boxed{\text{By the inductive step.}}$$
$$= 8 \times (3^{2k} + p)$$

Therefore $3^{2(k+1)} + 7$ is divisible by 8, and so if the result holds for $n = k$ it holds for $n = k + 1$. Since the result holds for $n = 1$, by induction $3^{2n} + 7$ is divisible by 8 for all natural numbers n.

Example 11.15

A recurrence relation is defined by $a_n = 2a_{n-1} + 3$ for $n \in \mathbb{N}$, with $a_0 = 1$.

Prove that $a_n = 2^{n+2} - 3$ for all $n \in \mathbb{N}$.

Consider $n = 1$

$a_1 = 2a_0 + 3 = 2 \times 1 + 3 = 5$, and also $2^{1+2} - 3 = 8 - 3 = 5 = a_1$, so the result holds for $n = 1$.

Assume true for $n = k$

Assume $a_k = 2^{k+2} - 3$

Consider $n = k + 1$

Aim: to show that $a_{k+1} = 2^{(k+1)+2} - 3$

$$a_{k+1} = 2a_k + 3$$
$$= 2(2^{k+2} - 3) + 3$$
$$= 2^{k+3} - 6 + 3$$
$$= 2^{k+3} - 3$$
$$= 2^{(k+1)+2} - 3$$

Therefore if the result holds for $n = k$, it holds for $n = k + 1$. Since the result holds for $n = 1$, by induction $a_n = 2^{n+2} - 3$ for all $n \in \mathbb{N}$.

Exercise 11E

1 Use induction to show that $1 + 3 + 5 + 7 + \ldots + (2n - 1) = n^2$ for all $n \in \mathbb{N}$.

★ 2 Prove that $3^n > 4n$ for all integers $n > 2$.

★ 3 Prove by induction that $3n^5 + 7n$ is a multiple of 5 for all $n \in \mathbb{N}$.

4 Show that $3^{2n+1} + 2^{n-1}$ is a multiple of 7 for all $n \in \mathbb{N}$.

★ 5 By using proof by induction, show that:

$$\begin{pmatrix} 1 & 1 \\ 0 & 1 \end{pmatrix}^n = \begin{pmatrix} 1 & n \\ 0 & 1 \end{pmatrix} \text{ for all } n \in \mathbb{N}.$$

6 If A is the matrix

$$\begin{pmatrix} 2 & a \\ 0 & 1 \end{pmatrix}$$

where a is some real number, prove by induction that:

$$A^n = \begin{pmatrix} 2^n & a\left(2^n - 1\right) \\ 0 & 1 \end{pmatrix}$$

for all natural numbers n.

7 Prove de Moivre's theorem, that is, prove that $(\cos \theta + i \sin \theta)^n = \cos (n\theta) + i \sin (n\theta)$ for all integers $n \geqslant 1$.

8 Use induction to prove the following.

a Given that $e^x e^y = e^{x+y}$, show that $(e^x)^n = e^{nx}$ for all $n \in \mathbb{N}$.

b Given that $\ln x + \ln y = \ln xy$, show that $n \ln x = \ln x^n$ for all $n \in \mathbb{N}$.

★ 9 A recurrence relation is defined by $a_n = 2a_{n-1} + 1$ for $n \in \mathbb{N}$, with $a_0 = 0$. Prove that $a_n = 2^n - 1$ for all $n \in \mathbb{N}$.

10 A recurrence relation is defined by $u_{n+1} = \dfrac{u_n}{u_n + 1}$ for $n \in \mathbb{N}$, with $u_0 = 1$.

Prove that $u_n = \dfrac{1}{n + 1}$ for all $n \in \mathbb{N}$.

★ 11 a Calculate the sum $\displaystyle\sum_{k=1}^{n} \frac{1}{k(k + 1)}$ for $n = 1, 2, 3$ and 4.

b Make a conjecture about the value of this sum for an arbitrary natural number n, and use induction to prove your conjecture.

12 a Calculate the sum $\displaystyle\sum_{k=1}^{n} \frac{k - 1}{k!}$ for $n = 1, 2, 3, 4$ and 5.

b Make a conjecture about the value of this sum for an arbitrary natural number n, and use induction to prove your conjecture.

★ **13** Prove the following identities using proof by induction.

a $\displaystyle\sum_{r=1}^{n} r(3r - 1) = n^2(n + 1)$ for all $n \in \mathbb{N}$

b $\displaystyle\sum_{r=1}^{n} \left(4r^3 + 3r^2 + r\right) = n(n + 1)^3$ for all $n \in \mathbb{N}$

c $\displaystyle\sum_{r=1}^{n} (-1)^r r^2 = \frac{(-1)^n n(n + 1)}{2}$ for all $n \in \mathbb{N}$

14 Prove this identity, which is sometimes referred to as Nicomachus's theorem:

$$\sum_{i=1}^{n} i^3 = \left(\sum_{i=1}^{n} i\right)^2 \text{ for all } n \in \mathbb{N}$$

> The inductive step here is quite tricky. It might help to write
>
> $$\sum_{i=1}^{k+1} i = (1 + 2 + \dots + k) + (k + 1)$$
>
> and use the identity from Example 11.11 to rewrite $1 + 2 + \dots + k$.

Direct proof

In a direct proof, we start with an initial assumption and prove a series of logical implications until we arrive at the desired conclusion. It is crucial that throughout the process we make no **additional** assumptions.

A number of steps may be required before a direct proof is complete. For example, if our initial assumption is a statement P, and the conclusion we wish to prove is Q, we might first prove that $P \Rightarrow R$ for another statement R, and then prove that $R \Rightarrow Q$. Since $P \Rightarrow R$ and $R \Rightarrow Q$ implies $P \Rightarrow Q$, this would be sufficient to prove the result.

Example 11.16

Prove that $4x < 25$ for $x \in \{1, 2, 3, 4, 5\}$.

$4 \times 1 = 4 < 25$

$4 \times 2 = 8 < 25$

$4 \times 3 = 12 < 25$ •————

$4 \times 4 = 16 < 25$

$4 \times 5 = 20 < 25$

> When asked to prove a result about a relatively small set, it is often easiest to use **proof by exhaustion**. We simply check the result holds in each case.

Since the result is true in all cases, $4x < 25$ for $x \in \{1, 2, 3, 4, 5\}$.

Example 11.17

Prove that if a and b are even integers, then their sum $a + b$ is also an even integer.

Aim: to show $a + b = 2m$ for some integer m.

> Identify what you need to show by using the definition of an even integer. Note that you are being asked to prove something about an infinite set (the integers), so proof by exhaustion is not an option.

Since a and b are both even, it must be the case that $a = 2c$ for some integer c, and $b = 2d$ for some integer d.

> It is important you use the initial assumption about a and b being even.

Therefore:

$a + b = 2c + 2d$

$\qquad = 2(c + d)$

> Since a and b may be different integers, you must use different letters when writing them as multiples of 2 (c and d in this case).

Since $c + d$ is an integer, $a + b$ is also an even integer (as required).

Example 11.18

If $m = 3k + 1$ for some integer k, prove that m^2 is **not** divisible by 3.

Aim: to show $m^2 = 3s + 1$ or $m^2 = 3s + 2$, where s is some integer.

> When counting up through the integers, after every integer which is divisible by 3 there are two integers which are not divisible by 3.

$m^2 = (3k + 1)^2$

> Use the initial assumption about m.

$\qquad = 9k^2 + 6k + 1$

$\qquad = 3(3k^2 + 2k) + 1$

> Remove the common factor of 3 from the first two terms.

Since $3k^2 + 2k$ is an integer, m^2 is not divisible by 3 (as required).

Exercise 11F

1 Use proof by exhaustion to prove each statement.

   a $3x - 4 < 9$ for $x \in \{0, 1, 2, 3, 4\}$.

   b $x^2 - x \geqslant 0$ for $x \in \{-2, -1, 0, 1, 2\}$.

   c If y is an odd integer in the set $\{1, 2, 3, 4, 5, 6, 7\}$, then y^3 is an odd integer.

   d If $z = \dfrac{x}{y}$ is a rational number with $x \in \{0, 1, 2\}$ and $y \in \{2, 3, 4\}$, then $z \leqslant 1$.

2 Let a and b be natural numbers. Prove these statements:

   a If a is odd and b is even, then $a + b$ is odd.

   b If a is odd and b is odd, then $a + b$ is even.

   c If a is even and b is even, then the product ab is even.

   d If a is even and b is odd, then the product ab is even.

   e If a is odd and b is odd, then the product ab is odd.

3   Decide whether each of the following statements is true or false. If true, give a proof; if false, give a counterexample.

   a   Let a, b and c be natural numbers. Then $a + b > a + c$ if and only if $b > c$.

   b   Let a, b and c be integers. Then $a - b < a - c$ if and only if $b > c$.

   c   Let a, b and c be natural numbers. Then $ab > ac$ if and only if $b > c$.

   d   Let a, b and c be integers. Then $ab > ac$ if and only if $b > c$.

★ 4   Prove that the square of any odd natural number is also odd.

★ 5   Prove that $m^2 + n^2 < (m + n)^2$ for all $m, n \in \mathbb{N}$.

6   Show that if $m = 3k + 2$ for some integer k, then m^2 is **not** divisible by 3.

★ 7   Prove that the product of any three consecutive natural numbers is divisible by 6.

8   Prove that the difference between the squares of any two consecutive odd numbers is divisible by 8.

9   Show that if f is an injective function with domain $\mathbb{R}$, then $2f$ is an injective function.

10 Let n be a natural number. For each of these statements, decide whether it is true or false. If true, give a proof; if false, give a counterexample.

   a   If n is a multiple of 4 then so is n^2.   b   If n^2 is a multiple of 4 then so is n.

11 Let A and B be non-zero 2×2 matrices with real entries. For each of these statements, decide whether it is true or false. If true, give a proof; if false, give a counterexample.

   a kA is a non-zero matrix for all $k \in \mathbb{R}$, $k \neq 0$.   b $A + B$ is a non-zero matrix.

   c AB is a non-zero matrix.   d $\det(A) \neq 0$ e $A^2 \neq 0$

12 There are often different ways to prove the same result. Example 11.11 used induction to prove that

$$1 + 2 + 3 + \ldots + n = \frac{n(n + 1)}{2} \text{ for all } n \in \mathbb{N}.$$

Find a direct proof of this statement.

> Try adding the sum $1 + 2 + 3 + \ldots + n$ to itself with the order of the terms reversed.

Indirect proof

Some statements can be difficult to prove directly. There are some alternative proof techniques which we can use if a direct proof looks hard to produce. This section looks at two such methods, namely **proof by contrapositive** and **proof by contradiction**.

Proof by contrapositive

The **proof by contrapositive** method relies on the fact that for statements P and Q, the statements $P \Rightarrow Q$ is equivalent to its **contrapositive** statement $\neg Q \Rightarrow \neg P$. So instead of assuming P and trying to prove Q, we can assume that Q is false and prove that this implies P is also false. In certain situations it can be much easier to prove this directly than the original statement.

It is helpful to show why $P \Rightarrow Q$ is equivalent to $\neg Q \Rightarrow \neg P$. This is an 'if and only if' statement, so there are two directions to prove:

1 $(P \Rightarrow Q) \Rightarrow (\neg Q \Rightarrow \neg P)$

   Since P implies Q, the only way that Q can be false is if P is also false. Therefore $\neg Q$ implies $\neg P$.

2 $(\neg Q \Rightarrow \neg P) \Rightarrow (P \Rightarrow Q)$

   Since $\neg Q$ implies $\neg P$, the only way that $\neg P$ can be false if is $\neg Q$ is also false. Hence $\neg\neg P$ implies $\neg\neg Q$. But $\neg\neg P = P$ and $\neg\neg Q = Q$, therefore P implies Q.

Example 11.19

Prove that if n^2 is an even integer then n is also even.

To prove the contrapositive, we must show that n **not** being even implies n^2 is **not** even. ●———

> Both the original assumption and conclusion are negated, and the direction of the implication is reversed.

Since any integer which is not even must be odd, a simpler way of expressing what we want to prove is that n being odd implies n^2 is odd.

Assume that n is odd. Therefore $n = 2k + 1$ for some integer k.

So:

$n^2 = (2k + 1)^2 = 4k^2 + 4k + 1$

$= 2(2k^2 + 2k) + 1$

and since $2k^2 + 2k$ is an integer, this implies that n^2 is odd. Since the contrapositive statement is true, the original statement also holds.

Example 11.20

Suppose that x and y are real numbers. Prove that if $x + y$ is irrational, then at least one of x and y is irrational.

We'll prove the contrapositive statement. We need to show that if both x and y **are** rational, then $x + y$ is also rational. ●———

> Remember that $\neg(P \text{ or } Q) = P$ and Q.

Assuming both x and y are rational, write: $x = \dfrac{m}{n}$ and $y = \dfrac{r}{s}$ ●———

for some integers m, n, r and s (with $n, s \neq 0$).

> This is the reason why it is easier to prove the contrapositive statement in this example. It is much easier to prove a result about fractions than arbitrary irrational numbers.

Then:

$x + y = \dfrac{m}{n} + \dfrac{r}{s}$

$= \dfrac{ms}{ns} + \dfrac{nr}{ns} = \dfrac{ms + nr}{ns}$

Since $ms + nr$ is an integer, and ns is a non-zero integer, this shows that $x + y$ is rational. Since the contrapositive statement is true, the original statement also holds.

Exercise 11G

★ 1 Prove that if m^2 is an odd natural number, then m is also odd.

2 For $x \in \mathbb{Z}$, show that if $x^2 - 2x + 4$ is even, then x is even.

★ 3 Suppose that x and y are real numbers, with $y \neq 0$. If the following expressions describe irrational numbers, in each case prove that at least one of x and y is irrational.

a $x - 2y$ b $3xy$ c $\dfrac{x}{y}$

★ 4 Prove that if the product of two positive real numbers x and y is greater than 64, then at least one of x and y is greater than 8.

5 Suppose that x and y are real numbers.

Prove that if $-0.25 < xy < 0.25$, then either $-0.5 < x < 0.5$ or $-0.5 < y < 0.5$ (or both).

6 A complex number is called **algebraic** if it is the root of some polynomial with integer coefficients, i.e. a polynomial of the form

$$a_n x^n + a_{n-1} x^{n-1} + \ldots + a_1 x + a_0$$

where $a_0, a_1, \ldots, a_n$ are integers.

If a complex number is not algebraic, we say it is **transcendental.** Examples of transcendental numbers include π and e, the base of the natural logarithm.

Using proof by contrapositive, show that if x is a transcendental number, then $2x$ is also transcendental.

Proof by contradiction

Proof by contradiction is also an example of an indirect proof technique. It makes use of the fact that a mathematical statement must be either true or false.

Suppose we are trying to prove some statement P. We begin by **assuming** that P is **not** true, so we assume $\neg P$.

Then, with no further assumptions and using only logical reasoning, we use this to prove a **contradiction.**

This means that we show some statement which we know to be true is also false.

Since this clearly doesn't make sense, it must be the case that our initial assumption of $\neg P$ was incorrect. Therefore, P must hold.

Example 11.21

Use proof by contradiction to show that if f is an injective function, then $2f$ is an injective function.

> Recall that a function f is called injective if, for all x and y in the domain of f, $f(x) = f(y)$ if and only if $x = y$.

For contradiction, assume that $2f$ is **not** an injective function.

> Make sure you clearly state the initial assumption.

Therefore, there exist x and y in the domain of f such that $x \neq y$ but $2f(x) = 2f(y)$. Dividing both sides of this equation by 2, we have $f(x) = f(y)$.

Hence there exist x and y such that $x \neq y$ but $f(x) = f(y)$, which means that f is not injective. This contradicts the fact that f is injective.

> Write down where the contradiction occurs.

Therefore the initial assumption was false, and $2f$ is injective.

Alternatively we could have used direct proof in Example 11.21 (see Exercise 11F Q9). Examples 11.22 and 11.23 illustrate situations when direct proof is not a sensible option.

Example 11.22

Prove that $\sqrt{2}$ is not a rational number.

For contradiction, assume that $\sqrt{2}$ **is** a rational number. Then we may write $\sqrt{2} = \frac{m}{n}$ where m and n are integers with no common factors, and $n \neq 0$.

> It is the 'no common factors' statement that you will eventually contradict.

Squaring both sides: $2 = \frac{m^2}{n^2}$ and so $2n^2 = m^2$.

> Look at Example 11.19 if you're not convinced by this last statement.

Hence m^2 is even, and therefore m must also be even.

Since m is even, we may write $m = 2r$ for some integer r.

Then $2n^2 = m^2$
$$= (2r)^2$$
$$= 4r^2$$
$$= 2(2r^2)$$

which implies that $n^2 = 2r^2$, so n^2 is even. Thus n is also even.

We have shown that both m and n are even, so have 2 as a common factor. However, in the second sentence of the proof we chose m and n to have no common factors. This is a contradiction.

> Write down where the contradiction occurs.

Therefore, our initial assumption was false, and $\sqrt{2}$ is not a rational number.

> Since you made no other assumptions, the only possibility is that the initial assumption was false.

Example 11.23

Prove that there are an infinite number of prime numbers.

For contradiction, suppose there are only a finite number of prime numbers.

> Assuming a finite set and producing a contradiction is often easier than using direct proof to show the set **is** infinite. This is why proof by contradiction is a good choice for this example.

Say this number is k. Then we can list the prime numbers as:

$p_1, p_2, p_3, \dots, p_{k-1}, p_k$

Now take the product of all these prime numbers, and then add 1. We can write:

$x = p_1 p_2 p_3 \cdots p_{k-1} p_k + 1$

Since x will be a natural number, it must itself be divisible by at least one of the primes in our list. Without loss of generality, assume that p_1 divides x, so we may write $x = p_1 y$ for some natural number y. But now we have:

$p_1 y = p_1 p_2 p_3 \cdots p_{k-1} p_k + 1$

which can be rearranged to give:

$1 = p_1 y - p_1 p_2 p_3 \cdots p_{k-1} p_k$

$\quad = p_1 (y - p_2 p_3 \cdots p_{k-1} p_k)$

which implies that 1 is divisible by p_1. This is a contradiction, since 1 is not divisible by any prime. Hence there are an infinite number of prime numbers.

Exercise 11H

1. Prove that if m^2 is an integer which is divisible by 3, then m is also divisible by 3.

> You may wish to use Example 11.18 and Exercise 11F Q6.

★ 2. Assuming that $\sqrt{2}$ and $\sqrt{6}$ are irrational numbers, prove that $\sqrt{2} + \sqrt{3}$ is an irrational number.

3. Prove that $\sqrt{3}$ is not a rational number.

> Use a similar technique to Example 11.22, along with Q1 above.

4. Prove by contradiction that $\log_2 3$ is irrational.

★ 5. Prove that $\sqrt{2} + \sqrt{5} < \sqrt{15}$.

★ 6. Given that π is a transcendental number, use proof by contradiction to show that 3π is a transcendental number.

> See Exercise 11G Q6.

7. Modify the proof in Example 11.23 to prove that there are an infinite number of primes of the form $4n + 3$, where n is a natural number.

8 An island is inhabited entirely by Knights and Knaves. Knights always tell the truth, while Knaves always lie. When visiting this island, you meet a person who says to you 'I am not a liar'. Prove that this person is not an inhabitant of the island.

9 Example 11.22 proved that $\sqrt{2}$ is irrational, and Question 3 above required proving that $\sqrt{3}$ is irrational.

Generalise this by proving that if $a \in \mathbb{N}$ is not a square number, then $\sqrt{a}$ is irrational.

a Let $m \in \mathbb{N}$ with m^2 divisible by a. Use the fundamental theorem of arithmetic (see Chapter 10) to prove that m is divisible by a.

b Example 11.19 used proof by contrapositive to show that if m^2 is divisible by 2, then m is divisible by 2. Why would it be difficult to use proof by contrapositive in part **a**?

c By following the strategy shown in Example 11.22, and using part **a**, prove that if $a \in \mathbb{N}$ is not a square number, then $\sqrt{a}$ is irrational.

d Now suppose that a is a square number, and we try to use the same proof to show that $\sqrt{a}$ is irrational. Identify the point in your proof from part **c** where this strategy fails.

Chapter review

1 Write these statements using logical notation.

a $f(x) > 0$ for all $x \in \mathbb{R}$.

b $9a \neq 0$ for some $a \in \mathbb{Z}$.

c If b divides 16, then $2b$ divides 32.

d $\dfrac{2}{r} \neq 0$ if and only if $r \neq 0$.

2 Determine the negation of each statement, and write it using logical notation.

a $x = -4$

b For all $x \in \mathbb{R}$, $x = -4$.

c There exists $x \in \mathbb{R}$ such that $x = -4$.

d $x < -1 \cdot 8$ or $x > 3 \cdot 55$

e $y = 0$ or $(\exists z \in \mathbb{Q}: yz = 1)$

3 Suppose that $f: A \to B$ is a function, with A and B both containing exactly k elements, where k is a natural number. Let P be the following statement.

P: If f is injective, then f is surjective

a Prove the statement P.

b Write down, then prove, the converse of P.

c Hence write down two equivalent statements regarding the function f.

4 Prove that $2^n > 3n$ for all integers $n > 4$.

5 Use induction to prove that $\displaystyle\sum_{i=1}^{n}(3i - 2) = \frac{n(3n - 1)}{2}$ for all natural numbers n.

6 Suppose that $f(x) = x^2$ for all $x \in \mathbb{R}$. Prove that $f^n(x) = x^{2^n}$ for all $n \in \mathbb{N}$. (Here $f^n(x)$ denotes the composition of f with itself n times.)

7 Prove the following statements directly.

a If 5 divides x, then 5 divides $3x$.

b If x is even, then $x^2 + 3x - 1$ is odd.

c If a, b and c are natural numbers, then $2a - b > 2a - c$ if and only if $b < c$.

8 For each of these statements, decide whether it is true or false. If true, give a proof; if false, give a counterexample.

a For 2×2 matrices A and B, if $AB = 0$ then either $A = 0$ or $B = 0$.

b If A is a 2×2 matrix and k is a non-zero real number, $kA \neq 0$ implies $A \neq 0$.

9 Use proof by contrapositive to prove these statements.

a If $3x + y$ is an irrational number, then at least one of x and y is an irrational number.

b If the product of two positive real numbers x and y is greater than 121, then at least one of x and y is greater than 11.

10 Use proof by contradiction to show that $\sqrt{6}$ is an irrational number. (You may use result from earlier in this chapter).

11 Prove by contradiction that $\log_4 5$ is irrational.

- I can express simple mathematical statements using logical notation.
  ★ Exercise 11A Q2 ★ Exercise 11B Q1, Q6

- I can determine the negation of a simple mathematical statement.
  ★ Exercise 11C Q1, Q2

- I can provide counterexamples to disprove mathematical statements.
  ★ Exercise 11D Q1, Q5

- I can prove results using mathematical induction. ★ Exercise 11E Q2, Q3, Q5, Q9, Q11, Q13

- I can use direct proof methods to prove results. ★ Exercise 11F Q4, Q5, Q7

- I can prove results using contrapositive statements. ★ Exercise 11G Q1, Q3, Q4

- I can prove results using proof by contradiction. ★ Exercise 11H Q2, Q5, Q6